AF615276

Computers in Reciprocating Engines and Gas Turbines

Conference Organizing Committee

IMechE Conference Transactions

Third International Conference on

Computers in Reciprocating Engines and Gas Turbines

9–10 January 1996

Organized by the
Combustion Engines Group of the
Institution of Mechanical Engineers

Co-sponsored by the JSME, CIMAC, and SAE
with the support of SNECMA and Perkins Technology

IMechE Conference Transaction 1996 – 1

Published by Mechanical Engineering Publications Limited for
The Institution of Mechanical Engineers, London.

First Published 1996

ISSN 1356-1448
ISBN 0 85298 984 9

A CIP catalogue record for this book is available from the British Library

Printed and bound in Great Britain by Antony Rowe Ltd, Chippenham, Wiltshire

Contents

C499/004/96

Assessment and application of a new spray wall impaction model

A P WATKINS BA, MSc, DIC, PhD, MSAE and **K PARK** BSc, MSc, PhD
Department of Mechanical Engineering, UMIST, UK

SYNOPSIS

This paper describes a new mathematical sub-model for the calculation of a diesel spray impacting onto a wall. Comparison with experiments over a wide range of engine-like conditions, some employing non-orthogonal grid systems, shows that the new model gives good results for wall spray shape and development. Comparison with PDA data on drop velocities and sizes within a wall spray gives reasonably good agreement. Finally the computer code has been employed to consider the possible attributes of engine designs which employ spray impaction.

NOTATION

H	Height from wall
k	Energy loss coefficient
r	Radius
V	Velocity
We	Weber number

Greek

β	Impaction angle
χ	Transverse velocity coefficient

Subscripts

a, b	After, before impaction
film	Film front
scattering	Random transverse velocity
stable	Stable We range
unstable	Unstable We range

Superscripts

n	Normal to wall
t	Transverse to wall

1 INTRODUCTION

Automotive direct-injection diesel engines of small displacement are often fitted with injection equipment operating at high pressures. This is to ensure that all the fuel is injected in the short time available near TDC of compression. These diesels are often running at higher speeds than medium or large conventional diesels, thus the time available for the injection process is very short. As a consequence, the fuel spray will usually impact onto the piston surfaces, unless the flow is highly swirled to prevent it. This may lead to unburned hydrocarbons at the end of the combustion process, due to the relatively slow evaporation and mixing in the vicinity of a relatively cool wall. Steps need to be taken in the design process of such engines to avoid this situation occurring. In this paper the possible involvement of computational fluid dynamics

(CFD) in this process is addressed. In particular it looks at the need to develop sub-models which express the physics of the impaction of diesel sprays onto solid surfaces. The paper then goes on to suggest by example how such sub-models may be used in the design process of the engine.

Spray models developed for use in CFD have the major advantage that they are based on the physics of the fluid flow phenomena, and are therefore universally applicable. Most models of sprays which have been developed for engine use are based on a statistical approach, whereby the spray is divided into a large but finite number of drop parcels, the discrete droplet model (DDM). The properties of the spray are fully accounted for by sampling enough such parcels. This approach has many advantages in terms of computational efficiency. However, it has the disadvantage that certain processes require specific sub-models to describe them. One such process is the impaction of sprays onto solid surfaces.

Naber and Reitz (1) proposed two impingement models as alternatives to the Stick model, in which droplets that reach the wall stick to the wall at the impingement location and continue to vaporize. In the model called Reflect, droplets that reach the wall rebound with the same velocity magnitude. In the Jet model, an incident droplet leaves tangential to the surface without any change of velocity magnitude, in analogy with a liquid jet. They reported that the best predictions were obtained using the Jet model, when cases were examined and compared with experiment of sprays impacting onto flat surfaces at relatively large distances from the injector. Impaction on the walls of a diesel engine was also examined. This model was applied by Naber et al (2) to simulate normal spray impaction onto the land on a cut off pip in a piston bowl set very close to the injector. The spray failed to disperse after impaction and the penetration of the spray was greatly over predicted. They modified the impaction model, by adding a randomly selected normal rebound velocity, however the penetration soon after impaction onto the wall was still over predicted.

Other modifications of the Jet model, e.g. Wakisaka et al (3), have assumed the conservation of kinetic energy at impact on the surface. Treating the velocity of droplets after impaction in this manner may result in over prediction, because the energy loss of impaction on the wall is not considered.

Wang and Watkins (4) proposed an impaction model based on the approach Weber number of the droplets. In this model the energy loss coefficient was introduced to calculate the velocity of the drop after impaction. In the high Weber number region, the droplet incident normally on the wall loses all the kinetic energy after impaction, as in the Stick model. This caused under prediction in both wall spray radius and height. So they introduced a modification to the grazing collision model of O'Rourke and Bracco (5). The direction of the velocity after the extended grazing collision calculation is varied towards the thin regions of the spray. The wall spray height is improved greatly, but the wall spray radius predictions became worse.

None of the above models attempt to represent all the seven possible outcomes of drop impaction, depending on the wall conditions and the impact conditions of the drop, as identified by Bai and Gosman (6). Major classifications of the outcomes can be made in terms of the wall temperature, particularly whether it is above the boiling temperature of the fuel, whether the wall is dry or wet from previous impactions, the roughness of the wall, etc. Bai and Gosman themselves concentrate on low wall temperature conditions.

In this paper a new approach for the spray impaction model is presented, which is based on the model of Wang and Watkins (4) but extended to include other physical features of drop impaction onto solid surfaces, and it is anticipated that the extended grazing model may be dispensed with. The basic data employed to develop this model is that of Wachters and Westerling (7). This was for conditions in which the wall temperature was above the boiling temperature of the drops. Thus the model developed considers two main regimes, depending on the approach Weber number of the drops. These are the rebound regime and the boiling induced breakup regime. However, there is insufficient data available to fully describe the latter regime. Recourse to data obtained on cool walls has therefore had to be made, which compromises the model somewhat, and testing of the model is difficult because most data is for cool wall conditions.

Despite these problems, the new model has been subjected to considerable testing. These tests are given in full by Park (8). Here there is space only to report on a small subset of these tests. The application of this wall impaction model to the design of novel engine shapes is reported in (9).

2 MATHEMATICAL MODEL

The gas phase is modelled in terms of the Eulerian conservation equations of mass, momentum, energy and fuel vapour mass fraction, and turbulent transport is modelled by the k-ϵ turbulence model. The droplet parcel equations of trajectory, momentum, mass and energy are written in Lagrangian form. Each droplet parcel contains many thousands of drops assumed to have the same size, temperature, velocity components, etc. The actions of the gas phase on the liquid phase are accounted for through the shear terms in the liquid phase momentum equations and the heat transfer terms in the liquid phase energy and fuel mass conservation equations. The effects of the liquid phase on gas phase are given as sources or sinks in mass, momentum and energy equations in the gas phase. All these equations are expressed in terms of non-orthogonal grid systems.

The transport equations are discretised in the finite volume manner. The Euler implicit method is used for the transient term and the hybrid upwind/central difference scheme is used to approximate the convection and diffusion terms. The non-iterative but implicit PISO solution scheme is used to treat the velocity-pressure linkage.

The wall impaction model is founded on experiments conducted on individual droplets by Wachters and Westerling (7). These experiments show that the behaviour of droplets after impaction is divided into that normal to and tangential to the wall, and also depends on the Weber number of the drop before impaction. They show the deformation characteristics of a drop impinging on a hot wall. The shape of the approaching drop is nearly spherical. Shortly after the impact the drop undergoes a large deformation and a continuous film spreads out all around a compact central dome. At the end of the spreading of the film, at lower Weber number it shrinks into a drop again and rises up from the surface and for high Weber number it disintegrates in the film state, due to boiling of the fuel next to the wall. However their photographs show that much of the film forms into small drops and these rise up from the surface.

The energy loss is accounted for by the velocity relation of the droplet before/after impaction over the whole range of Weber number

$$\frac{V_a^2}{V_b^2} = 1 - k\cos^2(\beta) \quad (1)$$

Here V_a, V_b are the velocities of the drops after and before impaction respectively, β is the angle between the approaching drop and the normal to the surface, and the coefficient k is obtained from the relationship between the normal velocity components at arrival and at departure taken from the experimental results of Wachters and Westerling (7). The curve of We_a versus We_b tends to level out at $We_b = 80$, so here for $We_b > 100$ a constant value of $We_a = 1.0$ is employed to give a small rebound velocity to drops which form from the breakup of the liquid film. On the other hand the velocity component parallel to the surface does not change during the impaction.

For the stable region $We < 80$, the velocity component of the drop after impaction normal to the surface is given directly from equation (1). As the Weber number increases ($We_b > 80$), the droplet becomes unstable and breaks up into a large number of little droplets. Their velocities after impaction can be modelled by introducing a random instability factor and their direction by random selection. The impacting drop is broken up into many smaller drops. Each of these new drops could be treated separately. However this would involve an enormous increase in the storage, and CPU, requirements of the computer program. Instead the parcel is broken into two new parcels, that are assigned scattering velocities (see below) of opposite sign, but equal magnitude.

Naber and Farrell (10), also investigating drop impaction on surfaces heated above the boiling temperature of the liquid, showed that the number of droplets atomized from an impingement increased linearly with increasing impinging Weber number. However, their experiments only covered the range of We_b up to 120. We have here assumed that the linear variation continues to much higher values of We_b. For the test cases included in this paper, the impaction We_b ranges from about 500 to about 5000. This linear variation of droplet number is also adopted by Bai and Gosman (6), for impactions on cool walls.

The velocity components of the droplets of the two parcels normal to the surface after impaction are given by

$$(V_a^n)^1_{unstable} = (V_a^n)^2_{unstable} = (V_a^n)_{stable} R_{xx} \tag{2}$$

where $(V^n_a)_{stable}$ is the velocity component normal to the surface that is calculated as in the stable region through equation (1) and R_{xx} is a random variable uniform in [0,1].

The velocity components parallel to the surface after impaction are given as

$$(V_a^t)^{1,2}_{unstable} = V_b^t \pm V^t_{scattering} \tag{3}$$

where $V^t_{scattering}$ is the scattering velocity which is given randomly as a fraction of the velocity V_{film} of the front of the liquid film spreading out around the central dome. The direction of the drop motion after impact is also selected randomly in $[0, 2\pi]$. V_{film} is given by differentiating the equation of the radius of the film, deduced by Park (8) from the data of Wachters and Westerling (7), as

$$V_{film} = 0.835(3.096 - 2\chi)\, V_b^n \tag{4}$$

where the coefficient χ is a function of V^n_b. Park found that $\chi = 1$ gives the best results of spray wall penetration for low values of V^n_b. For high velocities $\chi = 1.28$ is used.

As explained in (8), these equations need to be adjusted to account for the use of non-orthogonal grids. This will not be gone into here. The new wall impaction model does not affect the free spray, the modelling of which has been subjected to many tests, as in e.g. Yule and Watkins (11).

3 TEST CASES

3.1 Wall spray shapes

To test the wall impaction sub-model a number of cases have been simulated. Park (8) shows 11 cases. Eight of these are testing the model when the spray impacts onto a flat surface and then spreads across it. Varied here are the injection pressures (13.8 - 19.6 MPa), the nozzle diameter (0.2 - 0.3 mm), the injection duration (1.2 - 2.4 ms), and the ambient gas pressure (1.0 - 2.9 MPa) and temperature (300 - 773 K). The cases can also be divided into two categories; these are impaction normal to the wall (5 cases), and at an angle to the wall (3 cases). The final three cases deal with very high velocity normal injection onto the land of a cut-off pip very close to the injector. For all the cases the simulation results given by the new model without or with the extended grazing model of Wang and Watkins (4) (denoted by New/New+Col), are compared with previous models and also experimental results.

A 35 x 35 line grid is used in the normal impaction cases. It is difficult to claim grid independence for spray induced flows, because of the danger of flooding of small grid cells by the liquid, and the gas phase equations breaking down. However the grid is thought to be fine enough to capture most of the spray and gas phase phenomena. For these simulations an orthogonal grid system is employed. For the drop phase a step size of 7μs is used and 10 drop parcels are introduced randomly throughout the time step. Thus 2000 drop parcels are introduced in each of the simulations. For these calculations this is sufficient to obtain statistically independent shapes for the wall sprays. A Rosin-Rammler size distribution is assumed for the spray at the nozzle exit, with a Sauter mean diameter (SMD) of 25μm.

Figure 1 shows results from a typical simulation. The details of the particular test case are: distance from injector nozzle to wall = 24 mm, injection pressure = 13.8 MPa, injection duration = 1.4 ms, nozzle diameter = 0.3 mm, ambient gas pressure = 1.5 MPa, and temperature = 300 K. This simulates one of the experiments of Saito et al (12). The third simulation shown in Figure 1, denoted by 'old', uses the wall impaction model of Wang and Watkins (4), but without the extended collision model. Clearly the latter gives a very poor resolution of the wall spray, which was why the extended collision model was developed. Both the New and New+Col models give a reasonable structure to the wall spray, with the concave region joining the free spray with the main part of the wall spray formed into a head vortex.

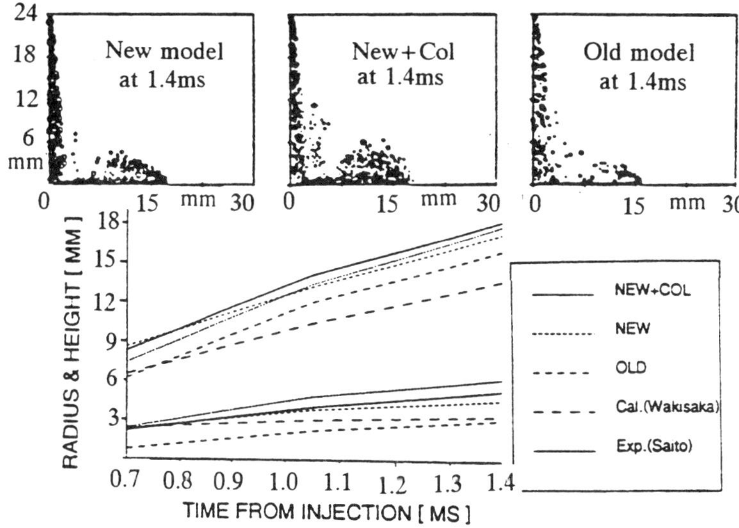

Fig 1 Comparison of spray penetrations

To distinguish between them, we need to compare the wall spray radius (the penetration along the wall) and the wall spray height or thickness, with experimental data. This is also shown in Figure 1. The spray penetration along the wall is predicted closely by both the New and New+Col models. They are clearly superior to the Old model. Also included in this Figure are the results obtained by Wakisaka et al (3), for this test case. The New model provides a much better description of the spray penetration. Similar conclusions can also be drawn with regard to the spray height. But here it is possible to distinguish between the results obtained using the New model and New+Col. The latter spreads the spray too far away from the wall.

The same general conclusions can be drawn from all the normal impaction test cases studied in (8). The angled impaction cases, with impaction angles of up to 30°, and using non-orthogonal grids, also confirm these conclusions. The New model provides much the best comparison of wall spray radius and height with experimental data. In these cases, the use of the extended collision model provides much worse results, with a large overprediction of the spray height and a consequent large underprediction of the spray radius.

The final three test cases detailed in (8) provide a very different test for the wall impaction models. The spray is impacted at high speed on the land of a cut-off pip in the centre of a bomb simulation of a piston bowl. Three test cases were examined experimentally and computationally by Naber et al (2). These varied only in terms of the injection pressure, which lay in the range 83 - 116 MPa. The other parameters were kept constant: wall distance = 6.34 mm, nozzle diameter = 0.406 mm, gas pressure = 3.2 MPa and gas temperature = 760 K.

Figure 2 compares the predicted evolution of the spray penetration after impaction on the land, with the experimental data (injection pressure = 83 MPa). The details of the wall impaction model appear to be less important, presumably because the spray has a shorter time in contact with the wall. Thus the differences between the New, New+Col and Old models are relatively small. What is important here is the way that the kinetic energy in the spray after impaction is calculated. All of these three models calculate it in essentially the same way. Naber et al's model clearly does not. The initial penetration is greatly overpredicted by their model. Later the spray tip velocity is too small, so that the penetration gradually tends towards the experimental value. However, it is only towards the end of spray injection that reasonable agreement is obtained.

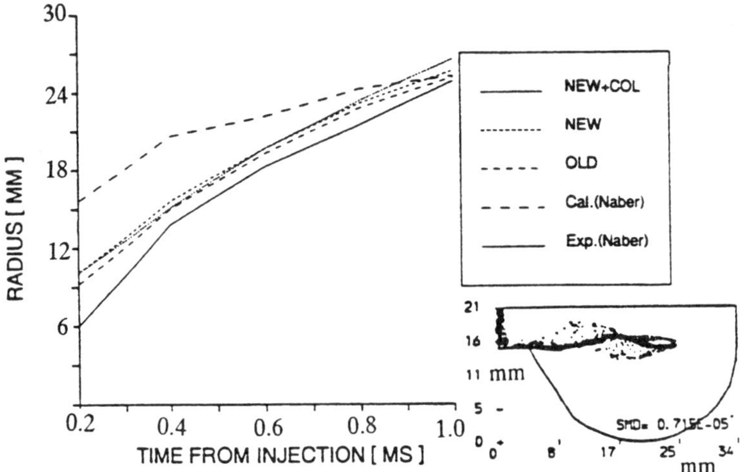

Fig 2 Comparison of spray penetration
(Inset: New model predictions at 1ms)

3.2 Drop velocities and sizes

In the experiments above the techniques used provide an outline image of the spray, usually by photographic means. However, they provides no details of the internal structure of the wall spray, although some information can be gathered by studying the temporal evolution of the sprays. To get a large amount of data concerning the internal structure requires techniques such as those based on lasers. Even then, because of problems of optical access to the main body of the spray, these techniques are only currently useful in relatively dispersed sprays. This means, for example, that the gas pressure must be low to allow the spray to disperse sufficiently after wall impaction.

Arcoumanis and Chang (13) used phase Doppler anemometry (PDA) to study the impaction of diesel sprays onto a heated or unheated flat plate at room conditions. The evolutions with time of the velocities at a number of points in the spray were examined, as were the transverse velocities and Sauter mean diameters of drops passing through a larger number of positions. The unheated plate case has been simulated, using the same wall impaction models as above. A full set of results is given in (8). Here again there is space for only a few of these.

Figure 3 compares with experimental data the predicted evolution in time of the drop velocities at three locations in the spray, using the New impaction model. H denotes the distance from the plate, and r the radius from the centre of the impaction site. Initially the spray moves radially outwards away from the impaction site. However, a head vortex soon develops and the spray first moves outwards away from the wall, then back towards the impaction site, and finally ends up moving almost directly downwards towards the wall. Thus once the head vortex has passed through, these points are in the entrainment part of the spray. Although there are substantial differences in detail between the simulated results and the experiment, both in terms of velocities and time of evolution, the predictions at r = 10 mm, H = 3 mm bear out this evolution. Agreement at the second point, r = 10 mm, H = 5 mm, is almost as good, except that the final condition is not quite right. The flow is being entrained at an angle approximately 30° to the correct line.

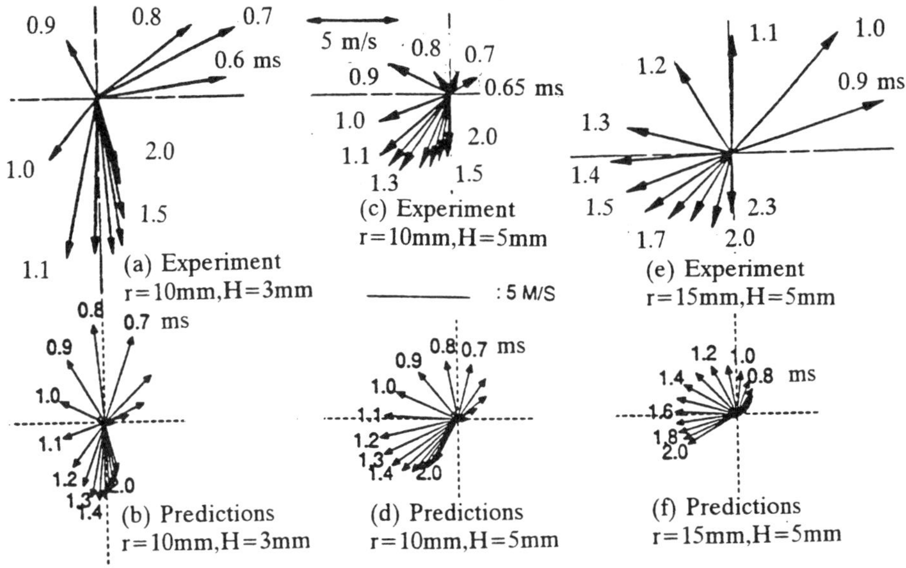

Fig 3 Comparison of spray velocities

The experimental results at the third point, r = 15 mm, H = 5 mm, indicate larger velocities than at the second point. It is not clear why this should be the case, unless they are caused by increased injection velocities. However, these should show up in the other points also. The predictions do not show this increase, rather they indicate a decrease in magnitude. There are substantial differences also in the final 'steady' position of the velocity vector, although there may still be some changes occurring after 2.0 ms, when the simulation was halted.

Some of the underprediction in drop velocities is thought to be because of the modelling of the normal component of drops induced by boiling. This does not match the corresponding velocity component in the breakup regime on cool walls, which is the correct regime for the present test case. Both the data of Mundo et al (14) and Nagaoka et al (15) indicate normal drop velocity components after impaction which are an order of magnitude greater than predicted by the current model for boiling induced breakup.

The predicted Sauter mean diameter of the drops as they pass through one of the above points is compared with experiment in Figure 4. For these calculations up to 4000 drop parcels are used, however the oscillations in the predictions indicate that insufficient resolution of the spray in terms of spray parcels has been obtained. Unfortunately computer constraints did not allow more parcels to be employed. The drop sizes are substantially underpredicted at this point, but appear to follow the correct trend. As shown in (8), the agreement with drop sizes is variable, with some points having an overprediction of drop sizes. However, it is clear that the impact model is producing roughly the correct sizes of drops after impaction.

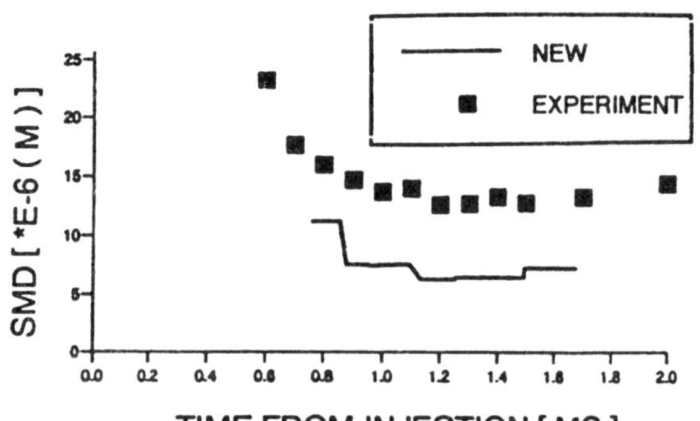

Fig 4 Comparison of Sauter mean diameter at r=10mm,H=3mm

4 ENGINE DESIGN

The computer code, complete with the wall impaction sub-model, has been used (8) and (9) to consider details of diesel engine design which deliberately use wall impaction as a means of breaking up the spray, instead of swirling the flow. One such design, due to Ogura and Lin (16), uses small impaction sites attached to the cylinder head. The detailed shape of the part onto which the spray is impacted is crucial to the subsequent engine performance, as demonstrated in (16). We have simulated one of these shapes, to see whether our model is capable of reproducing the correct spray shapes after impaction. Figure 5 shows a comparison between the outline shapes of experimental and predicted sprays. Reference (16) gives no temporal evolution for the spray, so here we have merely chosen a time at which our predictions best match the photographs. However, the results indicate that the new impaction model does faithfully reproduce the effects caused by the particular shape (shown inset on Figure 5) chosen for the impaction site.

Finally, we have further investigated (9) a design first suggested by the authors in (17). This provides for a number of cut-off pips in the piston bowl. One of these is in the centre, whilst others are placed symmetrically around the bowl periphery. Figure 6 shows predicted results of spray and fuel vapour mass fraction in a plane through one of the sprays injected onto a peripheral pip, and the spray injected onto the centre pip. The sprays combine so that the fuel vapour fills a large fraction of the available space, without requiring the ambient air to be swirled. Stoichiometric conditions have almost been reached in substantial volumes. All of the spray is directed away from the chamber walls, so that it would be unlikely that unburnt hydrocarbons would exist after combustion in this system.

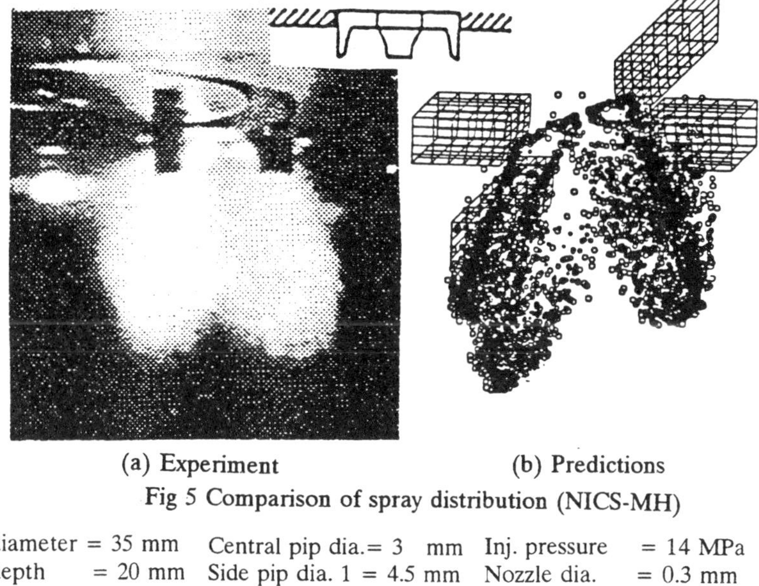

(a) Experiment (b) Predictions

Fig 5 Comparison of spray distribution (NICS-MH)

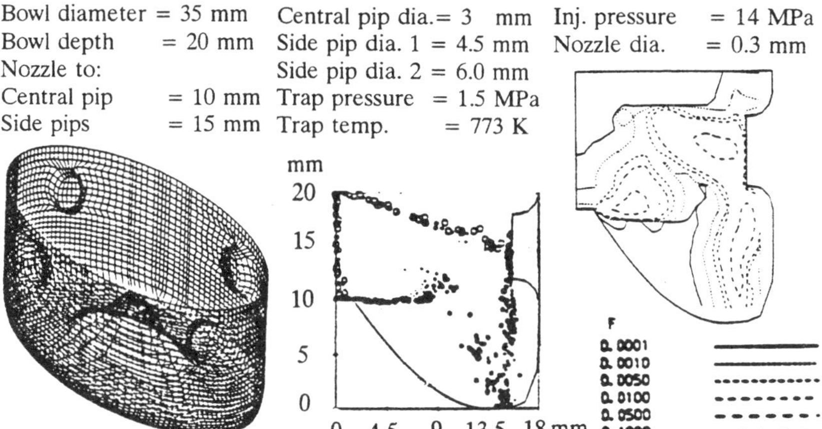

Fig 6 Spray and fuel vapour mass fraction distributions at 1 ms

5 CONCLUSIONS

For both normally and angled impacting sprays the new model gives predictions of wall spray shapes and penetration rates which are superior to earlier wall impaction models. It is able to handle very well the impaction of very high velocity sprays, despite the fact that these test cases are for cool walls and the model is built on hot wall data.

Comparisons with PDA data on the internal structure of a wall spray indicate that many of the flow features are predicted in a reasonable manner. These include the temporal evolution of the spray at discrete points, and the Sauter mean diameter of the drops. Errors are thought to be significant here because of the underprediction of normal velocity components of the broken up drops.

The code has been used to investigate design details of engines employing spray impaction. Preliminary indications are that the code, with the new wall impaction sub-model, does provide a means by which such engine designs could be assessed.

REFERENCES

(1) NABER, J.D. and REITZ, R.D. Modelling engine spray/wall impingement, SAE 880107, 1988.

(2) NABER, J.D., ENRIGHT, B. and FARRELL, P. Fuel impingement in a direct injection diesel engine, SAE 881316, 1988.

(3) WAKISAKA, T., YOSHIDA, S. K., ISSIKHI, Y. and SHIMAMOTO, Y. A study on spray models for numerically analysing fuel spray behaviour, Proc. 11th Symp. on Internal Combustion Engines, 1993, 241-246, JSME/JSAE, (in Japanese).

(4) WANG, D.M. and WATKINS, A.P. Numerical modelling of diesel wall spray phenomena, Int. J. of Heat and Fluid Flow, 1993, 14, 301-312.

(5) O'ROURKE, P.J. and BRACCO, F.V. Modelling of drop interactions in thick sprays and a comparison with experiment, Stratified Charge Automotive Engine Conf., London, 1980, IMechE, 101-116.

(6) BAI, C. and GOSMAN, A.D. Development of methodology for spray impingement situation, SAE 950283, 1995.

(7) WACHTERS, L.H.J. and WESTERLING, N.A.J. The heat transfer from a hot wall to impinging water drops in the spheroidal state, Chemical Engineering Science, 1966, 21, 1047-1056.

(8) PARK, K. Development of a non-orthogonal-grid computer code for the optimization of direct-injection diesel engine combustion chamber shapes. Ph.D. Thesis, University of Manchester Institute of Science and Technology, 1995.

(9) PARK, K. and WATKINS, A.P. An investigation of combustion chamber shapes for small automotive direct injection diesel engines employing spray impaction, Accepted by Proc. I. Mech. E., J. of Automobile Engg., 1995.

(10) NABER, J.D. and FARRELL, P. Hydrodynamics of droplet impingement on a heated surface, SAE 930919, 1993.

(11) YULE A.J. and WATKINS, A.P. Measurement and modelling of diesel sprays, Atomization and Sprays, 1991, 1, 441-465.

(12) SAITO, M., HASHIMOTO, A., NAGAE, M., SENDA, J. and FUJIMOTO, H. Characteristics of diesel spray impinging on a flat wall, Trans. JSME B(Japanese), 1991, 57, 3973-3980.

(13) ARCOUMANIS, C. and CHANG, J.C. Flow and heat transfer characteristics of impinging transient diesel sprays, SAE 940678, 1994.

(14) MUNDO, C., SOMMERFELD, M. and TROPEA, C. Droplet-wall collisions: deformation and breakup process on a flat surface, Int. J. Multiphase Flow, 1995, 21, 151-173.

(15) NAGAOKA, M., KAWAZOE, H. and NOMURA, N. Modelling fuel spray impingement on a hot wall for gasoline engines, SAE 940525, 1994.

(16) OGURA, M. and LIN, B. A new multi-impingement-wall head diffusion combustion system (NICS-MH) of a d.i. diesel engine, SAE 940196, 1994.

(17) PARK, K., WANG, D.M. and WATKINS, A.P. A contribution to the design of a novel direct injection diesel engine combustion system - analysis of pip size, Appl. Math. Modelling, 1993, 17, 114-124.

C499/050/96

Experimentally validated multi-dimensional simulation of mixture formation and combustion in gasoline engines

R TATSCHL, H FUCHS, and **W BRANSTÄTTER**
AVL List GmbH, Austria

SYNOPSIS

A computational fluid dynamics and combustion analysis code has been used to investigate aspects of SI engine mixture formation and combustion. Studies have been performed for a two-valve cylinder head configuration with a disc-shaped combustion chamber and for a production line four-valve pent-roof head under varying engine operating conditions. Wherever available, numerical results were compared with experimental data, in order to assess the predictive capability of the adopted computational method.

The flow pattern and fuel/air mixture composition fields at ignition time, especially in the four-valve engine, are found to be strongly dependent on the engine operating conditions. The results also demonstrate that the complex three-dimensional flow structure present at the time of ignition onset persists throughout a significant portion of the combustion process and exerts enduring influence on the flame propagation characteristics. The comparison of the calculated results with the experimental data shows good agreement with respect to mixture composition patterns, flame shape and propagation characteristics, as well as concerning integral quantities, such as combustion delay and burn duration for the engine configurations and operating conditions considered.

1 INTRODUCTION

Modern engine development and design is mainly focused on reduced fuel consumption and pollutant emissions, simultaneously aiming at shorter development times and a decrease in the number of prototypes required. Fuel consumption and pollutant emissions of SI engines are intimately related to the combustion process which is governed by the flow and the mixture composition patterns developed during intake and compression. Hence, there is considerable interest to gain a detailed insight into the governing processes and to get a deeper knowledge of the influence of engine design and operation parameters on the flow field and on the mixture composition distribution at the time of spark onset, and thus on flame propagation.

Today's CAE tools used for the design of SI engine combustion systems provide extraordinary insight into the details of port and in-cylinder air flow, fuel/air mixing and combustion. The need for a fast combustion system development, however, puts the focus onto items which can only be dealt with by means of numerical air flow and combustion analysis, such as the three-dimensional space and time resolved simulation of the impact of mixture composition, mean flow convection and turbulence intensity on early flame kernel development right after spark ignition, or the flow/flame interaction during the main combustion period. In this connection numerical methods used to simulate the flow, mixture formation and combustion processes in engines must meet the requests for validated precision and reliability.

It is the aim of this paper to describe the application of a thermofluids analysis code to the investigation of SI engine mixture formation and combustion and to validate the results by comparison with experimental data. The flow, mixture formation and combustion properties were studied numerically and experimentally in a disc-shaped combustion chamber of a single-cylinder research engine and in a modern production line four-valve pent-roof head under varying engine operation parameters.

2 ANALYSIS OF FLOW, MIXTURE FORMATION AND COMBUSTION

2.1 Modelling methodology

The in-cylinder flow and charge mixture composition evolution were obtained through simulation of the complete gas exchange and compression strokes. An unsteady gas dynamic and cycle calculation method was used to provide the initial and boundary conditions for the multidimensional simulation under the operating conditions of interest. This procedure enabled incorporation of all data pertinent to the operating engine, including the pressure wave oscillations in the intake system.

The CFD method applied in the present work is the thermofluids analysis software package FIRE. A number of recent publications have described the FIRE code system and its application to engine related flow, mixture formation and combustion simulation (1-6). Hence, only a brief description of the mathematical framework and the numerical solution method is provided.

Mathematical framework

FIRE solves the density-weighted, ensemble-averaged conservation equations of mass, momentum and enthalpy, in addition to the transport equations of the k-ε turbulence model. In the case of chemically reacting flows, additional transport equations are solved in order to capture the mixture formation process during intake and compression and the change in mixture composition due to chemical reaction during the combustion process. In the general case of inhomogeneous, partially-mixed combusting flows, additional differential transport equations for a mixture fraction 'f', an optional residual gas mass fraction 'g' and a reaction progress variable 'c' are solved, in order to determine the charge composition, thermodynamic state and heat release rate. The conservative scalar variable 'f' is equivalent to the total (burned and unburned) fuel mass fraction and thus a measure of the mixing between the air and the fuel, irrespective of the combustion process. The reaction progress variable 'c' is equivalent to the reaction product mass fraction that occurs such that either all the fuel or all the oxidant is depleted (or both for stoichiometric mixtures). The reaction progress variable 'c' is bounded by the values of zero and unity corresponding to the unburned and burned states, respectively, regardless of the equivalence ratio.

Combustion model

The hydrocarbon oxidation during engine combustion is expressed by a one-step irreversible combustion reaction in which fuel and oxygen are converted into the combustion products carbon-dioxide and water. The determination of the mean reaction rate which is governed by turbulent mixing and/or chemical kinetic effects, depending on the combustion regime, is the primary objective of combustion models for turbulent reactive flows (7). In the present study a combustion model based upon solving a full transport equation for the probability density function distribution (PDF) of the reaction progress variable 'c' is used (3-6). The solution of the PDF transport equation provides a complete statistical description of the scalar quantity 'c', thus fully accounting for the high non-linearities in the chemical rate expressions. The ability of the PDF combustion model to simultaneously account for both finite rate chemistry effects and the impact of turbulence on flame propagation obviates the need for any prior assumptions as to whether one or the other of the two processes governs the mean rate of reaction.

Solution method

The partial differential transport equations governing mean fluid motion and charge mixing are recast into the curvilinear non-orthogonal form and transformed to a contracting / expanding coordinate system, so to enable their solution on body-fitted computational grids with moving boundaries. The differential transport / conservation equations are discretised adopting the finite volume approach. The temporal discretisation is Euler implicit, in order to ensure unconditional numerical stability, for approximation of the spatial derivatives a hybrid central / upwind or optional higher order differencing schemes are used (8). This practices lead to coupled algebraic equation systems solved iteratively on a SIMPLE based pressure-velocity coupling procedure. The PDF transport equation is solved adopting a Monte Carlo simulation approach (9) in which the continuous probability density function p(c) is represented by an ensemble of notional elements located at the cell centers of the finite volume computational grid. In order to advance p(c) in time, the notional elements are transported across physical and composition space according to the processes of convection, diffusion, chemical reaction and mixing, which are simulated sequentially by means of an operator splitting technique.

2.2 Experimental investigations

Mixture formation and combustion properties were analysed in the optically accessed combustion chamber of a single cylinder research engine. A quartz-glass piston-liner arrangement provided optical access to the combustion chamber. For the two-valve head with an excentrically mounted spark plug optical access was given for a height of about 28 mm below the cylinder head surface. For both the two-valve and the four-valve head the central part of the piston covering some 60 % of the cylinder bore diameter was accessed through the quartz-glass piston window.

The fuel delivery during the intake stroke was visualised by planar Laser-Induced-Fluorescence from aromatic fuel components (10, 11). A KrF excimer laser (λ = 248 nm, pulse duration 15 ns) delivered the excitation flash, triggered by a signal from the crank-shaft encoder. Two perpendicular cylindrical lenses formed the light sheet with a thickness less than 1 mm. A gated and intensified CCD camera perpendicular to this plane gathered the fluorescent intensity through a 105 mm photographic lens. An edge filter at 290 nm in front of the lens took off the laser line and scattered light. The same detection system as for the LIF experiments was used for the flame studies. In this case the trigger unit was directly connected to the control station of the CCD camera. All images were statistically averaged in order to minimise effects from cycle-to cycle fluctuations.

The visualisation of flame propagation, the measurement of the gas velocities and the conventional pressure measurements provided the data for the analysis of turbulent combustion under the influence of engine speed, load and air-fuel ratio. Local, cycle resolved gas velocities were obtained by means of Laser-Doppler-Anemometer (LDA) measurements, flame front velocities were derived from paired flame photographs taken at a certain time interval within a single engine cycle. Air-fuel ratio was measured with a oxygen sensor in the exhaust pipe, cylinder gas pressure was measured with a quartz pressure transducer.

3 RESULTS

3.1 Cylinder filling and mixture formation

When fuel is injected long before inlet valve opening, considerable if not all the fuel sprayed onto the port walls and the back surface of the valve can evaporate prior to valve opening. This fuel stored upstream of the valve enters the cylinder with the accellerating gas flow after inlet valve opening. Fig. 1.1 shows the standard situation of filling the cylinder of a four valve production line engine with fuel which has evaporated in the inlet port and enters the cylinder during the early inlet phase. The vertical cross section shows the inlet jet flow governed by the geometry of the cylinder/port assembly. Part of this jet can propagate into the cylinder volume, whereas other parts are deflected by the cylinder wall and the piston surface. For the actual cross section, this results in accumulation of fuel below the inlet valve. With the ongoing inlet stroke an increasing amount of fresh air follows the earlier mixture of vapour and air. The mixing of fuel vapour, which has entered the cylinder just after inlet valve opening, with the continuing inflow of air further contributes to the process of charge homogenisation. Results in a horizontal cross sectional plane 5 mm below the cylinder head and a comparison with the measurements are given in Fig. 1.2.

3.2 Flame propagation

The inlet port of the two-valve research cylinder head is designed to generate minimum convective air motion in the disc-shaped combustion chamber. Hence, under non-firing conditions the gas velocities were found to be in the range of +/- 1.5 m/s at an engine speed of 1000 r/min throughout a large period of the compression and expansion stroke. In the firing mode, however, strong convective flow motion is induced by the volumetric expansion of the hot combustion products. As the combustion wave penetrates into the unburned mixture it acts as a propagating source of expanding gas. The observation of this phenomenon shows the accelleration of the cold gas ahead of the flame as the flame approaches and the deceleration and flow reversal as it passes over a fixed monitoring point in the combustion chamber. A comparison of measured and predicted data for an engine speed of 1000 r/min is given in Fig. 2, together with numerical results of the flame propagation process. Here the entire flame shows high symmetry in the radial direction with respect to the spark plug. In the axial direction, however, the propagation process is retarded, as can be seen in Fig. 3, which presents a sequence of flame photographs together with the calculated results. This retarded axial propagation appears right after ignition as the flame has to accelerate against the flow field induced by the upward moving piston.

On average, flame front velocities were found to be some 50 % above the peak gas velocities induced by the expansion of the hot combustion products. Their variation with air-excess ratio and further data on the influence of engine load (i.e. gas density) and engine speed (i.e. turbulence intensity levels) on flame front velocities are summarized in Fig. 4.1, 4.2 and 4.3, together with results from the experiments.

3.3 Interaction of flow, mixture composition and combustion

For a given fuel delivery strategy, the mixture formation and charge homogenisation process during intake and compression is governed by the interaction of the gas dynamics of the intake system and the combustion chamber / port arrangement that determine the characteristics of the intake generated flow and turbulence pattern. Since the turbulence time scales, and thus the intensity of turbulent mixing, are linearly related to engine speed the degree of charge homogenisation at the time of spark ignition does not significantly change with engine operation parameters, as shown in Fig. 5. The distribution of the remaining charge-inhomogeneity, however, depends on the evolution characteristics of the complex three-dimensional flow pattern. As can be seen in Fig. 5 the mixture composition distribution at the time of spark ignition significantly differs for the two engine operation points under consideration, clearly reflecting the impact of the flow field which is governed by the remnant of the induction process.

The turbulent flame speed usually found in engines is of the order of 10 - 30 m/s, depending on engine operation parameters. In a convective flow field the flame speed is superimposed upon the net gas velocities. This leads to a flame propagation which is determined by the turbulent flame front propagation and the directed convective charge motion. In multivalve engines port design aiming at increased tumble-air motion to obtain increased turbulence levels and thus faster burn rates must consider its effect on flame convection. This is demonstrated by examples given in Fig. 6, showing the influence of engine operation parameters, and hence of the mean flow field, on the evolution characteristics of the flame front.

3.4 Integral quantities

In addition to space- and time-resolved data providing detailed insight into the complex interaction processes of the flame front with the remnant of the intake generated mixture composition, flow and turbulence fields that determine flame propagation in SI engines, the numerical simulation also provides information on cylinder-averaged global data. Besides cylinder pressure and temperature traces, the mass-weighted cylinder-mean progress variable (i.e. normalised cumulative heat release), as shown in Fig. 7.1, may be extracted, in order to assess the characteristics of the combustion process. Fig. 7.2 shows calculated combustion delay and burn duration data for different four-valve engine configurations and comparison with measurements.

Analysis of integral combustion quantities may also be used for engine operation (ignition timing) optimisation with respect to combustion efficiency and thus minimised fuel consumption. Fig. 7.3 shows the variation with spark timing of the time delay between the appearance of peak values of pressure and temperature, which represents a measure of the matching of the burn characteristics of the engine and the ideal constant volume combustion process. Hence, the ignition timing that provides the shortest delay between pressure and temperature peaks leads to the combustion process with the best thermodynamic efficiency and consequently minimised fuel consumption.

4 SUMMARY

CFD modelling was used to investigate the flow, mixture formation and combustion process in two different SI engine configurations under varying operating conditions. Wherever available, computed results were compared with experimental data.

The comparison of the calculated results with the experimentally obtained cycle-averaged data shows good agreement with respect to flame shape and propagation characteristics. Cumulative heat release and flame propagation speed reveal to be in good correlation with the measurements over a range of operating conditions.

For the four-valve engine the study revealed several notable aspects of the intake generated flow pattern, mixture composition distribution and flame propagation:

- The mixture composition field and the flow pattern at ignition time were found to be strongly dependent on the engine operation condition.

- The degree of mixture inhomogeneity at the time of spark onset (for a given fuel delivery strategy) is not sensitive to the engine operation parameters.

- The flow structure present at ignition time persists throughout a significant portion of the combustion process and strongly interacts with the turbulent flame front exerting enduring influence on its propagation characteristics.

ACKNOWLEDGEMENTS

Parts of this work have been funded by the Styrian Government, the Austrian Research Foundation (FFF) and by the German Forschungsvereinigung Verbrennungskraftmaschinen (FVV). Their financial support is gratefully acknowledged.

REFERENCES

(1) BACHLER, G., BRANDSTAETTER, W., STEFFAN, H., WIESER, K., "Three-Dimensional Simulation of Flow, Mixture Formation and Combustion", Conference: The Working Process of the Internal Combustion Engine, University of Technology Graz, 1989.

(2) TATSCHL, R., BRANDSTAETTER, W., "Multidimensional Calculation of Spark Flame Initiation by Adopting a Generic Hydrocarbon Kinetic Scheme", In: Computational Methods in Applied Sciences, pp. 283-293, Elsevier Science Publishers, 1992.

(3) TATSCHL, R., BRANDSTAETTER, W., "The Application of a Hybrid Finite-Volume / Monte Carlo PDF Method to Engine Combustion Simulation", NATO Advanced Study Institute on Unsteady Combustion, Instituto Superior Tecnico, Lisboa, 1993.

(4) TATSCHL, R., "Multidimensional Calculation of Engine Combustion Using a Monte-Carlo PDF Method", Fifth Int. Conf. on Numerical Combustion, Garmisch-Partenkirchen, September 29 - October 1, 1993,

(5) TATSCHL, R., WIESER, K., REITBAUER, R., "Multidimensional Simulation of Flow Evolution, Mixture Preparation and Combustion in a 4-Valve Gasoline Engine", COMODIA 94, July 11 - 14, 1994, Yokohama.

(6) CARTELLIERI, W., CHMELA, F., KAPUS, P., TATSCHL, R., "Mechanisms Leading to Stable and Efficient Combustion in Lean Burn Gas Engines", COMODIA 94, July 11 - 14, 1994, Yokohama.

(7) BORGHI, R., "Turbulent Combustion Modelling", Prog. Energy Combust. Sci., Vol. 14, pp. 245-292, 1988.

(8) SCHIFFERMÜLLER, H., "Higher Order Differencing Schemes and their Application to Multi-Dimensional Flow Problems", PhD Thesis, Technical University Graz, 1993.

(9) POPE, S.B., "A Monte Carlo Method for the PDF Equations of Turbulent Flow", MIT-EL 80-012, 1980.

(10) WINKLHOFER, E., PHILIPP, G., FRAIDL, G., FUCHS, H., "Fuel and Flame Imaging in SI Engines", SAE 930871, 1993.

(11) WINKLHOFER, E., FUCHS, H., FRAIDL, G.,"Optical Research Engines - Tools in Gasoline Engine Development?" IMechE Seminar on "Measurement and Observation Analysis of Combustion in Engines", 1994.

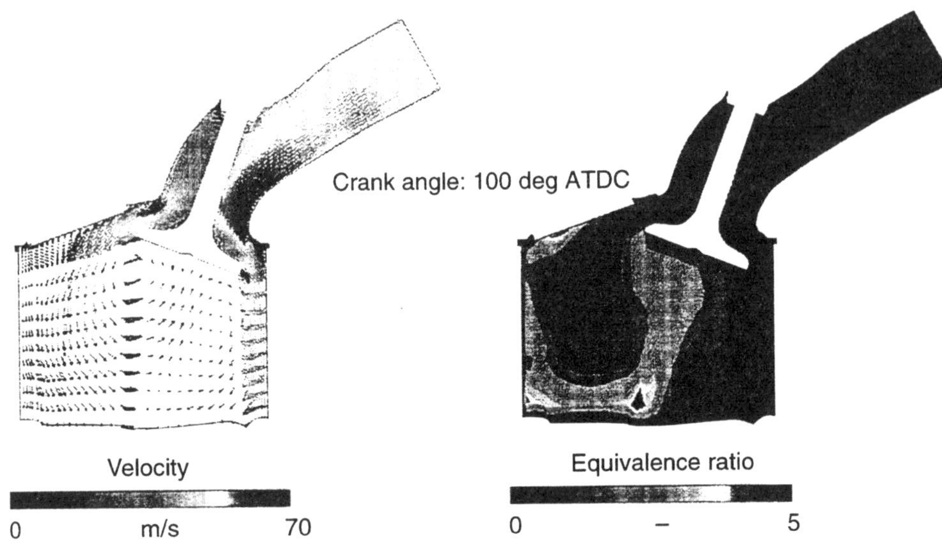

Fig. 1.1 Calculated velocity pattern and mixture composition during intake

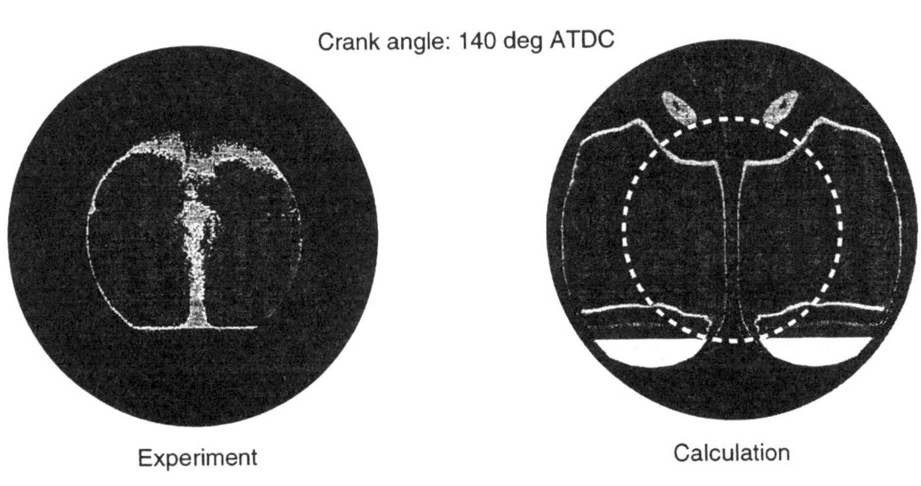

Fig. 1.2 Measured and calculated mixture composition pattern 5 mm below cylinder head

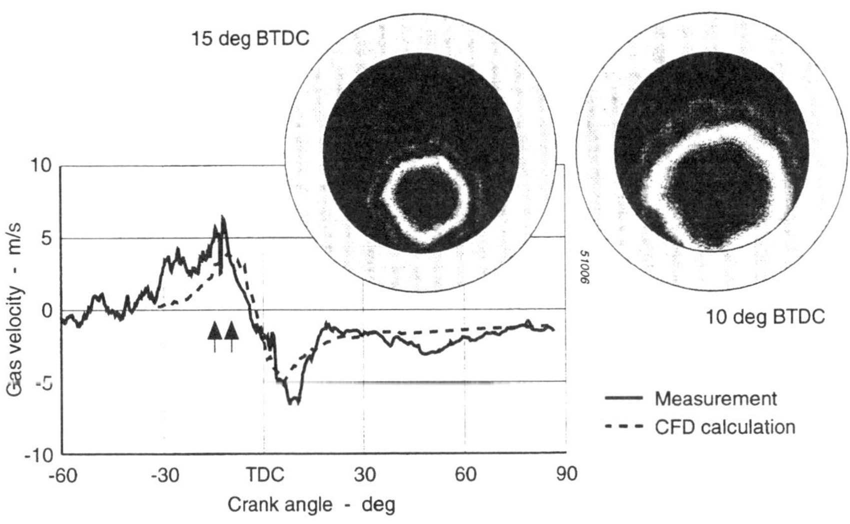

Fig. 2 Local gas velocity under the influence of flame propagation

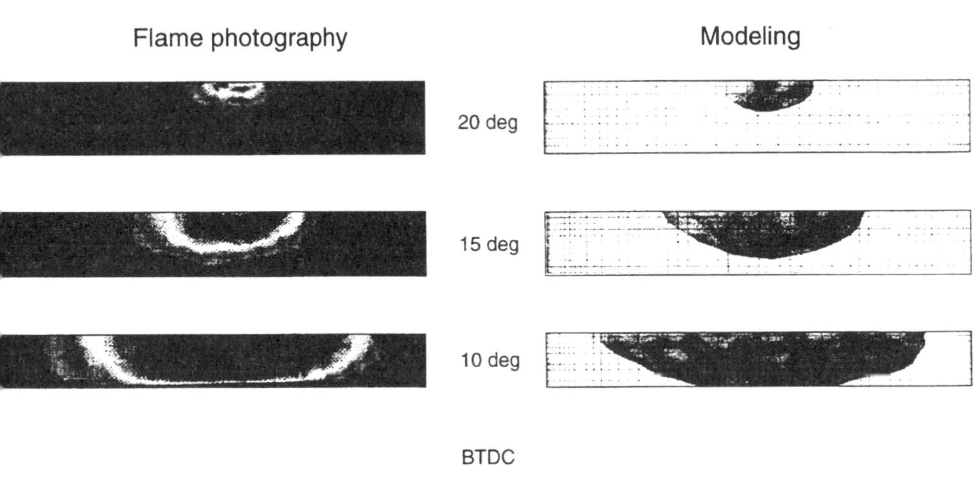

Fig. 3 Flame propagation: Cycle-averaged photographs and CFD results

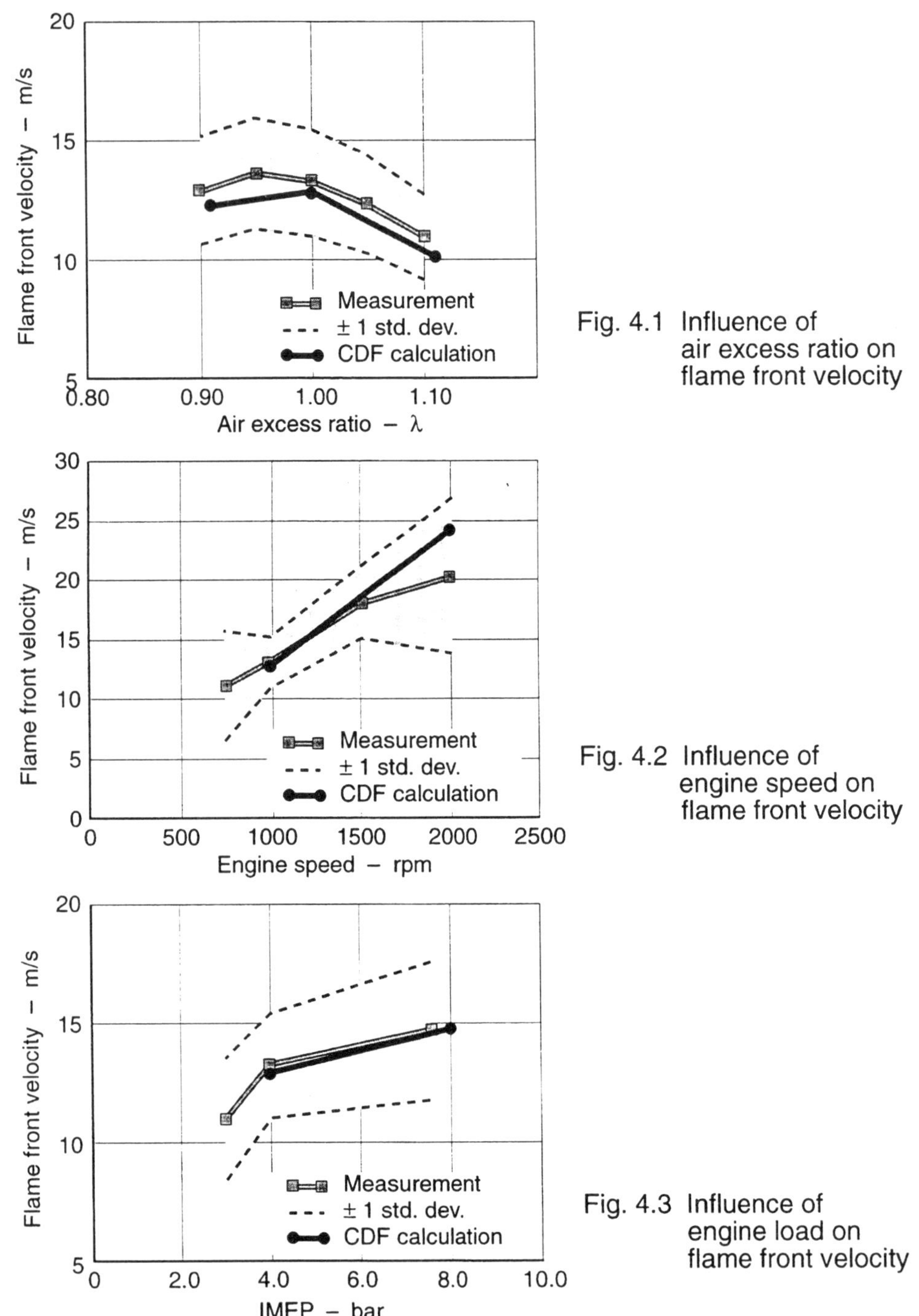

Fig. 4.1 Influence of air excess ratio on flame front velocity

Fig. 4.2 Influence of engine speed on flame front velocity

Fig. 4.3 Influence of engine load on flame front velocity

Fig. 4 Flame front velocity; influence of engine operation parameters

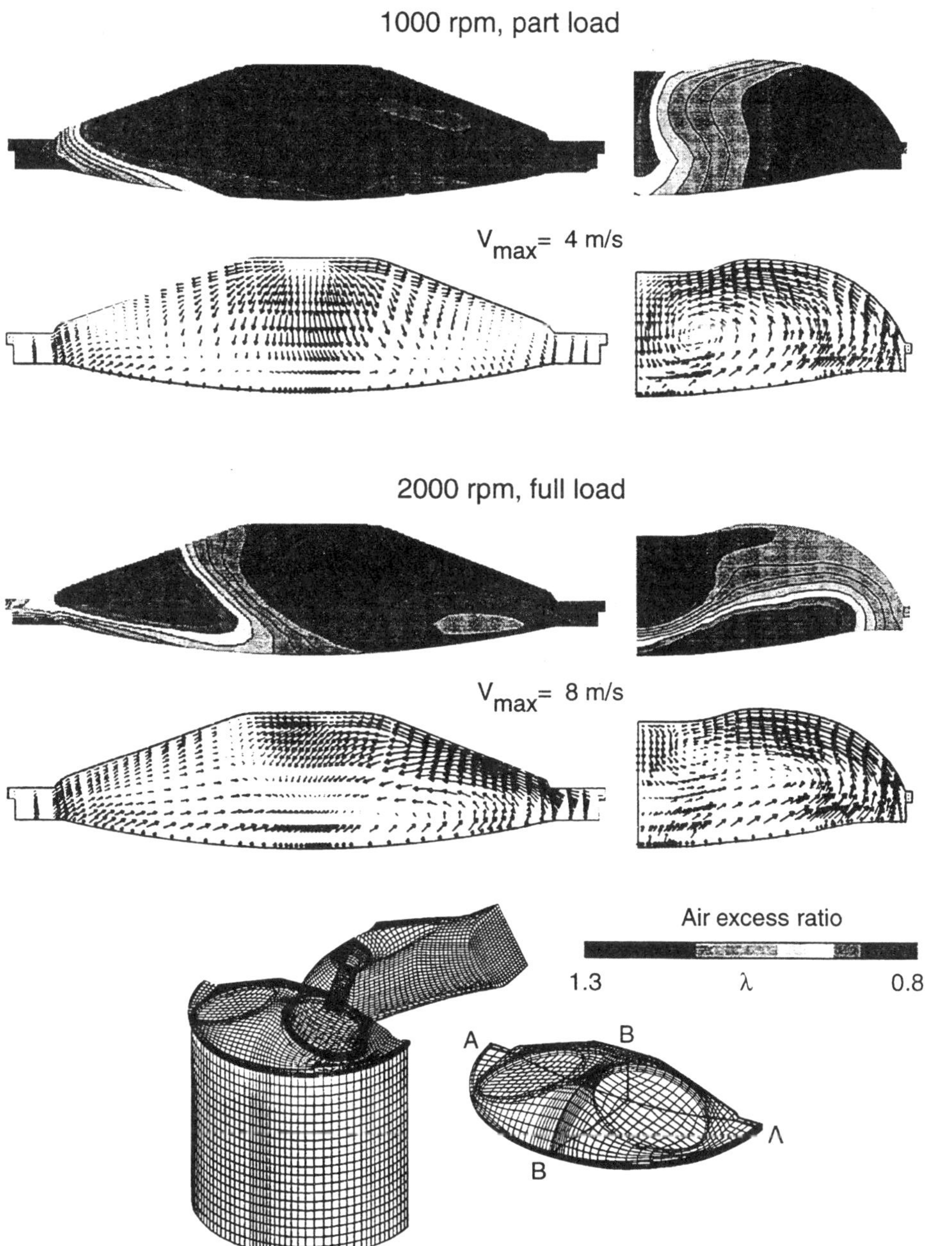

Fig. 5 Velocity pattern and mixture composition distribution at spark timing; influence of engine operation parameters

1000 rpm, part load

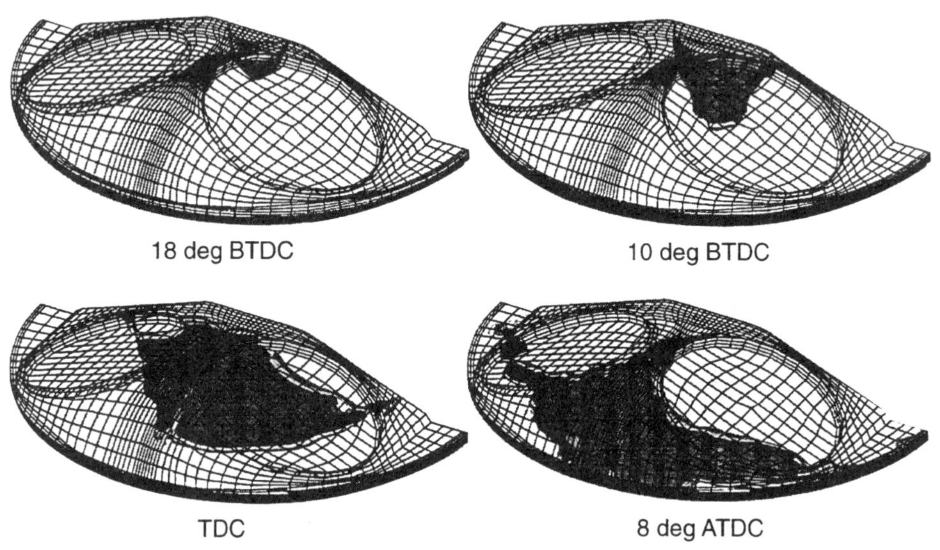

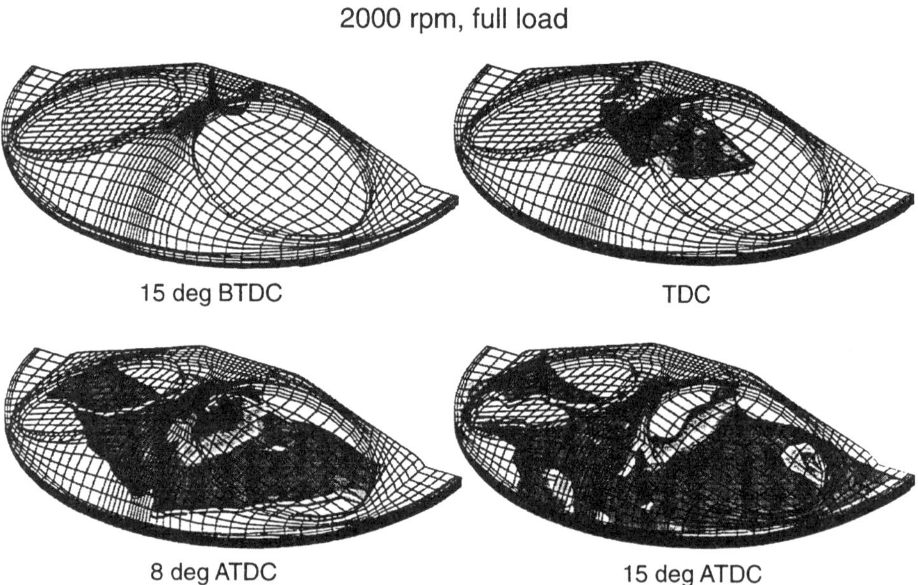

Fig. 6 Flame front evolution in a 4-valve engine; influence of engine operation parameters

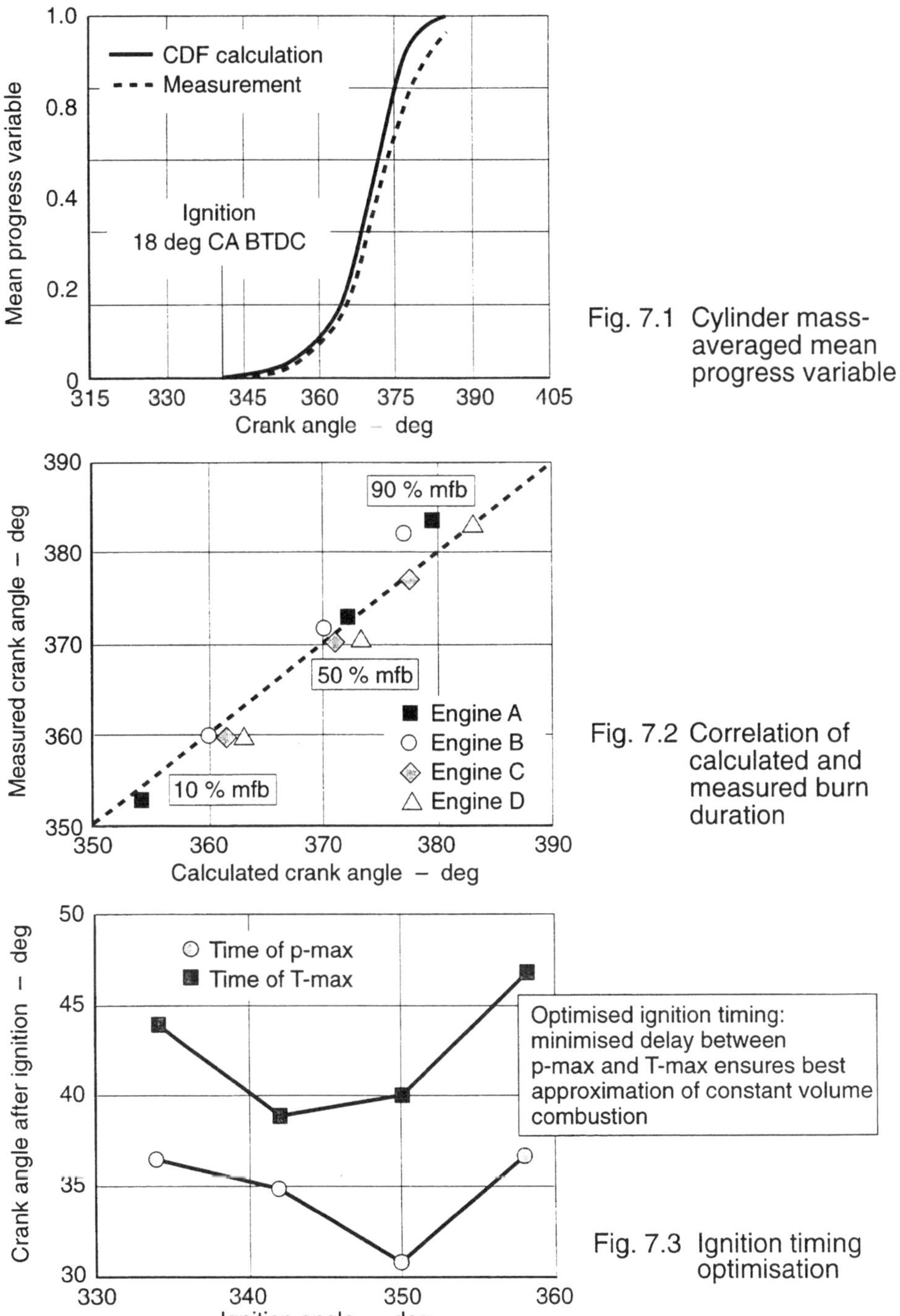

Fig. 7.1 Cylinder mass-averaged mean progress variable

Fig. 7.2 Correlation of calculated and measured burn duration

Fig. 7.3 Ignition timing optimisation

Fig. 7 Integral values; combustion process assessment and ignition timing optimisation

C499/053/96

Identification of flow phenomena in fuel injection systems for diesel engines from dynamic measurements

H OFNER and **W EGARTNER**
AVL List GmbH, Austria

SYNOPSIS

A method is introduced to identify flow resistance due to friction, cavitating flow and fluid properties in fuel injection systems. For this procedure dynamic measurements are used and mathematical models are required for simulation of the physical phenomena. A strategy is described how to obtain optimum access to the different phenomena and how the mathematical part of the identification can be solved. The practical application of the method is demonstrated by an example. Success was achieved for identification of flow losses and fluid properties. The attempt to identify a cavitating state in injection nozzles failed because of simplifications in the physical models.

NOTATION

a	speed of sound (m/s)
K.....	cavitation factor defined by eg. (5), (6)
E	compressibility of liquid (N/m^2)
f....	cross sectional flow area of line (m^2)
g....	general function for time dependant measurements or calculated results
M ..	equation (11); theoretically infinity, reasonable results are achieved with $M \approx 20$
p	pressure (N/m^2)
p_v ...	vapour pressure (N/m^2)
R.....	friction force related to unit mass, equ. (9) (N/kg)
r	radius of the circular tube (m)
s	specific entropy (J/kg/K)
ssq....	"sum of squares", equ. (13)
T	temperature (K)
t.....	time (sec)
w ...	fluid flow velocity (m/s)
w_o ...	fluid flow velocity at narrowest cross section of orifice (m/s)
x.....	spatial coordiante (m)

ρ ... density of liquid phase (kg/m³)

ν .... kinematic viscosity (m²/s)

ζ .... friction factor

$\zeta_{2\text{-}ph.}$ friction factor which characterizes the particular state where a gas-liquid zone fills up the whole cross sectional flow area of the orifice's contraction.

ω_n... roots of Bessel function of order zero; equation (11)

1 INTRODUCTION

Computational methods are commonly used for layout and analysis of fuel injection systems but also gain importance as analytical tools in support of engine tests. By such methods physical effects can be shown up which by experimental techniques could hardly be assessed. This leads to a deeper understanding of a system's behaviour and as a consequence e.g. instabilities can be revealed or even hardware failures can be detected.

System instabilities and damage are often caused by cavitation. On top of this, cavitation in fuel injection nozzles of DI diesel engines contributes to the fuel spray break up and thus even influences the combustion process. Therefore one central motivation to start the research activities as described lay in the need to identify cavitation. This aim has not been reached yet, however a feasible method is clearly demonstrated. Success was achieved for identification of flow losses and fluid properties. The method has already been used within the field of injection systems for alternative fuels /5/.

Within this paper "identification" means a method for determination of appropriate mathematical models for simulation of fluid flow and engine parts motions. The procedure starts with the application of measured data as boundary conditions to a system of equations which unambigously define the fuel flow and the parts' motions in the system. In a second step the calculated results are compared with some "observations", which are additonal measurements. The identification procedure achieves agreement between calculated results and observations by variation of the model parameters and extending the physical models.

One of the most severe problems thereby is to distinguish between particular phenomena which in real systems usually overlap. Consequently one effect may be overemphasized to the disadvantage of some other, which, depending on the kind and quality of the applied measurements, often leads to ambigous results.

Driven by the need for a comprehensive identification methode, AVL in cooperation with the Institute of Mathematics, Technical University of Graz, inititated activities aiming at mathematical methods of parameter identification to fuel injection systems /1/. The early stages of this project revealed the necessity of improving the physics of the models. This covered the frictional losses in unsteady flows, cavitation in orifices and the thermodynamic state properties of fluids. Furthermore, strategies to achieve access to the specific phenomena at different locations in an injection system had to be developed.

In this paper a general strategy to approach the problem is introduced. Subsequently it will be demonstrated by an example how different flow phenomena can unambigously be extracted from measurements. The pyhsical flow models, which are of central importance throughout this paper, will be explained and their limits of validity will be shown by means of the example.

2 FUNDAMENTALS FOR THE IDENTIFICATION OF FLOW PHENOMENA

2.1 Physical models for fuel flow

2.1.1 Basic assumptions

The major job of an identification procedure as described in the introduction is carried out within the mathematical domain of the system equations. From the mathematical point of view the parameter identification represents an optimization problem for achieving optimum agreement between calculated and measured results. Consequentlay the physical models must unambigously mirror the physical phenomena which are identified and therefore represent the keypoint of the whole procedure.

Modelling of the fluid flow in hydraulic systems, like fuel injection systems for diesel engines, has been the subject of various publications during recent decades /e.g. 16, 17, 18/. These references are given as an example only, they refer to several other sources in their parts but still give no complete list. It has been the demand of this paper to start out with the fundamentals as published in this literature. Based on these the models of elementary parts of hydraulic systems such as orifices, valves and pipes are introduced in a general form and emphasis was layed on pointing out the assumptions which are usually made for reasons of simplifying and accelerating their numerical solution. Flow phenomena like frictional losses and cavitation are discussed in more detail.

2.1.2 Frictional losses and cavitation in orifices and valves

Flow losses due to friction

With fuel injection systems for reciprocating engines considerable attention is usually directed to dynamic effects, in spite of that commonly used flow simulation methods for orifices are identical to those for steady state flow hydraulics. Basically, if dealing with one-dimensional flow models, the Bernoulli equation is applied neglecting the term which considers the inertia effects. This assumption is certainly valid if, by definition, these components expand over short distances (a quantitative evaluation of this simplification was made in /3/ for a diesel injection nozzle).

The energy losses caused by friction are expressed in terms of a factor ζ which is related to a specific kinetic energy. If differences in geodesic heights are also excluded (this is part of the demand for short distances!) the Bernoulli equation yields:

$$\left(\frac{p}{\rho}+\frac{w^2}{2}\right)_{in}=\left(\frac{p}{\rho}+\frac{w^2}{2}\right)_{out}+\zeta\cdot\frac{w_o^{\;2}}{2} \tag{1}$$

The friction factor ζ can be related to any velocity w_o within the orifice but in most cases the one at the narrowest cross section is useful. For practically all applications of orifices and valves in diesel injection systems, the specific kinetic energies $w^2/2$ on the input and output side are small compared to the specific energies due to pressure p/ρ and thus can be neglected. Furthermore, if an average value for the density ρ is introduced and considered constant, equation (1) reduces to

$$p_{in}-p_{out}=\zeta\cdot\rho\cdot\frac{w_0^{\;2}}{2} \tag{2}$$

This formula is also commonly used for determining of ζ from steady state flow measurements.

However, one has to be aware that the condition

$$\zeta = const. \tag{3}$$

is valid only if the flow is turbulent and a single phase fluid exists in the orifice (e.g. no cavitation occurring). Otherwise ζ becomes a function of the flow velocity and fluid viscosity for a laminar and transient flow, or it depends on both the actual static pressure and the vapour pressure of a liquid if cavitation occurs.

Cavitating flow

The local fluid velocities in orifices can achieve very high values, e.g. at the flow around sharp edges. This sometimes causes the static pressure to drop even below the vapour pressure which leads to the formation of gas bubbles. Experiments of a visualized steady state flow of diesel fuel in an orifice have shown that according to the high turbulence in an orifice's contraction the gas bubbles intensively mix with liquid to apparently homogeneous two-phase zones (/4/, some visualized flows have been published in /5/).

The gas-liquid zones have little influence on the orifice's resistance factor ζ as long as a pure liquid flow is dominant at their contraction. In this case equation (2) can still be applied. But, as soon as the gas-liquid zone fills up the whole cross sectional area of an orifice, it's flow characteristics suddenly change and become independent of the back pressure p_{out}.

Such an effect can clearly be seen in figure 1 which shows the characteristics of a commonly used multi-hole VCO (valve covered orifice) injection nozzle. It should be noted that here equal symbols refer to certain input pressures p_{in}. Consequently the pressure drop was varied by changing p_{out} while p_{in} each time was kept at its constant state. The figure shows that many experimental points correlate very well with equation (2), but for each input pressure one particular state exists where the mass flow rate cannot be increased any more by increasing the pressure drop. Exactly at these transitions from the parabola of the Bernoulli equation into the vertical sections, a zone of gas-liquid flow has filled up the whole cross sectional flow area of the orifice's contraction as has been shown by the flow visualization method stated above /5/. These points, which are independant from p_{out}, show a proportional behaviour of p_{in} to the squared mass flow rate. Consequently they can be very well correlated by

$$p_{in} - p_v = \zeta_{2-ph} \quad \rho \ \frac{w_o^{\ 2}}{2} \tag{4}$$

The vapour pressure p_v is insignificant within diesel injection systems and can be considered zero. It is introduced here with respect to keeping to a thermodynamic consistancy.

A formula similar to equation (4) has been published by /14/ who based their considerations on studies given in /15/. Equation (4) is of a high practical importance for all kinds of nozzles which are used for injecting liquid into a gaseous atmosphere. As soon as the flow state described by equation (4) is reached within the spray holes of an injection nozzle, the break up of the injected spray instantaneously improves (short distances of the spray hole lengths are assumed here, therefore it is considered that once the gas-liquid flow state is formed at the hole entrance, it also discharges into the gas). The high content of vapour makes this two phase fluid compressible like a gas and as a consequence complex gasdynamic processes take place which emphasize the spray break up. This phenomenon has been already published by AVL in /5/, where the tendency to forming this two phase state was investigated for Dimethyl Ether and compared to that for diesel fuel. Concerning the influence of a cavitating flow on the spray break up, similar conclusions were drawn in /6/.

For the quantitative evaluation of an orifice's tendency to forming a gas-liquid state as described above, a dimensionless "cavitation factor" K is introduced which normalizes the pressure drop of an orifice by the inlet pressure minus vapour pressure p_v:

$$K = \frac{p_{in} - p_{out}}{p_{in} - p_v} \qquad (5)$$

This definition is unusual as most of the authors dealing with cavitation relate the pressure drop to $(p_{out}-p_v)$, e.g. /7/, although the form of equation (5) can also be found in literature, e.g. /8/. This latter form is prefered here because it leads to the most simple modelling of the cavitation effects described.

The experiments (see figure 1) have shown that the transitions between the flow phenomena described by equations (2) and (4) happen spontaneously at certain states. This means that at these transition points both the equations (2) and (7) are valid, and consequently a mathematical expression can be derived for their determination by equalizing these equations.

If w_o is expressed from both the equations (2) and (4) and these are equalized, a simple transformation leads to

$$\frac{p_{in} - p_{out}}{p_{in} - p_v} = \frac{\zeta}{\zeta_{2-ph}} = K_{lim} \qquad (6)$$

This is a very meaningful equation showing that the borderline between the two flow regimes can be expressed in terms of the pressures or by the resistance factors. In other words, if. e.g. the resistance factors are known, equation (6) defines the active flow regime for all pressures p_{in} and p_{out} of an orifice. Consequently, the following relationship can be written for a partially cavitating orifice:

$$p_{in} - p_{out} = \zeta \cdot \rho \frac{w_o^2}{2} \quad if \quad \frac{p_{in} - p_{out}}{p_{in} - p_v} \leq \frac{\zeta}{\zeta_{2-ph}}$$

$$p_{in} - p_v = \zeta_{2-ph} \rho \cdot \frac{w_o^2}{2} \quad if \quad \frac{p_{in} - p_{out}}{p_{in} - p_v} > \frac{\zeta}{\zeta_{2-ph}} \qquad (7)$$

In the case of transient conditions this relationship is easy to handle if the resistance factors are known from steady-state flow measurements and $w_o(t)$ is calculated for given pressures p_{in} (t) and p_{out} (t). In this case the conditions on the right hand side of equation (7) unambiguously define which model equation is valid. Far more complicated is the solution of the inverse problem which means that ζ and ζ_{2-ph} are to be identified from given pressures and velocities. Here the unknown parameters themselves determine which model equation has to be used. However, equation (7) indicates the possibility to unambiguously identify the cavitation phenomena described above which is of a high technical importance.

2.1.3 Unsteady flow in lines

Within this paper a "line" indicates any hydraulical connection with a constant cross sectional flow area. This means that also bores and chambers in engine parts (e.g. nozzle holder, nozzle) are modelled in the same way as lines. At sudden changes of flow area individual line models are combined. Continuous changes of flow area over long distances practically never occur in fuel injection systems.

Within lines, both the inertia effects of a fluid and its compressibility must be considered. Usually the modelling is based on the laws of mass and momentum conservation. The energy conservation law can be neglected as no heat transfer to the environment is considered within one analysed event (e.g. one single fuel injection cycle). Furthermore no phase changes of the fuel are taken into account, except cavitation, which implies locally the evaporation of small amounts of liquid. This "line cavitation" frequently appears in pipes of diesel injection systems and is caused by short negative pressure peaks ("suction waves") which drop below the liquid's vapour pressure. Although the effect of the "line caviation" is similar to that of the "flow cavitation" described in chapter 2.1.2, it is strictly distinguished between these phenomena as they are caused by different physical mechanisms. As a consequence they also require different methods for modelling.

The compressibility of a fluid is defined by its thermodynamic behaviour with respect to pressure and density. If handling liquids, it is advantageous to introduce a state equation in terms of the adiabatic speed of sound

$$a^2(p,T) = \frac{\partial p}{\partial \rho}\Big|_{s=const} \tag{8}$$

The use of this equation leads to the following form of the conservation laws of mass and momentum:

$$\begin{aligned} &\text{mass:} && p_t + a^2 w \rho_x + a^2 \rho w_x = 0 \\ &\text{momentum:} && w_t + w w_x + 1/\rho\, p_x = R \end{aligned} \tag{9}$$

The indices t and x indicate the partial derivations versus the independant variables x (spatial coordinate) and t (time). The flow velocity w, pressure p and density ρ depend on x and t and are defined by equations (8) and (9) including a thermodynamic relationship for the speed of sound a (p, T). R, on the right side of the momentum equation in (9), is the specific force (related to unit mass) due to friction. In case $R=0$, a loss free fluid motion is considered.

Several simplified approaches for friction damping of pressure waves can be found in literature which use empirical parameters and do not take into account the nonstationary increase of friction. One physical profound function which avoids these simplifications is given in /11/. It is derived from the two-dimensional Navier Stokes equations and considers laminar flow. Consequently the only parameter used for quantification of flow losses due to friction is the fluid's viscosity

$$R(x,t,\nu) = \frac{\nu}{f}\int_o^t w_\tau(x,\tau)D(t-\tau)d\tau \tag{10}$$

D is a damping function which depends on the geometry of the line's cross section For a circular form it was obtained in /11/:

$$D(t) = -4\pi\sum_{n=1}^{M} exp(-\omega_n^{\,2}\ \nu t / r^2) \tag{11}$$

The assumption of a laminar flow is no severe simplification if dealing with diesel injection systems. E.g. in /11/ it is stated that within pulsating flows a laminar flow regime is existent up to much higher Reynolds numbers than known from stationary flows. The high acceleration/deceleration forces acting on the fluid particles hinder the onset of a turbulent exchange of momentum. It was confirmed within this work that equation (10) excellently models the frictional losses.

The classical method for numerical solution of equations (9) is the method of characteristics (e.g. described in details for equation (9) in /11/). As this leads to unacceptably long computation times, other algorithms were considered and a new method that drastically accelerates the solution was developed and published in /2/. However, some simplifications within the equations (9) are necessary if this new method is applied. First, both the speed of sound a and the fluid's compressibility E defined by

$$E = a^2 \cdot \rho \tag{12}$$

must be considered constant. As a consequence the density ρ is also constant. Furthermore, the convective terms ($a^2 w \rho_x$ in mass conservation and $w\ w_x$ in momentum conservation) are considered zero. These simplifications are often found in technical publications and sometimes the authors even pass over them without any comments. For all results published in this paper, these simplifications were applied and the accelerated solution method /2/ was used.

It is anticipated here that the assumption of a constant velocity of sound led to a systematic error at the calculation of the fluid velocities from measured pressure traces. This is described in detail in chapter 3.

2.2 System measurements

Besides the model, the measurements within a system are of fundamental importance for the identification of physical phenomena. Their kind, quality and location must be carefully considered in order to obtain maximum of information.

Generally it is assumed for this work that only measuring devices are applied which usually are available on modern test beds for engine development and fuel injection pump testing. This means that pressures, motions of engine parts (e.g. injector needle lift) and accumulated quantities (e.g. injected fuel quantity per engine cycle) are usually measured. As the transducers and sensors cannot be positioned at any desired locations within the system, injection line pressures and the injector's needle lift trace are usually the only data available from engine tests. On the fuel injection pump test rig, sac hole pressure or the injector's back pressure in the equipment into which the fuel is injected (e.g. a Bosch-tube) and the fuel quantity per cycle are additionally recorded.

If parameters are identified from measurements, the experimental data must overdetermine the model's variables. This means that besides the boundary- and initial conditions, which determine the model equations, additional data is required. This latter data is called the "observations" as it represents the measurements used for comparison with the calculated results.

2.3 Identification procedure

The "observation", which is the measured data used for comparison with calculated results, should contain a maximum of information on the different phenomena. However, it must be considered that this data reflect an overlap of several effects which are difficult to distinguish. If we think of e.g. the flow resistance within a system of orifices and lines, the losses are modelled by ζ, equation (2), and R, equation (10). These are two completely different functions with respect to their numerical treatment. Within the previous chapters it was already strictly distinguished between the "orifice type" equations, eq (2), and the "line type" equations, eq (9). Therfore it is obvious to also look for an identification strategy which splits the system equations into individual groups of "orifice type" and "line type" equations and to carry out the identification procedures within these subsystems. As a consequence the measuring adapters must be placed correspondingly, the boundary conditions must be defined for all subsystems and care must be taken to keep to physically consistent parameters for the whole system.

Once the system equations are set up and the boundary and initial conditions are defined by measurements, the pressures and velocities can be calculated at any location within the system. The next step provides the comparison between the "observations" and the calculated results. The quantitative deviation between measured and calculated traces is expressed by the sum of n squared differences ssq between them at the timesteps i along the time dependant functions $g_{calc}(t)$ and $g_{meas}(t)$:

$$ssq = \sum_{i=1}^{n} (g_{calc,i} - g_{meas,i})^2 \qquad (13)$$

The calculated function $g_{calc}(t)$ changes due to the variation of model parameters, e.g. fluid properties, resistance factors etc. As a consequence also ssq changes. Optimum agreement between measured and calculated results is achieved if ssq is a minimum. For the automatic search of the ssq minimum, which usually depends on an arbitrary number of parameters, a Levenberg Marquardt algorithm is used /13/. This was taken from a mathematical program library /12/ and adapted for the problems under discussion.

The ssq minimization may lead to ambiguous results for the model parameters. This most times reveals that physical models must be improved.

3 EXAMPLE

3.1 System description

For demonstration of the identification procedure a simplified system is used. Figure 2 shows schematically an injection tube with a nozzle holder and an injection nozzle mounted to the right end of it. The aim of this example is to identify the frictional losses for a non stationary flow in the tube and the losses and cavitation in the injection nozzle.

It is outlined in the sketch of figure 2 that the non stationary flow was created by a diesel fuel injector which was fixed to the left end of the tube. This injector was driven by a fuel injection pump on the test rig and injected diesel fuel periodically into the tube in order to produce pressure and velocity pulses. The system was equipped with pressure transducers at positions 1 and 2 along the tube, in the nozzle chamber at position 3 and downstream the injection nozzle at position 4. The injection nozzle used was the same VCO nozzle as shown in figure 1. The needle spring of the nozzle was removed and the needle lift was fixed at 0.27 mm, thus the spray holes formed the only contraction of the injection nozzle.

From the position 4 a second tube recirculated the fuel by an adjustable pressure valve into the tank. This pressure valve allowed the residual pressure in the measuring equipment to be adjusted to the desired value.

Figure 2 shows the measured pressure traces at positions 1, 2, 3 and 4. It clearly can be seen how the pressure pulse, created at position 1, travels along the tube and that the fuel is partly throttled by the orifice and partly reflected back into the tube.

Considering the models for the flow in pipe, nozzle holder and nozzle, the pressure traces at positions 1 and 4 could be applied as boundary conditions and as a consequence the model equations would be unambiguously defined. But, as described in chapter 2.3, it is advantageous to split the system into subsystems which exclusively contain "line type" or "orifice type" equations. This can be realized by introducing the measured pressure trace at position 3, right before the nozzle orifice.

In this case both the tube plus nozzle holder and the injection nozzle can be calculated independantly. The system of "line type" equations includes the injection pipe, the nozzle holder's long holes passage and the nozzle chamber which are three individual "lines" with different diameters and lengths. Within the "orifice type" system the nozzle sprayholes represent only one individual orifice. The redefined boundary conditions for the subsystems are the pressures at positions 1 and 3 for the lines and those at positions 3 and 4 for the nozzle orifice.

3.2 Identification of flow losses in lines

As described in chapter 2.1.3, the model used for calculation of flow losses is based on the assumption of laminar flow, eq. (10). As a consequence exclusively the viscosity determines the fluid's losses of momentum due to friction.

Considering the line model as derived from the conservation laws given in equations (9), all model parameters represent thermodynamical data of the fluid. The velocity of sound a determines it's "compressibility" and is defined by equation (8). The density ρ refers to the residual pressure in the line which must be defined as initial condition. Any changes in ρ due to pressure are defined by the velocity of sound using equation (8).

According to these considerations it gets clear that the identification of flow losses means to identify fluid properties. These are in particular the viscosity and the velocity of sound.

Within the example under discussion both the pressures at positions 1 and 3 are used as boundary conditions for the subsystem which includes the line models. On top of these the identification procedure requires an "observation", which also can be provided by a measured line pressure trace, e.g. the one at position 2 (see figure 2).

The next step covers the comparison between "observation" and simulated pressure trace at position 2. This is made in terms of calculating *ssq* from equation (13) using the measured line pressure for the generalized function g_{meas} and the calculated pressure for g_{calc}.

Figure 3 shows how the *ssq* depends on both the fluid's velocity of sound and its viscosity. A characteristic minimum is loacted at a = 1355 m/s, ν = 2.6 cs. It is interesting to note that the identified properties at the *ssq* minimum are within the range of measured fluid properties published in /9/.

A comparison between the calculated pressure using the identified parameters and the measured pressure trace at position 2 is given in figure 4.

Once the model parameters are identified, the variables - pressure and velocity - can be determined at any location within the system. One location of major interest is the injection nozzle, right upstream the needle seat. If the velocity at this location is multiplied by the relation between the flow areas of nozzle spray holes and nozzle chamber (this corresponds to the continuity equation for incompressible fluids) the velocity in the spray holes can be determined.

The figure 5 shows the calculated velocity function in the spray holes. Here the full lines refer to the velocity as calculated from the identified line parameters, the other lines demonstrate the changes due to variations of the viscosity and the velocity of sound. The viscosity strongly amplifies or diminishes the velocity function and as a consequence the fuel quantity, which discharges from the tube within the time period analysed, is strongly influenced.

Figure 6 shows the calculated quantities per event versus fuel viscosity. As the quantity is known from measurements on the fuel injection pump test rig, this diagram offers a second possibility to determine the fluid's viscosity. It can be seen that the viscosity determined from figure 6 very well corresponds to the value determined from the *ssq* minimization shown in figure 3.

Variations of the velocity of sound distort the velocity function in a different way. Here the fuel quantity per event is hardly influenced but the peak velocities. This latter influence is of high importance with respect to the identification of the flow resistance and cavitation in the nozzle spray holes.

3.3 Flow losses and cavitation in nozzle orifice

As described in chapter 2.1.2 the flow losses of an orifice can be modelled by a resistance factor ζ. Furthermore a cavitating state can be identified by a similar factor $\zeta_{2\text{-}ph}$ which indicates that a gas-liquid zone has filled up the whole cross sectional area of the spray holes. The idea is to identify both the factors ζ and $\zeta_{2\text{-}ph}$ as defined by equation (7) from the pressure traces at positions 3 and 4 (see figure 2). As a result the active flow regime in the nozzle spray holes can be determined.

Compared to the subsystem containing lines it is much more difficult to provide an "observation" within the nozzle orifice. The only value which can be measured with reasonable effort is the fluid quantity which discharges from the orifice within a certain time period, e.g. an injection cycle. However, even this cannot be measured at an engine but only at the fuel injection pump test rig. Furthermore, the quantity is only an integral value which does not unambiguously define the shape of the flow velocity function versus time but only quantifies its time integral (the trapped area below the velocity function).

At this point the results of the line analysis, in particular the velocity in the nozzle spray holes as shown in figure 5, can be brought in. If knowing both this velocity and the pressure drop across the orifice, the resistance factor ζ is easy to determine. Figure 7 shows the calculated velocity term of equation (2) $\rho w_o^2/2$ versus the pressure drop p_{in}-p_{out} for the nozzle spray holes. In this figure the symbols refer to a number of descrete points along the time dependant velocity function as shown in figure 5 for the identified line parameters.

It can be seen in figure 7 that at high velocities the points are shifted toward lower pressure drops. This tendency was observed at various examples and obviously originates from a systematic error, most probably from the assumption of a constant velocity of sound. It must be noted that the experimental data /9/ indicate an approximatly 7% increase in velocity of sound if the pressure is changed from 100 to 300 bar. This fact certainly has severe consequences on the velocity function as given in figure 5. In this figure the velocity shows a very sensitive reaction on variations of the velocity of sound.

The figure 7 clearly shows up a limit for the proposed method. The cavitating state in the nozzle holes could not be identified from this diagram. Furthermore the flow resistance factors show some scattering if correlated from data as shown in figure 7. The reason for that are simplifications made within the model for unsteady flows in lines.

4 CONCLUSIONS AND FUTURE ASPECTS

The proposed method can be used for identification of frictional losses in unsteady pipe flows. This was shown by means of an example. Considering the mathematical model for the flow in pipes the losses due to friction could not be handled independently from other parameters but had to be identified simultaneously with the liquid's compressibility. This was expressed in terms of the velocity of sound.

For orifices a model was introduced that shows up a cavitating state from their characteristics defined by pressure drop versus flow velocity. If dealing with steady state flows this state can be determined from flow tests. In the paper the attempt is demonstrated to identify the cavitation from dynamic measurements.

Here the method did not lead to unambiguous results. The flow velocity in the orifice, which was calculated from the flow dynamics of an adjacent line, distorted the orifice's characteristics. The reason was a systematic error which most probably originated from the assumption of a constant velocity of sound in the line model.

Based on these findings it got clear that the future activities have to concentrate on a more accurate modelling of fuels' thermodynamics and on the incorporation of these into the flow calculation models of injection systems. E.g. /10/ describes the thermic and caloric state properties of diesel fuel in a comprehensive manner and also includes its evaporation and solubility behaviour of gases. This program package primarily has been developed with respect to calculating cavitating flows but also the formation of gas cavities due to dynamical effects in lines. For the fuel flow calculation in injection systems a new solver algorithm is currently developed as the method used so far is based on the assumption of a constant velocity of sound.

Literature

/ 1/ W. Egartner: AVL internal work within doctoral thesis, not yet finished, started 1991

/ 2/ M. Kroller: Efficient Computation of a Mathematical Model for the Damping of Pressure Waves in Tubes of Circular Form, Numerical Methods for Partial Differential Equation, 11, 41-60 (1995), Copyright John Wiley.

/ 3/ T. Gruber: Bestimmung der Widerstandscharakteristik am Nadelsitz einer Einspritzdüse, Diplomarbeit, Fachbereich Physikalische Ingenieurwissenschaften, TU Berlin, Okt. 1993.

/ 4/ AVL internal R&D work on visualization of nozzle flow, Sept. 1992.

/ 5/ P. Kapus, H. Ofner: Development of Fuel Injection Equipment and Combustion System for DI Diesels Operated on Dimethyl Ether, SAE Paper 950062.

/ 6/ C. Soteriou, R. Andrews, M. Smith: Direct Injection Diesel Sprays and the Effect of Cavitation and Hydraulic Flip on Atomization, SAE Paper 950080.

/ 7/ W. Bergwerk: Flow Pattern in Diesel Nozzle Spray Holes, Proc. Instn. Mech. Engrs., Vol. 173, No. 25, 1959.

/ 8/ W.H. Nurick: Orifice Cavitation and its Effect on Spray Mixing, Trans of the ASME, Journal of Fluids Engineering, Dec. 1976, VV 681-687.

/ 9/ D. Laforgia, S. Suriano: Sound Velocity Measurements for Different Fuels Varying Thermodynamical Parameters, ICE-Vol. 22, Heavy Duty Engines: A Look at the Future, ASME 1994.

/10/ AVL internal R&D work on thermodynamic properties of diesel fuel, 1994.

/11/ K. Melcher: Ein Reibungsmodell zur Berechnung von Instationären Strömungen in Rohrleitungen an Brennkraftmaschinen, Bosch Techn. Berichte 4 (1974) 7.

/12/ S. Burton et.al.: Minpack (Program Library), Argonne National Laboratory, Implementation of Levenberg Marquardt Algorithm in LMDIF, March 1980.

/13/ D.W. Marquardt: Journal of the Society for Industrial and Applied Mathematics, Vol. 11, pp. 431-441 (1963)

/14/ T.R. Ohrn, D.W. Senser, A.H. Lefebvre: Geometrical Effects on Discharge Coefficients for Plain-Orifice Atomizers, Atomization and Sprays, Vol. 1, no. 2, pp. 137-153, 1991

/15/ D. Pearce, A. Lichtarowicz: Discharge Performance of Long Orifices with Cavitating Flow, Paper presented at the Second Fluid Power Symposium, Guilford, England 1971

/16/ H. Hiroyuki: Diesel Engine Combustion and its Modelling; Diagnostics and Modelling of Combustion in Reciprocating Engines, COMODIA 1985, Tokyo, Sept. 4-6, 1985

/17/ H. Ofner, D.W. Gill: A general Purpose Simulation Model for High Pressure Fuel Injection and other Mechanical-Hydraulic Systems, Paper Presentation C430/009, IMechE 1991

/18/ M. Marcic: Calculation of the Diesel Fuel Injection Parameters, SAE-Paper 952071

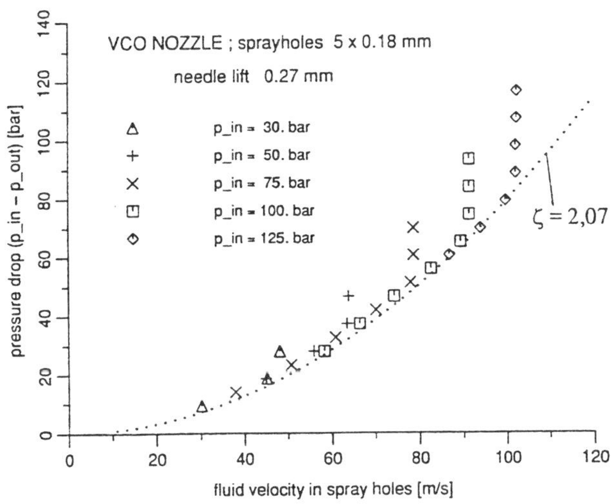

Fig. 1 Steady state flow characteristics of multi hole diesel injection nozzle

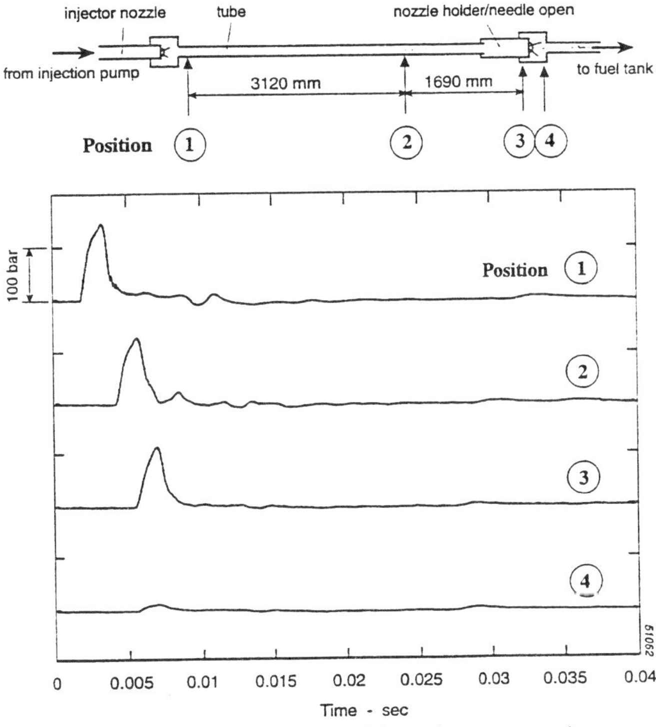

Fig. 2 Test equipment for identification of flow phenomena and measured pressure traces

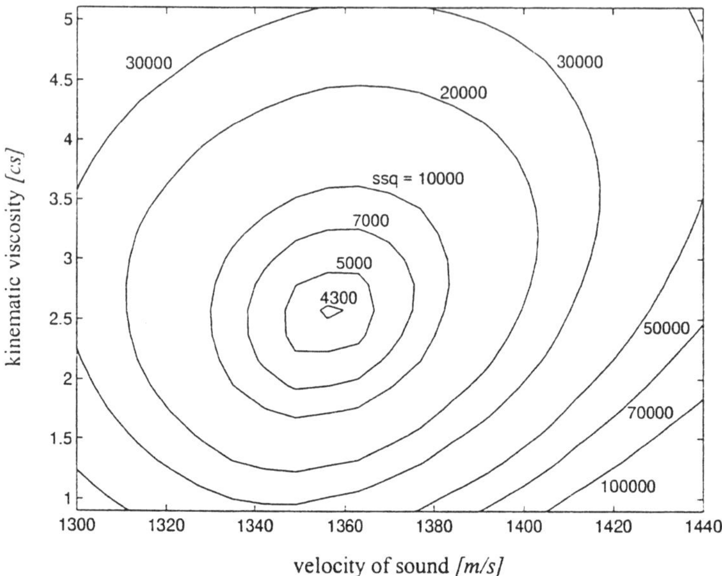

Fig. 3 The sum of squares function (equation (13)) versus velocity of sound and fluid viscosity; at the minimum the best agreement between measured and calculated pressures at position 2 (fig. 2) is achieved

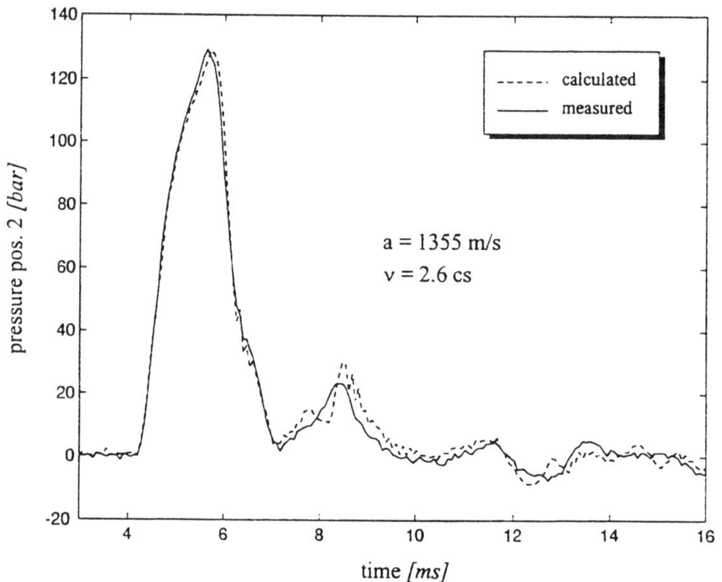

Fig. 4 Comparison of measured and calculated pressures at position 2 (see figure 2). For the calculation the parameters at the ssq-minimum are used (see figure 3)

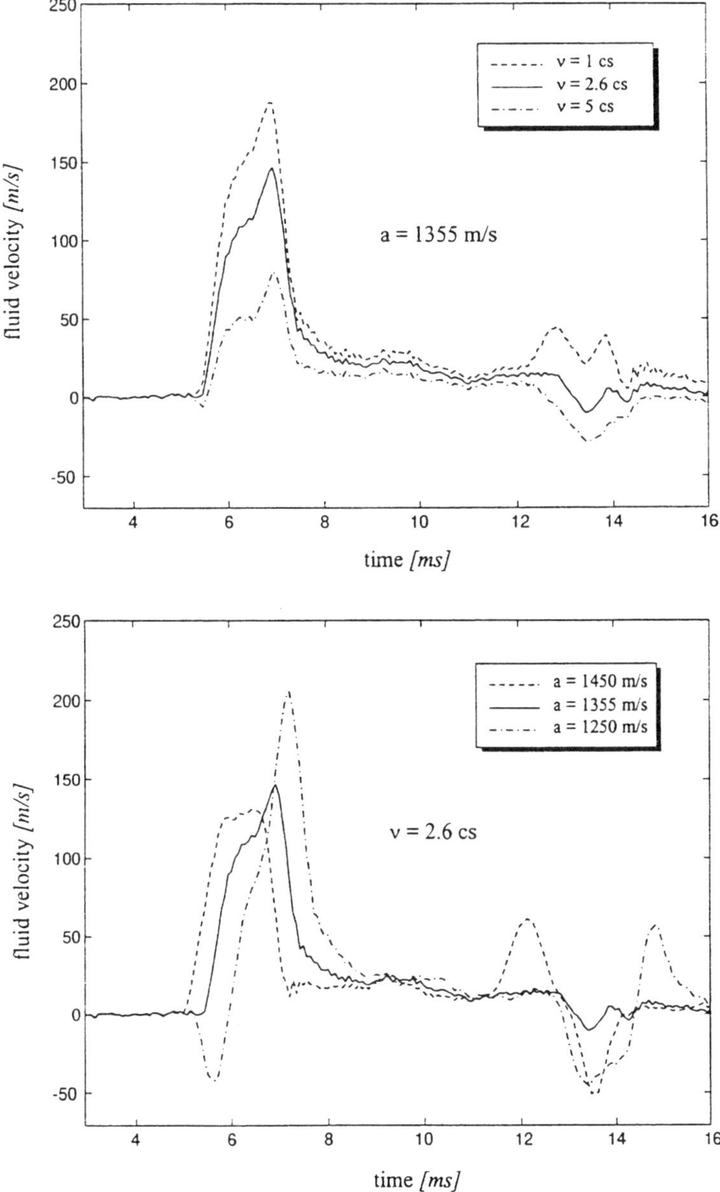

Fig. 5 Fluid flow velocity in nozzle spray holes calculated from line model. The diagrams show the influence of model parameters (ν, *a*) on the velocity function

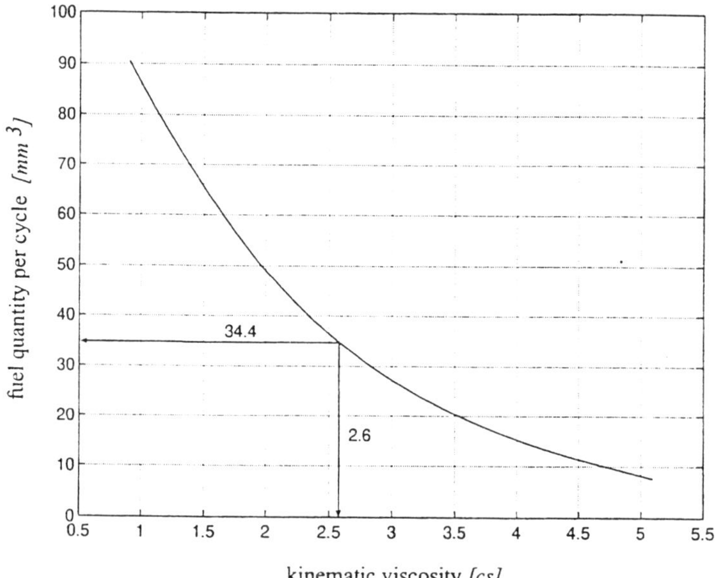

Fig. 6 Calculated fuel quantity per cycle versus fluid viscosity for identified velocity of sound (a = 1355 m/s); the measured quantity is 35 mm³

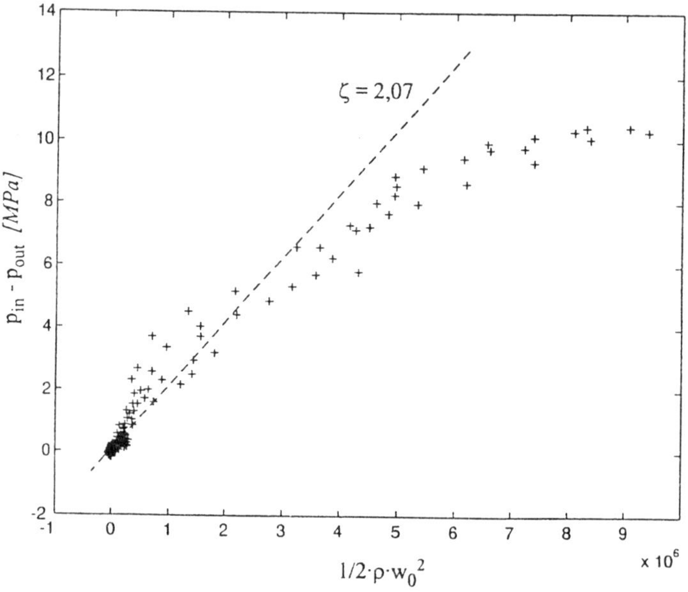

Fig. 7 Measured pressure drop of nozzle orifice (fig. 2; positions 3,4) versus velocity term of equation (2) $\rho w_o^2/2$. The same nozzle was used as specified in figure 1. The velocity was calculated from the line model using the identified parameters (ν = 2.6 cs, a = 1355 m/s, see figure 5)

C499/028/96

Computation of flow and heat transfer in rotating-disc systems

M WILSON, J X CHEN, and **J M OWEN** FRAeS, MASME, FIMechE
School of Mechanical Engineering, University of Bath, UK

SYNOPSIS

The flow in the internal cooling–air systems of gas turbines can be modelled using simple rotating–disc systems. In this paper, finite–volume methods and low Reynolds–number k–ε turbulence models are used to compute the flow and heat transfer between two confined discs with a radial outflow of cooling air. The two discs can rotate at different speeds and the systems can be characterised by the rotation ratio, Γ, of the slower to the faster disc. Computational results for axisymmetric, incompressible, steady flow are compared with data from purpose–built experimental rigs. The co–rotating (Γ=+1), rotor–stator (Γ=0) and contra–rotating (Γ=–1) disc systems discussed represent problems encountered in internal–air systems, and detailed understanding of the behaviour of these systems should lead to improvements in gas–turbine design with consequent improvements in efficiency.

NOTATION

a, b	inner and outer radius of disc respectively
C_m	moment coefficient ($=M/0.5\rho\Omega^2b^5$)
C_w	nondimensional flow rate ($=m/\mu b$)
C_p	specific heat capacity
G	gap ratio ($=s/b$)
k	turbulent kinetic energy, thermal conductivity
m	superposed mass flow–rate
M	moment on one side of disc
Nu	Nusselt number ($qr/k(T_{disc}-T_{ref})$)
Pr	Prandtl number
q	heat flux from disc to fluid
Re_ϕ, Re_t	rotational Reynolds number ($=\rho\Omega b^2/\mu$), turbulence Reynolds number ($=\rho k^2/\mu\varepsilon$)
r	radius
s, s_c	axial gap between discs, shroud clearance
S_r	swirl ratio of pre–swirl air ($=v_{\phi,p}/\Omega r_p$)
T	static temperature
U_τ	friction velocity
v_r, v_ϕ, v_z	mean–flow velocity components in r, φ, z directions
x	nondimensional radius ($=r/b$)
y,y^+	wall distance, nondimensional wall distance ($=\rho U_\tau y/\mu$)
z	axial co–ordinate
β	core rotation ratio ($=\omega/\Omega$)
ε	turbulence energy dissipation rate
Γ	ratio of angular velocity of slower to faster disc
λ_T, λ_L	turbulent flow parameter ($=C_wRe_\phi^{-0.8}$), laminar flow parameter ($=C_wRe_\phi^{-0.5}$)

μ	dynamic viscosity
ρ	density
τ_w	total wall shear stress
ϕ	tangential co-ordinate
ω	angular speed of inviscid core
Ω	angular speed of disc

Subscripts

d,s,p,b	disc, seal, pre-swirl, blade-cooling air
ref	reference value
t	turbulent

1. INTRODUCTION

In gas-turbine engines, a small percentage of the air is bled from the compressor to cool the nozzle-guide vanes, turbine blades and turbine discs. Cooling poses a design dilemma: too much cooling air is expensive and inefficient but too little can result in the failure of expensive components.

The flow and heat transfer associated with rotating turbine discs can be complex and difficult to compute, and reliable experimental data are needed to validate the computer codes being developed for this purpose. Rotating-disc systems, in which a plane disc rotates next to another rotating or stationary disc, provide simple models of the more complex geometries that occur in engines: these models can then be used to produce experimental data to validate codes.

It is convenient to classify rotating-disc systems using the parameter Γ, which is the ratio of the angular speed of the slower disc to that of the faster disc: for all cases $-1 \leq \Gamma \leq +1$. Fig 1 shows simplified models of three important cases: the rotating cavity, Γ=+1; the rotor-stator system, Γ=0; contra-rotating discs, Γ=−1. The flow structure associated with the rotor-stator system (which represents a turbine disc rotating close to a stationary casing) and the rotating cavity (which represents two co-rotating turbine or compressor discs) are discussed extensively by Owen and Rogers (1,2).

This paper presents an overview, using both new and published results, of some recent applications of elliptic solvers, incorporating low Reynolds number k-ε turbulence models, to the computation of flow and heat transfer in rotating-disc systems for Γ=+1, 0 and −1. The experimental rigs and computational models are described in sections 2 and 3, and comparisons between computations and data are made in section 4. Conclusions and possible directions for future developments are given in section 5.

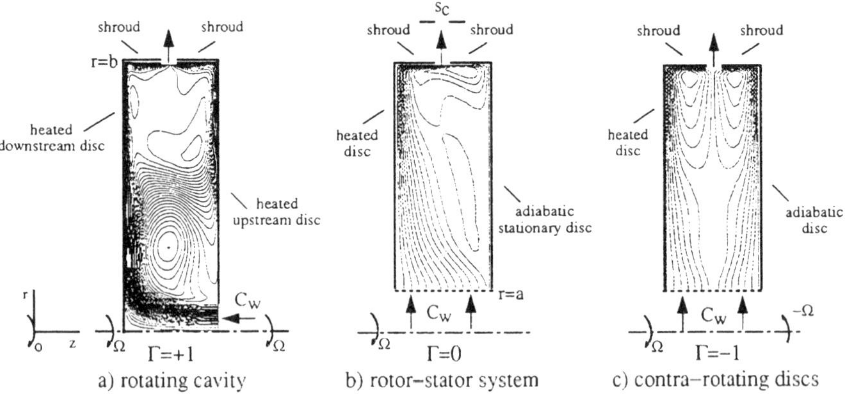

Fig. 1 Computational models and computed streamlines in rotating disc systems

2. EXPERIMENTAL RIGS AND SIMPLIFIED COMPUTATIONAL MODELS

2.1 Contra-rotating disc rig

The same experimental rig was used to make measurements of flow and heat transfer in both contra-rotating disc systems and for the rotor-stator configuration, the latter being the special case Γ=0. The apparatus was described by Gan et al (3) and Chen et al (4), and a brief summary is given here relevant to the computational models illustrated in Fig. 1b and Fig. 1c.

The rig comprised two discs, one of steel and one of transparent polycarbonate, which could be rotated independently up to 1500 rev/min. A peripheral shroud was attached to each disc using insulating material, giving an effective radius b=391 mm for the discs. The axial clearance s_c between the shrouds increased with speed and was estimated as 4 mm maximum, and the gap ratio for the cavity was G=0.12. A radial outflow of cooling air was supplied through gauze tubes of 100 mm diameter attached to the centre of the discs (a/b=0.128). For contra-rotating disc cases (Fig. 1c), a uniform radial flux was prescribed at $r=a$, based on the measured mass flow-rate and inlet air temperature. This was given inlet swirl to match the disc speeds, so that $v_\phi=\Omega a$ for $0 < z/s < 1/2$ and $v_\phi=-\Omega a$ for $1/2 < z/s < 1$. For the rotor-stator tests, the uniform radial flux at inlet was given zero swirl. Zero gradient conditions were applied at the axial clearance, except for the radial velocity which was prescribed from the superposed flow-rate and the area of the exit boundary. Low turbulence levels were associated with the inlet flow and did not affect the computed results, in view of the higher turbulence levels generated within the flow. No-slip conditions were imposed at solid surfaces.

Six thermocouples on the heated steel disc, located between x=0.2 and x=1.0, were used to provide fixed temperature boundary conditions, and six fluxmeters provided heat flux data for comparison with computational results. All other surfaces were assumed adiabatic. The transparent disc allowed LDA measurements to be made of radial and tangential velocity distributions between the discs. More details are given by Gan et al (3).

2.2 Pre-swirl rotor-stator rig

Wilson et al (5) gave a detailed description of an experimental rig used to study a direct transfer pre-swirl system. This comprised a rotor-stator system with a radial outflow of cooling air, separated by an inner seal from an axial cross-flow of cold air. A schematic diagram of the rig and the two superposed flows is given in Fig. 2.

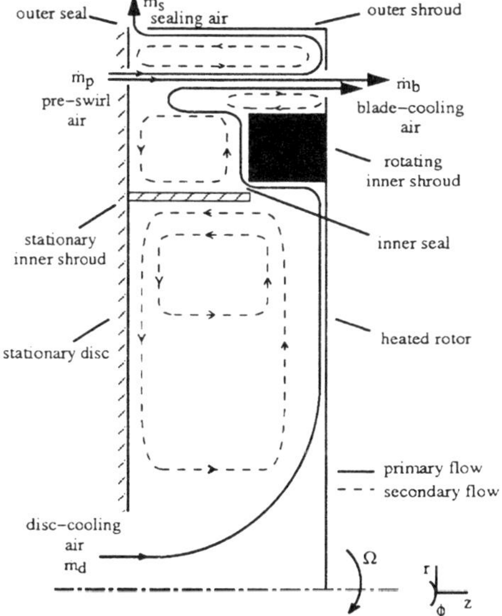

Fig. 2 Schematic diagram of pre-swirl rotor-stator system

The maximum rotational Reynolds number tested was $Re_\phi \approx 2.0 \times 10^6$ at 7000 rev/min, based on the outer seal centreline radius b=215 mm (G=0.103). The cross–flow entered the pre–swirl chamber, formed by the inner and outer seals, from sixty pre–swirl nozzles on the stationary disc (at a radius r_p=200 mm and angled at 20° to the tangential direction) and left through sixty blade–cooling holes on the rotor. In the computational model, the discrete pre–swirl nozzles and blade–cooling holes were represented by equivalent–area annular slots, giving a steady axisymmetric approximation to the real (3D time–dependent) system. Inlet velocities were deduced from measured mass flow–rates, and inlet temperatures were also known. The flow–rates in the experiment could be balanced so that $C_{w,b}=C_{w,p}$ and $C_{w,s}=C_{w,d}$, and this was used to divide the mass–flow leaving the system between the blade–cooling slot and the outer seal. The latter was repesented by an axial clearance between the rotor and the stationary outer shroud. Full details of modelling simplifications were given by Wilson et al (5).

The temperature of the heated disc was measured by six thermocouples located at $0.3 < x < 1$, and a polynomial fit to these data provided fixed temperature boundary conditions for the disc. Similar use was made of temperatures measured on the stator for $0.23 < x < 0.85$. Heat flux through the disc was measured by fluxmeters with radial locations at $0.65 < x < 1$. One of these was located under the inner shroud on the rotating disc (at x=0.86), and this measurement was used to distribute a constant heat flux around the surface of the shroud. It was not possible to make use of the measurement underneath the outer shroud in the simplified model.

Adiabatic surfaces were assumed wherever information was not available, including the stationary shrouds and the stator surface in the pre–swirl chamber. Computed results could be compared with tangential velocities and temperatures measured at eight radial locations on the mid–plane between the discs (z/s=0.5): this included one point inside the pre–swirl chamber at $x \approx 0.84$. The temperature of the air entering the blade–cooling holes was also measured.

The validation of models for flow and heat transfer in rotor–stator systems is clearly of importance in relation to the pre–swirl system described above. In an alternative configuration (now being studied experimentally), pre–swirl air flows radially outward between the heated disc and a rotating cover–plate mounted at the mid–plane (z/s=1/2). For this case, verification of codes against data for heated rotating cavities is of interest, and this is also described below.

3. COMPUTATIONAL METHOD

3.1 Finite–volume method

Numerical solutions have been obtained, using iterative pressure–correction methods, to discretised forms of the axisymmetric, steady, incompressible Reynolds–averaged Navier–Stokes and energy equations, in primitive variables, using hybrid–upwind differencing on a cylindrical–polar grid, and with staggered storage locations for the axial and radial components of velocity. In the $k-\varepsilon$ turbulence closure used here, turbulent diffusion is represented by a turbulent viscosity μ_t, computed from local values of k and ε, and with turbulent heat flux computed using a turbulent Prandtl number. Two different (but closely related) turbulence model formulations have been tested.

The rotating cavity and pre–swirl computations described below were carried out using a multigrid solver described by Gan et al (6) with the Launder and Sharma (7) low Reynolds number $k-\varepsilon$ turbulence model. The multigrid code was developed from that described by Vaughan, Gilham and Chew (8). Rotor–stator and contra–rotating disc computations were carried out using a single–grid solver with a low Reynolds–number $k-\varepsilon$ turbulence model developed by Morse (9). This code was described by Chen et al (4).

3.2 Turbulence models

Launder and Sharma (7) devised a low Reynolds number $k-\varepsilon$ turbulence model, for application to flow near a spinning disc, which includes terms in the turbulence transport equations representing the effect of molecular viscosity on turbulence structure near walls. The model also predicts transition from laminar to turbulent flow, and turbulence levels near the wall are reduced by damping of the turbulent viscosity μ_t, using an exponential function based on local turbulence Reynolds number, Re_t. Morse (9) found that the Launder and

Sharma model exhibited delayed transition to turbulence in free–disc and rotating cavity flows, and modified the model using a damping function based on dimensionless distance from the wall, y^+. The two models generally return similar results for cases where transitional effects are small.

Low Reynolds number models can give accurate predictions of flow and heat transfer over a range of rotating–disc problems, as described below, however approximately double the number of grid–points are required than would be needed to apply a high Reynolds number model with wall–functions. This is because the model needs to be applied in the laminar sub–layer very close to the wall, where the non–dimensional distance y^+ is less than unity.

4. COMPARISON BETWEEN COMPUTATIONS AND EXPERIMENTAL DATA

4.1 Rotating cavity with radial outflow (Γ=+1)

Computations were carried out (by I. Mirzaee, University of Bath) using the multigrid solver for comparison with tangential velocities, measured by Pincombe (10), on the radial mid–plane of a rotating cavity with $5.4\times10^5 < Re_\phi < 1.1\times10^6$ and $C_w \approx 2500$. A uniform axial flux at inlet was assumed (as illustrated in Fig. 1a) with $v_\phi/\Omega r=1$, and a 91 by 115 axial by radial grid was used. Fig. 3a shows the measured tangential velocities plotted against x^{-2}, illustrating that the data for each case follow a Rankine vortex behaviour:

$$v_\phi/\Omega r = A + Bx^{-2} \;\; ; \;\; A, B \text{ constant} \tag{1}$$

The computed results show that this behaviour is reasonably well captured by both the Launder and Sharma and Morse turbulence models (for laminar flow, the computed results agreed with the exact solution of the Ekman layer equations, for which A=1 and $B=\lambda_t/2\pi$).

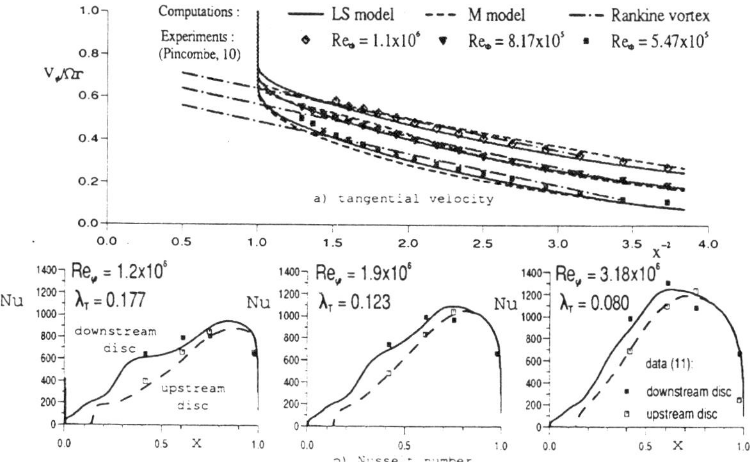

Fig. 3 Comparison between computations and data for a rotating cavity with radial outflow

Heat transfer computations were carried out (by H. Karabay, University of Bath), using the Launder and Sharma turbulence model, for comparison with the Nusselt number distributions reported by Northrop and Owen (11) for a rotating cavity with symmetrically heated discs, using the original data to prescribe inlet flow boundary conditions and to supply the temperature distributions for the two discs. Computational results for Nusselt number Nu are shown in Fig. 3b, in comparison with data for cases with $1.2\times10^6 \leq Re_\phi \leq 3.18\times10^6$ and $C_w \approx 13000$. The reference temperature used for the Nusselt numbers is the adiabatic disc temperature:

$$T_{ref} = T_d + 1/2 Pr^{1/3}\Omega^2 r^2/C_p \tag{2}$$

For cases with $\lambda_T < 0.2$, the peak Nusselt number on the downstream disc shows the end of the predicted source region, where all the flow has been entrained by the boundary layers: the non–entraining Ekman–type layer radially outward of the source region gives rise to reduced heat transfer.

The agreement between computations and data shown in Fig. 3b is reasonably good in most cases, although for $Re_\phi > 10^6$ the Nusselt numbers on the downstream disc are overpredicted in the outer part of the cavity, and the upstream disc values are slightly underpredicted. The agreement is similar to that obtained by Morse and Ong (12) who computed these cases using the Morse turbulence model.

4.2 Rotor–stator system with radial outflow (Γ=0)

A modified version of the Morse (9) turbulence model was used by Chen et al (4) to compute heat transfer cases for comparison with data from the rotating–disc rig described in section 2.1. A 92 by 76 axial by radial grid was used, contracted to the walls and with $y^+<0.5$ for application of the turbulence model. Grid refinement was also used around the mid–plane between the discs, in order to resolve the flow leaving at the shroud clearance (Fig. 1b). There was no significant difference between solutions obtained with this grid and one with double the number of points in each direction.

Fluid dynamics computations were verified by comparing predicted velocity distributions with measurements obtained at $Re_\phi=1.25\text{x}10^6$, C_w=6100 (λ_T=0.08) for G=0.12. The results are reproduced in Fig. 4 for radial locations with $0.6 < x < 0.85$ at which data were available. The peak radial velocity near the disc is slightly overpredicted, and the tangential velocity of the rotating core is underpredicted, but the general level of agreement is reasonable. The computed results at the lowest radius (x=0.6) are sensitive to the assumed inlet boundary conditions (section 2.1). The tangential velocity of the rotating core of fluid between the discs is reduced, relative to an enclosed rotor–stator system, due to the superposed flow with zero inlet swirl, and this increases the disc frictional moment.

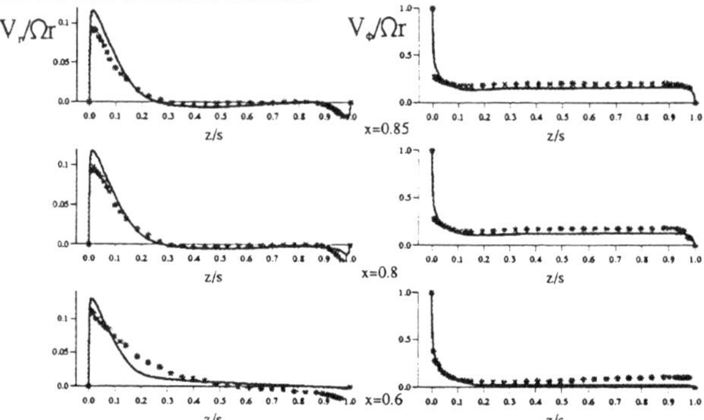

Fig. 4 Comparison between computations and data for a rotor–stator system with radial outflow $Re_\phi=1.25\text{x}10^6$, C_w=6100 (λ_T=0.08), G=0.12

For "wide gap" cases (with separate boundary layers on the discs), Daily et al (13) correlated the moment coefficients as:

$$C_m = C_m^*(1 + 13.9\beta^*\lambda_T G^{-1/8}) \ ; \ C_m^*=0.0510G^{0.1}Re_\phi^{-0.2} \ ; \ \beta^*=0.43 \qquad (3)$$

where C_m^* and β^* are the moment coefficient and core rotation ratio for the corresponding enclosed system (λ_T=0). The correlation gives a value $C_m=4.08\text{x}10^{-3}$ for the case described above, and the computations gave $C_m=3.80\text{x}10^{-3}$ (the Launder and Sharma model gave an almost identical result). The value of C_m^* is about half that for a free–disc, and C_m rises toward the free–disc value as $\lambda_T \rightarrow 0.22$, the free–disc entrainment rate. The effect of C_w on computed velocity distributions was studied parametrically by Chen et al (4).

Four heat transfer cases were computed at Re_ϕ=1.25x10^6, with C_w=2530, 3900, 6100 and 9680 (0.03 < λ_T < 0.13), using the same grid as above. The computed Nusselt numbers for the four cases followed the measured trend of increasing heat transfer with increasing C_w, and agreed reasonably well with the experimental data corrected for the effects of radiation, as described by Chen et al (4). The agreement improved as C_w increased, and it is thought that the radiation correction is more accurate at high C_w, where the ambient temperature should be better approximated by the superposed flow inlet temperature. The Nusselt numbers for the highest flow–rate case (λ_T≈0.13) were similar to those measured during free–disc tests (λ_T=0.22) away from the effects of the source region. Other results for heat transfer in a rotor–stator system are presented below.

4.3 Contra–rotating discs with radial outflow (Γ=–1)

Comparisons between computed and measured velocity profiles for Γ=–1, Re_ϕ=1.14x10^6, C_w=4026 (λ_T≈0.057) are shown in Fig. 5a. These computations were carried out by Chen et al (14) using a 92 by 78 grid (with y+<1 for the grid–node closest to the discs) and a modified Morse (9) turbulence model: many other comparisons were presented by Gan et al (3) and Gan et al (6) using the Launder and Sharma model, for which very similar results were obtained. The computations and data in Fig. 5a agree well, with the exception of the radial velocity near z/s=1 where measurements were difficult to make. The radial velocity distributions show recirculating flow in the mid–plane at x=0.6, indicating that these comparisons are made outside the source region (Fig. 1c). Gan et al (6) reported computations at higher values of λ_T which showed that the Launder and Sharma turbulence model tended to overpredict the radial extent of the source region (these comparisons with data were carried out at lower rotational Reynolds numbers, Re_ϕ≈3.8x10^5, where turbulent transition is more influential). Disc moment coefficients were not measured, but computed results showed that C_m was close to the free–disc value and insensitive to changes in superposed flow–rate. The results were also within about 2% of earlier predictions by Morse (15) for cases with 1.25x10^6 ≤ Re_ϕ ≤ 10^7.

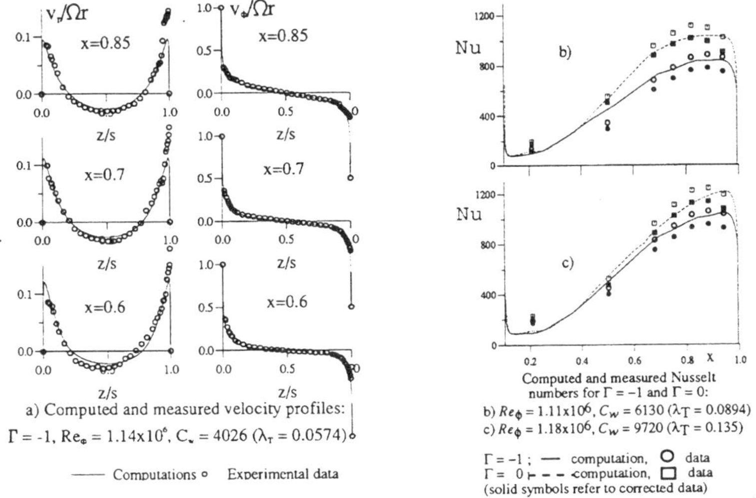

Fig. 5 Comparison between computations and data for a contra–rotating disc system with a radial outflow

Chen et al (14) compared computed and measured Nusselt numbers for a number of cases with 2x10^5 < Re_ϕ < 1.11x10^6 and 0.089 < λ_T < 0.51. The adiabatic disc was predicted to remain at the inlet air temperature until the end of the source region, x_e. Gan et al (3) applied the known free–disc entrainment rate to both discs to estimate this as:

$$x_e=1.37\lambda_T^{0.385} \qquad (4)$$

This agreed well with the predicted location in the cases considered by Chen et al (14). The temperature of the recirculating air in the mid-plane was found to be higher than that of the unheated disc.

Fig. 5b,c shows comparisons between computed and measured Nusselt numbers for both $\Gamma=-1$ and $\Gamma=0$ (the rotor-stator system) at equivalent test conditions. Reasonable agreement is obtained for $x>0.6$ (beyond the effects of the source region) with experimental data corrected for radiation effects. The heat transfer in the contra-rotating disc system is lower than that for the corresponding rotor-stator case, due to the recirculation of hot fluid in the non-rotating core flow between the discs. There is no recirculation in the rotating core for $\Gamma=0$.

4.4 Pre-swirl rotor-stator system (Γ=0)

The computed streamlines shown in Fig. 6a,b illustrate the flow patterns inside the pre-swirl chamber of the system described in section 2.2 and shown schematically in Fig. 2. The computations were carried out by Wilson et al (5) at $Re_\phi=1.23\times10^6$ and $C_{w,d}\approx2250$ ($\lambda_T\approx0.06$) using the Launder and Sharma turbulence model on a 115 by 323 grid: block obstructions within the grid represented the inner shrouds. Comparison between Fig. 6a and Fig. 6b shows the predicted effect of doubling the pre-swirl flow-rate, from $C_{w,p}/C_{w,d}=2.18$ to 4.52 (the swirl ratio also changes as a consequence, from S_r about 1 to S_r about 2). In Fig. 6a, the pre-swirl cross-flow causes toroidal recirculations above and below the cross-flow centreline. The disc cooling air flows radially outward from the inner system and into the pre-swirl chamber through the inner seal. The boundary layer remains attached to the disc as it flows around the rotating shroud, and is ingested into the blade-cooling slot at $x\approx0.93$. No reverse flow (from the pre-swirl chamber to the inner system) occurs at the inner seal in the predictions.

When the pre-swirl flow-rate is doubled (Fig. 6b), the disc-cooling air boundary layer separates from the top of the rotating inner shroud, and is ingested into the blade-cooling slot by entrainment into the under-side of the cross-flow. Chen et al (16,17) carried out three-dimensional laminar flow computations of a simplified pre-swirl system which showed that these "direct" or "indirect" routes for ingestion of disc-cooling air (Fig. 6a and Fig. 6b respectively) affected the temperature of the air at the blade-cooling holes.

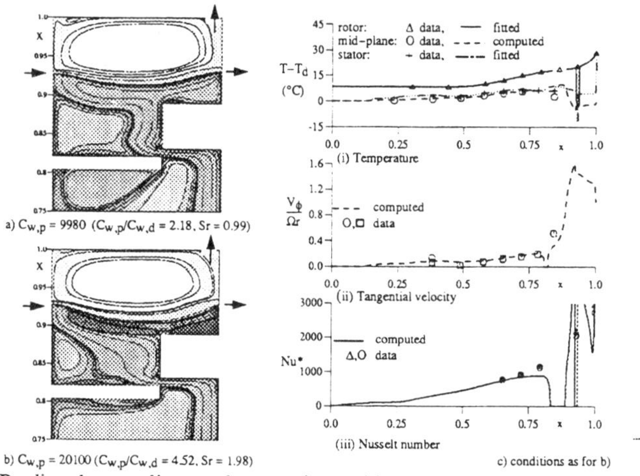

Fig. 6 Predicted streamlines and comparison with experimental data for a pre-swirl rotor-stator system, $Re_\phi=1.23\times10^6$, $C_{w,d}\approx2250$ ($\lambda_T\approx0.06$)

The two computations described above were carried out by Wilson et al (5) in order to make comparisons with measurements from two of the tests in an extensive experimental programme. Fig. 6c shows these comparisons for the case illustrated in Fig. 6b. Fig. 6c(i) shows the fitted temperatures used as boundary conditions on the rotor and stator, and the computed mid-plane air temperature. The mid-plane temperature is well predicted in the inner rotor-stator system, but there is insufficient data to validate properly the computation of the interacting flows in the pre-swirl chamber. The same comments apply to the computed mid-plane tangential velocities shown in Fig. 6c(ii). Fig. 6c(iii) shows that the measured Nusselt numbers in the inner system are slightly underpredicted. The high values for Nu for $x>0.82$ approximately (i.e. inside the pre-swirl chamber) are due to the impingement of the cross-flow on the disc: no direct comparison could be made between the computed results from the axisymmetric model and the measurement made between the blade-cooling holes on the disc at x=0.93.

The measured (and computed) Nusselt numbers in the inner system were found to be virtually identical for the two cases illustrated in Fig. 5a and Fig. 5b, suggesting that no reverse flow occurred at the inner seal. Similar computations, carried out at a lower disc-cooling air flow-rate ($\lambda_T \approx 0.03$), did show an effect of increased pre-swirl flow-rate on the Nusselt numbers in the inner system, but the computed results again showed no effect. These results were discussed in more detail by Wilson et al (5).

Computed values of air temperature rise, from pre-swirl inlet to blade-cooling exit, were found to be consistently underpredicted by about 30% (corresponding to around 2°C) compared with measured values, although the trends of the experimental data were correctly reproduced for the cases considered. The individual effects of geometrical and axisymmetric simplifications in the model, turbulence model deficiencies in predicting the recirculating, impinging flow in the pre-chamber, and sensitivity to assumed thermal boundary conditions have yet to be explored, and many more experimental data-sets are available for the further studies required to bring about improvements in prediction of the air temperature rise.

5. CONCLUSIONS

Low Reynolds number k–ε turbulence models have been shown to give sufficiently good agreement with experimental data for flow and heat transfer in different classes of rotating-disc systems with a radial outflow of cooling air, for rotational Reynolds numbers up to 3.8×10^6, although around 1.25×10^6 for most of the cases discussed. Other workers have found that more economical turbulence models, using mixing-length models or wall-functions near solid surfaces, do not give the same level of agreement with heat transfer data for the range of systems with $\Gamma=+1$, 0 and -1. The need to carry out computations at higher rotational Reynolds numbers for practical applications to gas-turbines (where $Re_\phi \approx 10^7$), to model flows with strong axial flow and recirculation such as the pre-swirl problem, and to compute systems involving radial inflow rather than outflow, all pose additional difficulties which will require further validation, and testing of alternative turbulence models. The work described in this paper should provide useful results to assess more sophisticated models, as well as the less computationally-demanding models which may be required for designs involving fully three-dimensional rotating-disc flows.

5. ACKNOWLEDGEMENTS

The experimental and computational work described above has been sponsored by EPSRC, European Gas Turbines Ltd,. Motoren und Turbinen Union and DRA. The rotating cavity computations were carried out at the University of Bath by Mr H. Karabay and Mr I. Mirzaee who are funded by the Turkish and Iranian governments respectively.

REFERENCES

(1) OWEN, J.M. and ROGERS, R.H. Flow and heat transfer in rotating disc systems: vol. 1, rotor-stator systems, 1989 (Research Studies Press, Taunton, U.K. and John Wiley, New York, USA)

(2) OWEN, J.M. and ROGERS, R.H. Flow and heat transfer in rotating disc systems: vol. 2,

rotating cavities, 1995 (Research Studies Press, Taunton, U.K. and John Wiley, New York, USA)

(3) GAN, X., KILIC, M. and OWEN, J. M. Flow between contra–rotating discs, ASME Paper 93–GT–286, 1993 (To be published in J.Turbomachinery)

(4) CHEN, J–X., GAN, X. and OWEN, J.M. Heat transfer in an air–cooled rotor–stator system, ASME Paper 94–GT–55, 1994 (To be published in J. Turbomachinery)

(5) WILSON, M., PILBROW, R. and OWEN, J.M. Flow and heat transfer in a pre–swirl rotor–stator system, ASME Paper 95–GT–239, 1995 (To be published in J. Turbomachinery)

(6) GAN, X., KILIC, M. AND OWEN, J. M. Superposed flow between two discs contrarotating at differential speeds, Int. J. Heat and Fluid Flow, 1994, 15, 438–446

(7) LAUNDER, B. E. and SHARMA, B. I. Application of the energy–dissipation model of turbulence to flow near a spinning disc, Letters in Heat and Mass Transfer, 1974, 1, 131–138

(8) VAUGHAN, C. M., GILHAM, S. and CHEW, J.W. Numerical solutions of rotating disc flows using a non–linear multigrid algorithm, Proc. 6th Conf. Num. Meth. Lam. Turb. Flow, 1989, (Pineridge Press, Swansea), 66–73

(9) MORSE, A.P. Application of a low Reynolds number k–ε turbulence model to high–speed rotating cavity flows, J. Turbomachinery, 1991, 113, 98–105

(10) PINCOMBE, J. R. Optical Measurements of the flow inside a rotating cylinder, D.Phil thesis, 1983, University of Sussex

(11) NORTHROP, A and OWEN, J. M. Heat transfer measurements in rotating disc systems part 2: the rotating cavity with a radial outflow of cooling air. Int. J. Heat Fluid Flow, 1988, 9, 27–36

(12) MORSE, A.P. and ONG, C.L. Computation of heat transfer in rotating cavities using a two–equation model of turbulence, J. Turbomachinery, 1992, 114, 247–255

(13) DAILY, J. W., ERNST, W. D. and ASBEDIAN, V. V. Enclosed rotating discs with superimposed throughflow, 1964, Dept. Civil Engng, Hydrodyn. Lab MIT Rep. No.64

(14) CHEN, J–X., GAN, X. and OWEN, J.M. Heat transfer from air–cooled contra–rotating discs ASME Paper 95–GT–184, 1995 (To be published in J. Turbomachinery)

(15) MORSE, A.P. Assessment of laminar–turbulent transition in closed disc geometries, J. Turbomachinery, 1991, 113, 131–138

(16) CHEN, J. X., OWEN, J. M. and WILSON, M. Parallel computing techniques applied to rotor–stator systems: fluid dynamics computations, Proc. 6th Conf. Num. Meth. Lam. Turb. Flow, 1993a, (Pineridge Press, Swansea), 899–911

(17) CHEN, J. X., OWEN, J. M. and WILSON, M. Parallel computing techniques applied to rotor–stator systems: thermal computations, Proc. 8th Int. Conf. Num. Meth. in Thermal Problems, 1993b, (Pineridge Press, Swansea), 1212–1226

C499/035/96

CFD developments for turbine blade heat transfer

J W CHEW BSc, DPhil, MASME, I J TAYLOR BSc, and J J BONSELL
Rolls Royce plc, Derby, UK

SYNOPSIS

In the present paper the use of 3D CFD methods in predicting the temperature field for turbine blades is considered. While CFD calculations resolving full details of the film cooling jets and internal flows are hardly practicable owing to computing and turbulence model limitations, CFD coupled with appropriate models of these effects is now showing promise. Modelling approximations are proposed and illustrated. The topics covered include film cooling, internal flows, coupling of the solid heat conduction with the convective calculation, and end wall heat transfer. The emphasis is on 3D temperature prediction through linked calculations. Comparisons of predictions with experimental data are included.

1. INTRODUCTION

In modern gas turbine engines the temperature of the gas leaving the combustor and entering the high pressure turbine is well above that which can be tolerated by the blade material. Component integrity is achieved through a combination of film and internal blade cooling. An example of a cooling system for a rotor is shown in fig. 1. Such systems involve complex internal cooling geometries with turbulators employed to enhance heat transfer and (particularly for the guide vane) may involve hundreds of film cooling holes. For guide vanes, impingement cooling of the inner surface is also sometimes used. The gas temperature entering the turbine is likely to vary both radially and circumferentially, depending on combustor design, and ejection of the cooling air may cause undesirable aerodynamic losses in the mainstream. Clearly the design and optimisation of the cooling system is a difficult and challenging problem.

Given the complexity of the problem and the known difficulties in predicting convective heat transfer it is not surprising that current design procedures rely heavily on

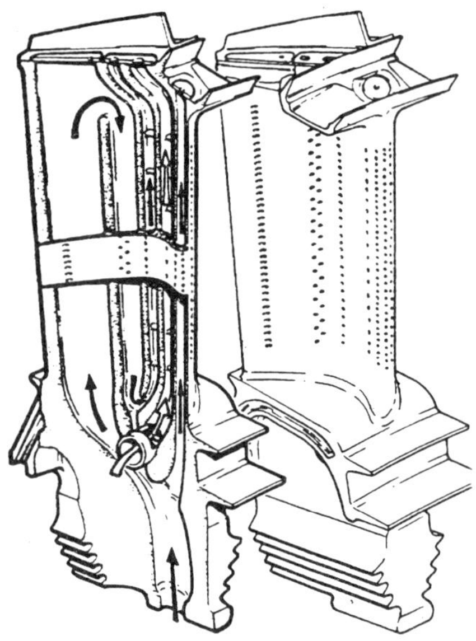

Fig. 1. Example of a cooled HP turbine blade.

experimental testing. However, it is now widely accepted that computational fluid dynamics (CFD) can be of assistance, and will play a more important role in the future. This optimistic view is supported in a recent review by Simoneau and Simon (1). These authors herald the dawning of a new era in gas path heat transfer prediction and suggest that major efforts in this field are now appropriate. It is principally advances in computing hardware that have improved the outlook over recent years.

In a system with multiple film cooling holes a CFD solution of the Reynolds-averaged Navier-Stokes equations with an eddy viscosity model of turbulence for a complete guide vane or rotor blade component is still beyond current capabilities. For example, consider flow through a single film cooling hole as studied by Leylek and Zerkle (2). These workers produced a CFD solution on a structured mesh of about 200 000 nodes. Even economising on grid points through the use of unstructured meshes it is clear that such grid resolution is impractical for a complete component. Of course, the issues of what degree of mesh resolution is required and what accuracy can be obtained using current turbulence models are contentious. The view put forward here is that while such conventional CFD calculations can and should be applied to elements of the system, such as individual cooling holes or turbulator configurations, a different approach is needed for whole component models.

The emphasis in this paper is on 3D component modelling. Typically, at present such models might be based on a finite element representation of the component allowing various heat transfer boundary conditions such as specified temperature, heat flux, or heat transfer coefficients and gas temperature.

The geometry definition may not include resolution of turbulators or film cooling holes, as the need to mesh such features increases the calculation cost considerably. The internal flow distribution may be specified from a 1D network analysis with assumptions regarding discharge coefficients, bend and entry losses and friction factors. Energy balances on the internal flow may then be included in the thermal model to calculated local gas temperatures; with further assumptions regarding heat transfer coefficients, the internal convective boundary conditions are then complete. The gas path boundary conditions might be based on heat transfer coefficients and gas temperatures derived either from CFD or empirical correlations. Other surfaces such as blade tips or inner surface of the annulus wall require further assumptions.

In an industrial environment it is highly desirable to have short elapse times for analyses and this frequently conflicts with the desire to implement more advanced predictive techniques. Thus, to gain acceptance, the advantages of using CFD methods will have to be clearly demonstrable. Thermal models may be used to provide temperature fields for stress calculations and to extrapolate from one running condition for which engine data is available to another condition. It is common practice for engineers to adjust the convective boundary condition assumptions to match experimental data. Ideally, use of CFD would eliminate the need for such adjustments, but this is unlikely to be the case in practice. For example, the inlet conditions to the CFD calculation will be subject to some uncertainty and may need modification in the light of measurements. Such issues must be considered in defining a design system.

In the next two sections of this paper CFD modelling of heat transfer in the main gas path and the internal cooling system are considered in more detail. Coupling of fluid and solid calculations are then discussed in section 4. Conclusions are then summarised in section 5.

2. GAS SIDE HEAT TRANSFER

2.1 Without Film Cooling

There is a considerable body of published work in this area. From their survey of the subject, Simoneau and Simon (1) conclude that although a truly reliable 3D predictive capability is not yet in hand, the tools to construct such a capability are available. Here we illustrate current capabilities by showing some recent results. These were obtained using a code based on the algorithm described by Moore (3) which has been applied to many turbomachinery flows (see, for example, Northall et al, 4). In this code the steady, Reynolds-averaged Navier-Stokes solutions are solved using a pressure correction algorithm. A mixing length model of turbulence has been employed with wall functions invoked at the near-wall points. Essentially these wall functions patch the solution to standard results for velocity and temperature profiles in the near-wall region; no special allowance for the effect of pressure gradient is included here.

Experimental heat transfer data on flat plates with streamwise pressure gradients representative of turbines are

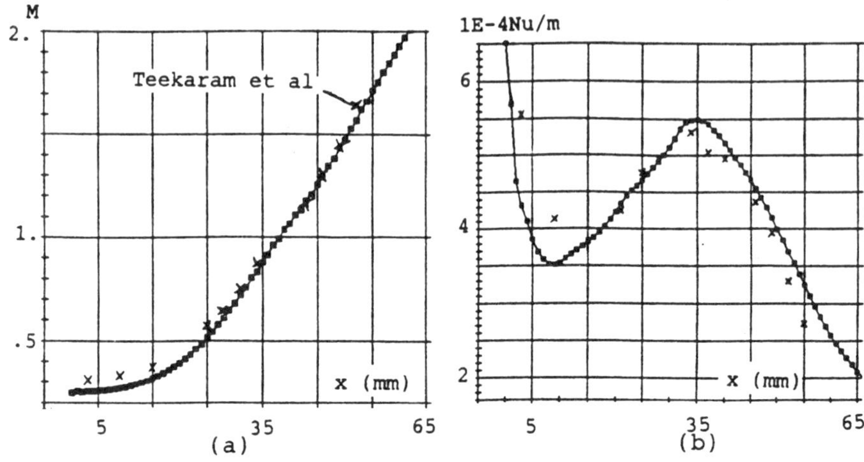

(a) Isentropic surface Mach number
(b) Nusselt number per unit length

Fig. 2. Comparison of calculated and measured results for flat plate heat transfer (× measurement, ▪-▪ CFD).

given by Teekaram et al (5). Fig. 2 shows a comparison of CFD results with heat transfer measurements and the surface Mach number distribution for one particular condition. Here the flow is subject to a strong favourable pressure gradient in the streamwise (x) direction and the inlet plane for the calculation coincides with a turbulent trip wire in the experiment. In the CFD model the tunnel test section was treated as two dimensional with inlet total pressure and temperature chosen to match test conditions. Setting the inlet boundary layer thickness to zero in the calculations was found to give the best agreement with the measurements in the inlet region. Tests with different calculation meshes and inlet assumptions showed some variation of the heat transfer in the vicinity of the inlet plane, but satisfactory insensitivity of the results over the bulk of the test section. It is apparent from fig. 2a that upstream of the tunnel contraction the computed Mach numbers are significantly less than the experimental values. This discrepancy may be associated with small nonuniformities in the inlet flow which would not be reproduced with the simplified CFD geometry. With lower velocities driving the heat transfer in the calculations than the experiment, an underprediction of heat transfer rate, as shown for the first two data points in fig 2b, might be expected.

From comparisons such as those in fig. 2 it has been concluded that for fully turbulent, 2D flow a predictive accuracy of about 10% is possible with a relatively simple turbulence model. The non-dimensional near-wall mesh spacing (y+=distance x friction velocity / kinematic viscosity) for the results in this figure is about 20. This is representative of the values currently achieved in 3D aerodynamic calculations with this CFD code. Numerical experiments indicate that y+ values of up to 100 may be used without severe degradation of accuracy, although there is clearly some sensitivity to mesh spacing. To achieve better accuracy and repeatability it seems that either more sophisticated wall functions are required or meshes resolving the laminar sublayer should be used.

The flat plate case considered above does, of course, represent a considerable simplification compared to engine conditions. The neglect of curvature is perhaps the most obvious approximation but may not be the most important. Tests against cascade data have indicated that, in fully turbulent regions, models such as that described above give reasonable results. The most difficult aspect of modelling single blade row examples is often the laminar-turbulent transition. Considerable research has been published and continues into this and other mechanisms that are expected to modify the heat transfer on engine blades. These include the effects of high freestream turbulence and unsteadiness due to wake passing from the upstream blade row. It is reasonable to expect progress in these areas in the next few years as more experimental data and increased computer power become available but, at present, there is a lack of reliable predictive methods.

Prediction of heat transfer to the end walls formed by the intersection of the airfoil with the hub and shroud puts further demands on the model. In the vicinity of these surfaces the flow is truly three dimensional. A horseshoe vortex is formed and the cross-passage pressure gradient leads to secondary flow skewing the boundary layer towards the suction surface. Current predictive capability for this region is illustrated in fig. 3. Here CFD heat transfer calculations are compared with Harvey and Jones'(6) measurements for a nozzle guide vane at conditions giving exit Mach and Reynolds numbers (based on true chord) of 0.89 and 2.6E6. The experimental data were obtained using thin film gauges with limited spatial resolution and so some smoothing of the surface heat transfer is likely in this presentation. Fully turbulent flow has been assumed in the calculation, as is consistent with the experimental evidence. Overall the CFD prediction is considered good. The general trends are reproduced by the model, and more detailed comparison indicates an accuracy level of about 20%.

One shortcoming of the endwall CFD calculation is that the predicted secondary flow is weaker than that observed. This is clearly shown in fig. 4 where limiting streaklines at the surface are compared with Harvey and Jones' results from oil flow visualisation. It is consistent with studies of swept wings (eg. Moreau, 7) that isotropic eddy viscosity models such as that used here tend to underestimate the cross stream velocity component. Adoption of turbulence model modifications applied to the swept wing problem might well give benefits for the present application. The skewed endwall boundary layer also has similarities to flow on a rotating disc. As explained by Virr et al (8) the need to resolve the secondary velocity profile may (depending on Reynolds number) require closer near wall mesh spacing than is normally considered adequate for use of wall functions. For the present example near-wall y+ values of about 20 were found adequate.

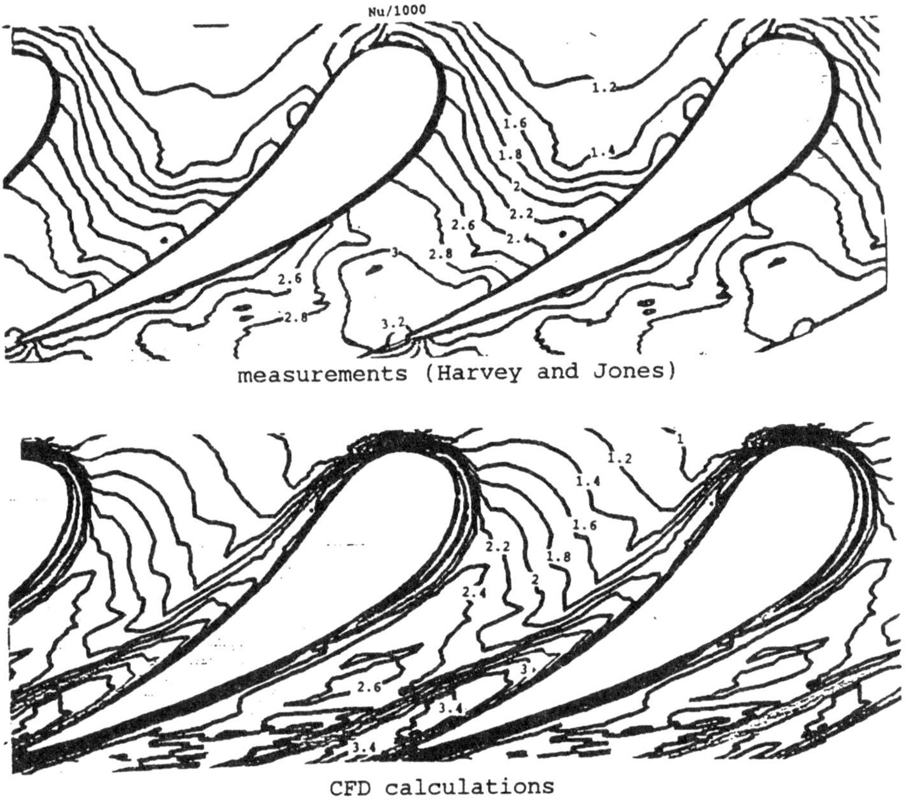

Nu/1000

measurements (Harvey and Jones)

CFD calculations

Fig. 3. Comparison of calculated and measured hub end wall Nusselt numbers (values give Nu/1000 based on true chord, and inlet gas and surface metal temperatures).

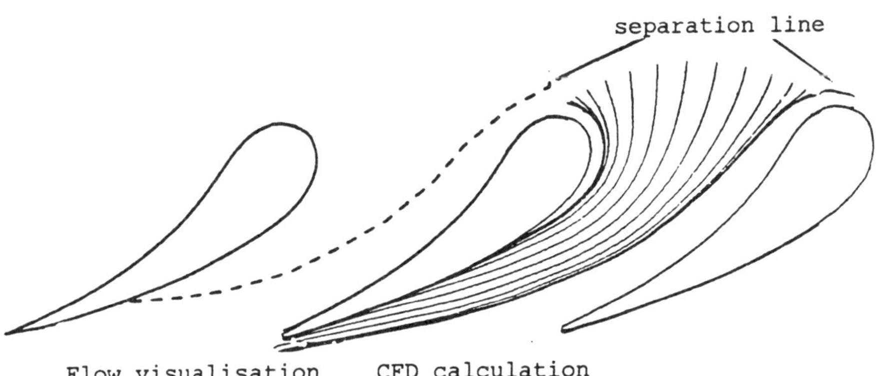

Fig. 4. Comparison of calculated and observed hub end wall streamlines.

2.2 With Film Cooling

Without film cooling todays engines could not operate at the temperatures they do. Hence modelling of this effect is essential. Until recently most attempts to develop design methods have used 2D, boundary layer CFD codes with film cooling effects averaged along the row of holes. Early examples are given by Herring (9) and Crawford et al (10). Developments have continued up to the present day. For example, Schonung and Rodi (11) have used information from 3D CFD studies of jets in cross flow in their 2D model. Norton et al (12) and Kusila et al (13) have based their approaches on modelling the coolant jet trajectory and interaction with the mainstream.

In recent years attention has been switching to 3D CFD codes and film cooling modelling in this context is now receiving attention. Garg and Gaugler (14) have presented full 3D CFD calculations for an airfoil section with up to 32 film cooling holes. Up to one million grid points were used in their calculations which, in some cases, differed considerably from experimental data. As yet, the open literature contains very few reports of more empirically based models being applied within 3D codes.

As a first step in representing film cooling within the 3D code described in section 2.1 a relatively simple model has been incorporated. The method used is similar to that of Crawford et al (10). Coolant is injected into the mainstream in a plane containing the film cooling row and approximately normal to the surface. The local injection rate is proportional to the local mainstream mass flux and is defined through specification of a mass shed ratio. The coolant may be injected some distance from the surface and is mixed with the mainstream so as to conserve momentum and energy. Since the effects of the film are averaged along the row of holes, the calculation mesh does not need to resolve each hole.

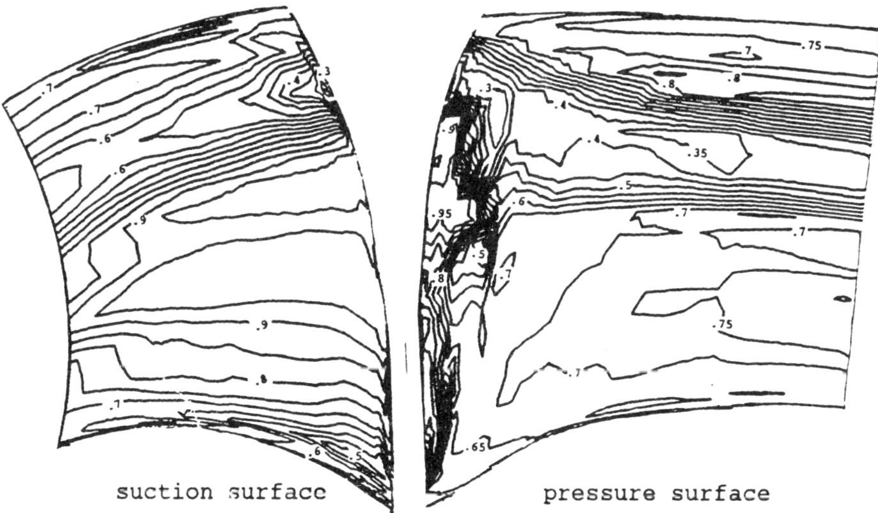

Fig. 5. Calculated adiabatic cooling effectiveness for a nozzle guide vane (based on gas-to-surface temperature difference / gas-to-coolant temperature difference).

Fig. 5 shows an example of application of this model for a film cooled nozzle guide vane. The particular interest in this investigation was in the three dimensional aspects of the results. The hot spot at about 85% height on the leading edge is due to a shortfall in coolant supply pressure at this point. This clearly produces a hot streak on the pressure surface. Other 3D effects are evident. Note, for example, the streaking on the suction surface. The model was successful in that it qualitatively reproduced many of the features of a thermal paint test and helped explain experimental results that had initially aroused surprise. Quantitatively, the results are sensitive to the choice of mass shed ratio. Not surprisingly, since no attempt was made to optimise the choice of mass shed ratio the calculated surface temperatures did not agree with data. However the potential for improvement of this method, building on previous work for boundary layer codes, is clear.

3. INTERNAL BLADE COOLING

Internal cooling plays an important part in maintaining metal temperatures, particularly for high pressure rotors. Modern designs usually include rib roughness elements for enhanced heat transfer. At present, correlations of experimental data are often used to estimate internal heat transfer rates. An interesting alternative is the integral-type method presented by Mayle (15). This method is claimed to correlate data for a range of conditions. It may be noted that, with modern experimental techniques, full coverage heat transfer measurements can be obtained for non-rotating ribbed ducts relatively quickly and efficiently (see, for example, Wang et al, 16).

There is a considerable body of work in the open literature concerning application of CFD to internal cooling flows. One example of a full 3D calculation of ribbed duct flow is given by Prakash and Zerkle (17). Their results are encouraging but they note difficulties with turbulence modelling in some parameter ranges. An example of a CFD calculation for a turbine internal cooling system is given by Dawes (18). The geometry studied includes pin fins and baffles, but no cooling holes. The accuracy of the calculation is unclear and the adequacy of the solution mesh could be questioned. However, there is no doubt that it will be feasible to use much finer meshes in the future and many workers see this as the best way forward. Another approach was tried by Cunha (19). This author treated the ribbed walls as rough surfaces using modified wall functions to represent the effect of the ribs.

While the CFD code used here is capable of detailed calculations of flow around roughness elements, high computing costs and turbulence model difficulties have limited its application to date. With these restrictions in mind an alternative approach that averages the effect of the ribs along the duct has been implemented. In this model effects of the ribs are represented by additional source terms in the governing equations. Similar methods have been used in boundary layer calculations for different types of roughness elements by other workers (eg. Christoph and Pletcher, 20, and Taylor et al

21).

Application of the rib model to fully-developed flow in a square cross-section duct with ribs on two opposing sides is illustrated in fig. 6. In these calculations the drag force due to the rib is assumed to be directed normal to the ribs in planes parallel to the ribbed surface. This drag is averaged over the duct length and distributed over that part of the cross-section into which the ribs extend. The magnitude of the drag is set equal to a coefficient times the dynamic head based on the local velocity component normal to the rib, and it is included in the usual momentum equations as a source (or body force) term. Comparison with Han et al's (22) friction factor data for various angle ribs is shown in fig. 6a. Here the results are for a fixed axial pressure gradient corresponding to the product of the friction factor and the Reynolds number squared equal to 4.23E7, the rib height is 0.063 times the duct side, and the pitch to height ratio is 10. The same drag coefficient was used in all these calculations. This simple model captures the experimental trend remarkably well for rib-to-duct axis angles of 45 to 90 degrees. At lower angles frictional drag is more significant (relative to form drag) and the model does not do so well.

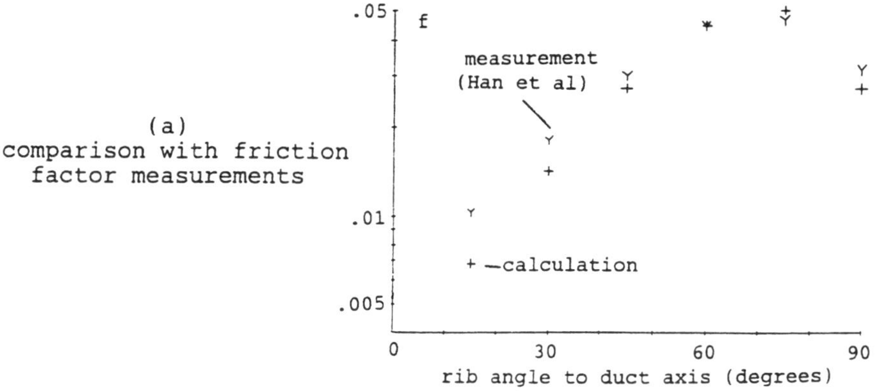

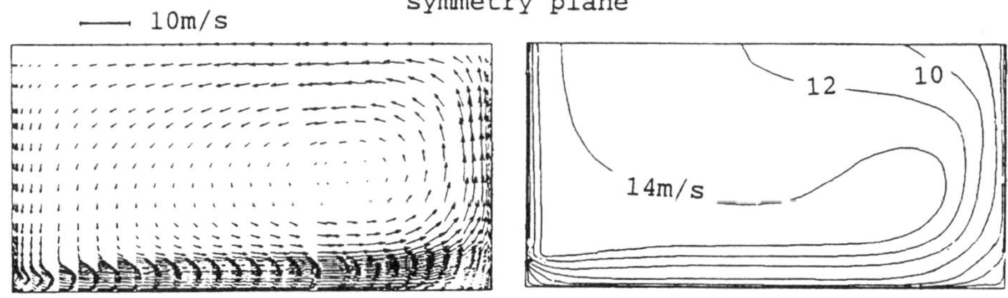

(b) secondary velocity vectors (c) axial velocity contours

Fig. 6. CFD calculations using the rib-roughness model.

As shown in figs 6b and 6c, the model predicts secondary flow and asymmetry of the axial flow due to angled ribs. It may therefore be useful in estimating heat transfer variations around the duct. It should be noted however that the diffusive mechanisms governing heat transfer may be more difficult to correlate than the form drag which dominates the results in fig 5a. This approach, if successful, is likely to rely heavily on correlation against measurements.

4. FLUID-SOLID COUPLING

During the design process component temperatures through a flight cycle may be required. In addition to being accurate an ideal model would be user-friendly and computationally quick, have the flexibility to allow for various types of boundary conditions, and allow the user to easily change and experiment with the boundary conditions. Of course the model must also link in with other parts of the system such as the geometry database and the stress calculations. The question arises as to how CFD calculations can most effectively be applied in a thermal model of the component. Attention here focuses on the gas path convective heat transfer; similar considerations apply to other surfaces.

An obvious approach would be to perform simultaneous time-marching calculations for the fluid and solid regions and exchange surface information at the end of each time step. Alternatively, one might combine the discretised equations for the two domains and so retain full coupling in a time-marching scheme. While these approaches would have very general applicability they do not take advantage of the physical nature of the present problem. For example, the time-scales for the gas motion are much shorter than those for the solid temperature changes, and so at any instant it is acceptable to treat the flow as steady. (At least, that is, while blade passing effects are neglected or are represented by a time-averaged model in the CFD calculation.) Further insight is given by consideration of the coupling between the gas energy equation and the other equations governing the flow.

It has long been recognised that, for situations in which the fluid properties are independent of the temperature field, the flow energy equation is linear in temperature. This fact underpins the use of heat transfer coefficients to characterise convective heat flux. For a linear equation, superposition of solutions for different thermal boundary conditions is possible. This enables the convective heat flux for complex thermal boundary conditions to be represented by a number of heat transfer coefficient-like parameters. Jones (23) has applied this methodology to the turbine situation. For 3D, flight cycle calculations Jones' method offers very good computational efficiency, but would involve approximations in modelling the effects of coolant injection temperature variations and non-uniform surface temperature distributions. In principle, the method can overcome these approximations, but it would then become cumbersome.

It is useful to estimate the influence of fluid property variations (due to surface temperature changes) on the surface heat flux. Typically, for flat plate heat transfer, a correction factor equal to the gas-to-wall temperature ratio raised to the power 0.4 is applied to a Nusselt correlation to account for this effect (see, for example, Teekaram et al (5). It follows that if surface temperature variations within ±100K then errors in heat flux introduced by neglecting property variations should be no more than about 4%. Compared to the flat plate one might expect more effect of buoyancy (and hence property variation) in the turbine. In the main gas path this is expected to be small. However, for rotor internal blade cooling flows there is evidence of buoyancy effects being significant (eg. Johnson et al, 24).

A flexible approach to coupling the fluid and solid calculations has been adopted in the present studies. These are based around the CFD code described earlier and a 3D finite element solver for the solid, both of which are well established in the company. The finite element solver supersedes an earlier 2D automatic analysis code described by Armstrong and Edmunds (25) and has been developed by those authors and their colleagues. CFD solutions are generated first and the discretised flow energy equation is stored in the form of coefficients at nodes on the CFD mesh. The finite element solver can then access the energy equation (which corresponds to a fixed flow field and fluid properties) and calculate convective heat fluxes for particular surface and gas inlet temperature distributions. Since this is a linear equation the solution is relatively quick. Heat fluxes from the fluid energy equation are then passed to the solid model in the form of heat transfer coefficients and gas temperatures. Iterating this procedure gives a solution satisfying the fluid energy equation and the solid model. If required, additional iterations updating the full energy equation through a CFD solution with revised thermal boundary conditions can be introduced so as to include the effect of fluid property variations.

In practice the main difficulty in implementing this approach has been in transferring surface information between the fluid and solid calculation meshes. For various reasons, including modelling approximations and component expansion at running conditions, geometries for the two models are not identical. This problem was overcome by defining a mapping that forces a number of discrete points on the two geometries to correspond. Normal projection between surfaces is then used to obtain corresponding surface points for mesh nodes, and the data on one surface can then be interpolated onto the other.

An example of rotor blade surface temperatures calculated using a CFD generated gas energy equation as the boundary condition on the airfoil is shown in fig. 7. The 3D effects that are evident are associated with the radially non-uniform mainstream gas temperature assumed at inlet, the secondary flow in the blade passage and the internal cooling model.

CONCLUSION

A move towards use of 3D CFD codes for modelling gas path is underway. Encouraging results have been obtained for representative problems using a relatively simple turbulence model and a 'row-averaged' film cooling model. Linking of the CFD calculation with a heat conduction solution for the blade

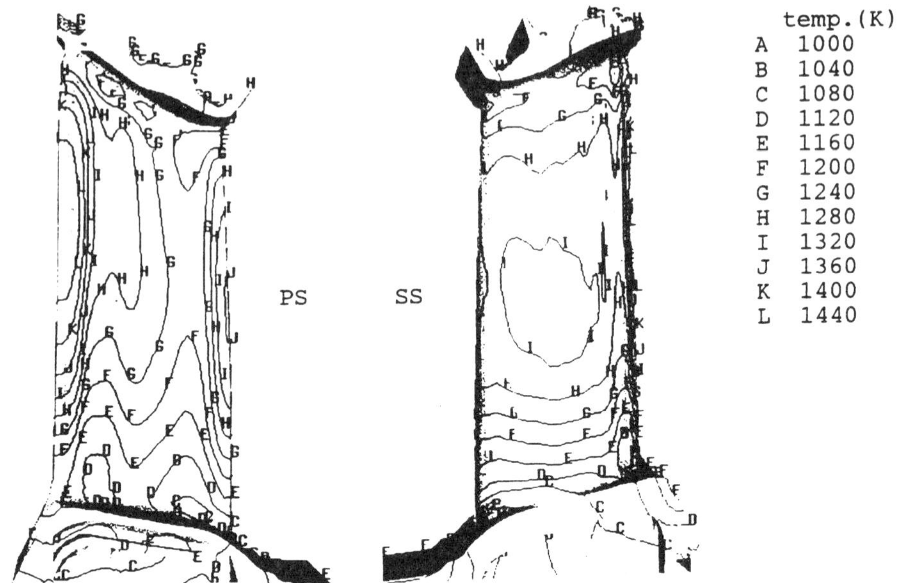

Fig. 7. Blade temperatures from coupled solid-fluid model.

has also been demonstrated. Work on use of CFD models for internal cooling is not so far advanced but is showing promise. Empirically based models of roughness elements offer an alternative to conventional CFD calculations.

With computer hardware continuing to improve it is inevitable that use of CFD will expand in the future. It will be feasible to improve component thermal models by use of CFD on non-gas path surfaces, by taking better account of the coupling between different elements of the system, and by improving geometry resolution. Nevertheless, turbulence model limitations and the complexity of the problem mean that heavy reliance will continue to be placed on experimental data for the foreseeable future. The importance of evaluating models against representative experimental data cannot be overemphasised.

ACKNOWLEDGEMENTS

We gratefully acknowledge collaboration with our colleagues in the Aerothermal Methods, Mechanical Methods and Turbine Engineering groups at Rolls-Royce and the Osney Laboratory, University of Oxford.

REFERENCES

1. SIMONEAU, R.J. and SIMON, F.F. Progress towards understanding and predicting heat transfer in the turbine gas path. Int. J. Heat and Fluid Flow, 1993, 14, 106-128.

2. LEYLEK, J.H. and ZERKLE, R.D. Discrete-jet film cooling: a comparison of computational results with experiments. ASME Gas Turbine Congress, 1993, paper 93-GT-207.
3. MOORE, J.G. Calculation of 3D flow without numerical mixing. AGARD Lecture Series No. 140 on 3D Computational Techniques Applied to Internal Flows in Propulsion Systems, 1985.
4. NORTHALL, J.D., MOORE, J.G. and MOORE, J. Three-dimensional viscous flow calculation for loss prediction in turbine blade rows. I. Mech. E. Conf. on Turbomachinery - Efficiency prediction and improvement, 1987.
5. TEEKARAM, A.J.H., FORTH, C.P. and JONES, T.V. Film cooling in the presence of mainstream pressure gradients. ASME J. Turbomachinery, 1991, 113, 484-492.
6. HARVEY, N.W. and JONES, T.V. Measurement and calculation of end wall heat transfer & aerodynamics on a nozzle guide vane in annular cascade. ASME Gas Turbine Congress, 1990, paper 90-GT-301.
7. MOREAU, V. A study of anisotropic effects in three-dimensional, incompressible turbulent boundary layers driven by pressure gradients. Ecole Polytechnique Federale de Lausanne, 1988, thesis 755.
8. VIRR, G.P., CHEW, J.W. and COUPLAND, J. Application of computational fluid dynamics to turbine disk cavities. ASME J. Turbomachinery, 1994, 116, 701-708.
9. HERRING, H.J. A method of predicting the behaviour of a turbulent boundary layer with discrete transpiration jets. ASME J. Engineering for Power, 1975, 97, 214-224.
10. CRAWFORD, M.E., KAYS, W.M. and MOFFAT, R.J. Full-coverage film cooling Part II: Heat transfer data and numerical simulation. ASME Gas Turbine Conf., 1980, paper 80-GT-44.
11. SCHONUNG, B. and RODI, W. Prediction of film cooling by a row of holes with a two-dimensional boundary layer procedure. ASME Gas Turbine Conf., 1987, paper 87-GT-122.
12. NORTON, R.J.G., FOREST, A.E., WHITE, A.J., EPSTEIN, A.H., SCHULTZ, D.L. and OLDFIELD, M.L.G. Turbine cooling design, Vol. 1, Wright Patterson Airforce report WRDC-TR-89-2109.
13. KUSILA, P., LEBOEUF, F. and PERRIN, G. Computation of a wall boundary layer with discrete jet injections. ASME Gas Turbine Congress, 1991, paper 91-GT-143.
14. GARG, V.K. and GAUGLER, R.E. Prediction of film cooling on gas turbine airfoils. ASME Gas Turbine Congress, 1994, paper 94-GT-16.
15. MAYLE, R.E., 1989. Pressure loss and heat transfer in channels roughened on two opposed walls. ASME paper 89-GT-86.
16. WANG, Z., IRELAND, P., JONES, T.V. and DAVENPORT, R. A colour image processing system for transient liquid crystal heat transfer experiments. ASME Gas Turbine Congress, 1994, paper 94-GT-290.
17. PRAKASH, C. and ZERKLE, R. Prediction of turbulent flow and heat transfer in a ribbed rectangular duct with and without rotation. ASME Gas Turbine Congress, 1993, paper 93-GT-206.
18. DAWES, W.N. The solution-adaptive numerical simulation of the three-dimensional viscous flow in the serpentine coolant passage of a radial inflow turbine blade. ASME J. Turbomachinery, 1994, 116, 141-148.
19. CUNHA, F.J. Turbulent flow and heat transfer in gas turbine blade cooling passages. ASME Gas Turbine Congress, 1992, paper 92-GT-239.

20. CHRISTOPH, G.H. and PLETCHER, R.H. Prediction of rough wall skin friction and heat transfer. AIAA J., 21, 509-515.
21. TAYLOR, R.P., COLEMAN, H.W. and HODGE, B.K. Prediction of turbulent rough-wall skin friction using a discrete element approach. ASME J. Fluids Engineering, 1985, 107, 251-257.
22. HAN, J.C., PARK, J.S. and LEI, C.K. Heat transfer enhancement in channels with turbulence promoters. ASME J. Engineering for Gas Turbines and Power, 107, 628-635.
23. JONES, T.V. Definition of heat transfer coefficients in the turbine situation. I. Mech E. Symposium on Turbomachinery: Latest developments in a changing scene, 1991, paper C423/046.
24. JOHNSON, B.V., STEUBER, G.D., YEH, F.C. and WAGNER, J.H. Heat transfer in rotating serpentine passages with trips skewed to the flow. ASME Gas Turbine Congress, 1992, paper 92-GT-191.
25. ARMSTRONG, I. and EDMUNDS, T.M. Fully automatic analysis in the industrial environment. Proc. second Int. Conf. on Quality Assurance and Standards in Finite Element Analysis, NAFEMS, 1989.

C499/038/96

Multigrid computation of three-dimensional turbulent flow in inlet 'S' ducts

Y ZHAO BSc, MSc, PhD, MAIAA and **Z M DING** BSc, MSc
School of Mechanical and Production Engineering, Nanyang Technological University, Singapore

SYNOPSIS

A general CFD code based on the multigrid/multiblock strategy has been developed to study 3-D turbulent flow in inlet S-shaped ducts for aircraft engines. An explicit five-stage Runge-Kutta time-stepping scheme with central or 3rd-order upwind TVD spatial discretization is used as the basic solver in conjunction with a multigrid/multiblock scheme. With proper data management, this solver can achieve high efficiency in terms of computing time and memory requirement even for complex 3-D flow problems, in comparison with existing implicit codes for such study. Closure of the Reynolds stresses is through the two-equation k-ϵ models. Computations were carried out to simulate flow in a non-diffusing S-duct. Results are compared with experimental data where possible to validate the code. It is found that the k-ϵ model outperforms the zero-equation model in predicting the secondary flow.

1 INTRODUCTION

S-shaped intake ducts are often used in aircraft engine systems. The purpose of a S-duct is to deliver airflow from the wing or fuselage to the engine compressor. Applications of S-ducts can be found in a number of commercial and military aircraft, such as Boeing 727, Lockheed Tristar L-1011, General Dynamics F-16 and McDonnell-Douglas F-18.

Flow in such S-ducts is normally subsonic, compressible and turbulent. Because of centreline curvature and/or cross-sectional area increase, the flow is highly complex and three-dimensional. The geometry of the duct gives rise to strong secondary flow and even flow separation. A good intake duct design should be able to achieve deceleration of the incoming flow without separation, minimal total pressure loss, and uniform flow with small secondary flow at the compressor face. Study of such inlet ducts have traditionally been done by experiment because the flow is too complex for numerical simulation.

With the advent of new generation of computers, Computational Fluid Dynamics (CFD) algorithms and turbulence models, numerical study of S-ducts is fast becoming possible. The parabolized Navier-Stokes (PNS) computation, which is based the assumption of negligible viscous diffusion in stream-wise direction, has been used by Towne and Anderson [1], Towne and Flitcroft [2], Vakili et. al. [3], Malechi and Lord [4], for incompressible and compressible flows in various S-ducts. The thin-layer Navier-Stokes approach (TLNS), which neglects viscous diffusion in directions parallel to the solid surface as well as cross derivatives in the viscous terms, has also been used by Jenkins and Loeffler [5] to study compressible flow in a highly offset diffuser. More recently, Harloff et. al [6] used the full Navier-Stokes (FNS) approach to study compressible flows within non-diffusing and diffusing

S-ducts. In the above work, space-marching was used to solve the PNS equations while the implicit Beam and Warming scheme was applied to solve the TLNS and the FNS equations. In these studies an algebraic turbulence model (zero-equation model) was normally used, except that of Malechi anc Lord [4] who employed the k-ϵ model in their numerical calculation.

Explicit Runge-Kutta time marching scheme with multigrid strategy was first proposed by Jame son et.al. [7] for efficient computation of two-dimensional transonic external inviscid flow. Thi scheme was later extended to calculate two-dimensional transonic viscous external flow based on th Reynolds average Navier-Stokes equations (RANS) by Martinelli [8]. The scheme was also furthe refined by Radespiel et. al. [9] and Vatsa et. al. [10] to compute three-dimensional viscous exter nal flow based on the TLNS assumption. An Algebraic (zero-equation) turbulence model, i.e., th Baldwin-Lomax model [11] was employed in all the above-mentioned applications. The efficienc and accuracy of the scheme were found to be excellent in all the external flow computations. Th potential of this scheme with a multigrid/multiblock strategy can be even greater if implemente on massively parallel computers. On the hand, the capability of this method for calculating inter nal flow is not well investigated. Compared with flow around airfoils and wings, flow in S-ducts i very complex and three-dimensional with strong secondary flow and possible large separation, wher viscous effects are dominant in the whole flow field in all the grid directions.

This paper describes the development and application of a general CFD program for high spee internal turbulent flow computations in the framework of a multiblock/multigrid strategy. An explic multistage time-stepping scheme is employed for the numerical solution of the three-dimension mass-weighted Reynolds averaged Navier-Stokes (RANS) equations with an k-ϵ turbulence mode Multigrid technique, together with implicit residual smoothing and local time stepping for all th equations, is employed to improve solution efficiency. Compared with the previous work, the prese study is the first known application of an explicit scheme to the FNS computation of compressib 3-D duct flow with a low Reynolds number k-ϵ model. Results show that the numerical method use is efficient and robust for calculating 3-D flow in S-ducts, and the k-ϵ model used can better predi the secondary flow patterns than the Baldwin-Lomax algebraic model.

2 GOVERNING EQUATIONS

The governing equations are the Reynolds averaged Navier-Stokes (RANS) equations in integr form:

$$\frac{\partial}{\partial t}\int_{\Omega} W dV + \int_{s} \vec{F}_c \cdot \vec{n_s} dS = \int_{s} \vec{F}_v \cdot \vec{n_s} dS + \int_{\Omega} S_t dV, \qquad ($$

where

$$\vec{F}_c = F_c\vec{i} + G_c\vec{j} + H_c\vec{k};$$
$$\vec{F}_v = F_v\vec{i} + G_v\vec{j} + H_v\vec{k};$$

$$W = (\rho, \rho u, \rho v, \rho w, \rho E, \rho k, \rho\epsilon)^T;$$

$$\vec{F}_c = \begin{bmatrix} \rho\vec{U} \\ \rho u\vec{U} + p\vec{i} \\ \rho v\vec{U} + p\vec{j} \\ \rho w\vec{U} + p\vec{k} \\ \rho H\vec{U} \\ \rho k\vec{U} \\ \rho\epsilon\vec{U} \end{bmatrix}; \vec{F}_v = \begin{bmatrix} 0 \\ \vec{\tau}_x \\ \vec{\tau}_y \\ \vec{\tau}_z \\ \bar{\bar{\tau}} \cdot \vec{U} - \vec{q} \\ (\mu_l + \frac{\mu_t}{\sigma_k})\vec{\nabla} k \\ (\mu_l + \frac{\mu_t}{\sigma_\epsilon})\vec{\nabla}\epsilon \end{bmatrix};$$

$$S_t = (0,0,0,0,0,P_k - \rho\epsilon + D_k, f_\mu c_1 \frac{\epsilon}{k} P_k - f_2 c_2 \frac{\rho\epsilon^2}{k} + D_\epsilon)^T.$$

The terms in the above equations are as follows:

$$\rho E = \rho e + \frac{1}{2}\rho(u^2 + v^2 + w^2); \bar{\bar{\tau}} = \vec{\tau}_x \vec{i} + \vec{\tau}_y \vec{j} + \vec{\tau}_z \vec{k};$$

$$\vec{\tau}_i = \tau_{ix}\vec{i} + \tau_{iy}\vec{j} + \tau_{iz}\vec{k}; \tau_{ij} = \mu_e(\frac{\partial u_i}{\partial x_j} + \frac{\partial u_j}{\partial x_i} - \frac{2}{3}\delta_{ij}\frac{\partial u_k}{\partial x_k});$$

$$\mu_e = \mu_l + \mu_t; \mu_t = \rho c_\mu f_\mu \frac{k^2}{\epsilon}; \vec{q} = -c_p(\frac{\mu_l}{Pr_l} + \frac{\mu_t}{Pr_t})\vec{\nabla} T;$$

$$P_k - \frac{\partial u_i}{\partial x_j}(\mu_t S_{ij} - \frac{2}{3}\delta_{ij}\rho k); S_{ij} = \frac{\partial u_i}{\partial x_j} + \frac{\partial u_j}{\partial x_i} - \frac{2}{3}\delta_{ij}\frac{\partial u_k}{\partial x_k};$$

$$D_k = -\frac{2\mu_l k}{y_n^2}; D_\epsilon = -\frac{2\mu_l \epsilon}{y_n^2} exp(-\frac{y_n^+}{2});$$

The low Reynolds damping terms introduced in [12] are employed, as follows:

$$f_\mu = (1 - exp(-0.0115 y_n^+); f_1 = 1; f_2 = 1 - exp(-Re_T^2);$$

here $Re_T = \rho\frac{k^2}{\mu_l\epsilon}$ is the turbulent Reynolds number.

The above system of equations are closed by the state equation of perfect gas:

$$p = \rho(\gamma - 1)e. \quad (2)$$

The various constants in the k-ϵ model are listed as follows:

c_μ	c_1	c_2	σ_k	σ_ϵ
0.09	1.35	1.80	1.0	1.3

The above equations are non-dimensionalized using a reference length and free stream flow conditions. Only the non-dimensional form of the equations are used in numerical calculations.

3 NUMERICAL METHODS

3.1 The Explicit Finite Volume Method

A cell-centered finite volume technique is employed to discretize the spatial operators in the governing quations in integral form (1). This reduces the governing equations to a set of ordinary differential quations which are, in turn, solved by an explicit multistage Runge-Kutta time-marching scheme roposed by Jameson et. al. [7]. The three-dimensional flow domain is divided into a large number f hexahedral subdomains or cells. The integral form accepts the existence of flow discontinuities n the flow domain, and the finite volume approach ensures the conservation of mass, momentum nd energy over all these small finite volumes while the average rates of change of W in them are alculated.

The viscous fluxes required on the cell interfaces contain first derivatives of flow variables, which re calculated using a local coordinate transformation from Cartesian coordinates (x,y,z) to the urvilinear coordinates (ξ,η,ζ) in the three grid line directions.

The calculation of source terms for the turbulence equations requires the values of the firs derivatives of flow quantities at a cell centre. They are calculated by a volume-weighted metho based on the Green's theorem.

The spatially descritized equations form a set of coupled ordinary differential equations whic are integrated in time by means of an explicit five-stage Runge-Kutta scheme. In this scheme, th dissipative terms and the turbulence source terms are frozen at their values in the second stag for the calculations in all other stages, i.e. these terms are only calculated twice in the multistag computation.

As an explicit scheme, the maximum time step is determined by a stability criterion of th numerical method. Although it is impossible to derive an analytical stability condition for genera three-dimensional problems, the Courant-Friedrichs-lewy (CFL) condition for hyperbolic equation can be used to derive such a condition by analysing a linearized model equation. This condition ca then be extended approximately to general three-dimensional problems.

3.2 Artificial Viscosity vs. Higher-order Upwind TVD Scheme

Theoretically artificial viscosity is not needed in the numerical solution of RANS equations, becaus of the existence of the physical viscosity. However, to guarantee that spurious oscillations do not aris and convergence is achieved, especially in the presence of strong gradients and the destabilizing effect of the source terms of the k-ϵ equations, some adaptive dissipation is still added to the numerica solutions. Therefore artificial viscosity is constructed by adaptively blending the second and fourt differences as proposed Jameson et. al. [7]. The dissipation of the fourth difference provides a lo background level of dissipation to prevent non-linear instability. The dissipation of second differenc is used in areas where there are shocks. The background dissipation will strengthen the fourth-orde numerical dissipation while the artificial dissipation of the second difference will turn the schem into first-order accurate when shocks are present. The artificial viscosity terms will be added t solutions in the form of dissipative fluxes at all the cell faces and scaled by the spectral radii of th flux Jacobian matrix in the corresponding grid line direction. Details can be found in [10].

Care should be taken to make sure that the amount of adaptive dissipation added should b maintained at minimum level in order to avoid spurious viscous-like effects. In principle, the tot amount of artificial viscosity should be at least 10^{-3} times smaller than the total physical dissipatio

On the other hand, a 3rd-oder upwind TVD scheme of Osher and Chakravarthy [15], based on th Roe's approximate Riemann solver [16], is also implemented to replace the artificial viscosity. Th advantage of using a upwind TVD scheme is that one needs not adjust any smoothing coefficients t achieve stable and accurate solution.

The 3rd-order TVD scheme is implemented by adding a correction to the centre-difference fluxe which is analogous to adding an artificial viscosity. But it is based on the flow physics of wa propagation and decomposition. The corrected flux on a cell face is calculated as

$$\bar{F}_{i+1/2} = F^{CD}_{i+1/2} + (D^1 + D^2)_{i+1/2},$$

here F^{CD} is the flux evaluated using the centre-difference scheme, D^1 is the first-order numerical flu based on the Roe's approximate Riemann solver approach [16] and D^2 defines the the higher-od correction term that upgrades the accuracy. A minmod slope limiter is used in D_2.

3.3 Convergence Acceleration

A multigrid strategy proposed by Jameson et. al. [7] is adopted here to speed up the convergen rate of the numerical solution. It combines the accuracy of the fine grid with the fast convergen of coarse grids, which requires less computing effort due to the use of less grid points and large tin steps for time-marching to steady state. In order to save computing time, the viscous diffusion tern are evaluated only on the finest grid. This treatment will not alter the accuracy of the solution since the coarser grids are used only to damp out the low frequency errors on the finest grid, and are being driven by the fine grid residuals.

Convergence to steady state solutions can also be substantially accelerated by using the maximum local time step to advance the equations at each cell. This is equivalent to introducing an artificial wave speed that goes faster as it moves towards the far-field.

In order to further increase the CFL number, a implicit residual smoothing scheme is employed. The residual at one point of the flow field is replaced by a smoothed or weighted average of the residuals at the neighboring points. The averaged residuals are calculated implicitly in order to increase the maximum CFL number, thus increasing the convergence rate. Normally this procedure allows the CFL number to be increased by a factor of 2 or 3. For cells with high aspect ratios, the constant smoothing coefficients are replaced by the variable ones based on the spectral radii in three directions. The implicit residual smoothing is applied to first, third and fifth stages only and it is found that this treatment does not affect the robustness of the scheme while computing time is significantly reduced with increased CFL number.

It has been found that the smoothing of multi-grid corrections for every grid level is also necessary to achieve faster convergence, specially when highly stretched grid is used. It is due to the fact that the use of trilinear interpolation of multigrid corrections with high cell aspect ratio can introduce strong oscillations in the solution.

4 BOUNDARY AND INITIAL CONDITIONS

At the solid wall, no-slip and no-injection boundary conditions are imposed, i.e., the zero normal fluxes of mass, momentum and energy are imposed. In addition, the solid surface is assumed to be adiabatic and the pressure gradient normal to the wall at the surface is considered to be zero. The turbulence kinetic energy and the gradient of ϵ at the wall are taken as zero.

The far field boundary conditions for the mean-flow equations are based on one-dimensional inviscid flow theory. By the introduction of Riemann invariants for one-dimensional inviscid flow normal to the boundary, non-reflecting boundary conditions are imposed. Flow quantities at the boundary are obtained in accordance with the condition that Riemann variables for the incoming and outgoing waves are constant along the normal direction.

For outflow, all the flow quantities are linearly extrapolated from the flow domain. whist the pressure is prescribed according to experimental measurement.

Uniform flow quantities based on the inlet or far field condition is prescribed in the flow field as the initial condition.

5 DATA MANAGEMENT

For complex 3-D problems, the multiblock/multigrid strategy used here requires much more memory than a single-block/single grid code if the conventional data storage method is employed without modification. In order to minimize memory requirement under the limitation of static arrays in the FORTRAN, all the flow field data are stored in two large 1-D arrays, or vectors at the main program level. One is the storage array U which contains flow quantities in all the blocks with the finest grid. The other is the working array UO which contains flow quantities of a particular block with all levels of grids. Temporary 3-D arrays are only used at subroutine levels for a certain block and a certain level of grid.

6 RESULTS AND DISCUSSION

First of all, quasi-one-dimensional flow in a converging-diverging (CD) nozzle is simulated by the program developed for the purpose of code validation. A number of flow conditions were considered, ranging from subsonic, transonic to supersonic flows. Here only the result of the transonic flow with Jameson's artificial viscosity is presented due to lack of space. In figure 1, the computed pressure, density and Mach number along the centre line of the nozzle are compared with the exact solution. As shown, the agreement is very good. Similar agreement is found with all the other cases. If the 3rd-order upwind TVD scheme is employed, the results are even better near the shock. The convergence history in figure 1 (D) shows that a reduction of residual in the order of 10^{-6} is achieved in less than 1000 iterations.

Figure 2 shows a comparison of computed pressure contours for the supersonic flow through a channel with an arc bump, using the artificial viscosity and the 3rd-order TVD scheme. It is found that for supersonic flow, Jameson's scheme becomes unstable and a very large second-difference coefficient has to be used. It is evident that the TVD scheme has a better performance in terms of stability and accuracy. In addition, there is no need to adjust coefficients for different flow conditions.

The non-diffusing duct introduced by Vakili et. al. [3] is used in this study. In this experiment the inlet Mach number is 0.6 and Reynolds number is 1.76×10^6 based on inlet diameter. The same boundary conditions and duct geometry are used for the computation here. Predicted flow quantities will then be compared with the corresponding measurements for validation and further flow analysis. Figure 3 shows the computational grid used. A two-zone grid is used in the computation with two levels of multigrid. The grid size is as follows: zone one (upstream straight section) $19 \times 49 \times 33$; zone two (entrance to S-duct upto the exit) $55 \times 49 \times 33$.

A comparison of total pressure recovery with the corresponding measurement on different cross sections of the S-duct is shown in Figures 4 (a), (b) and (c). The left-hand side of each graph i the experimental measurement whist the right-hand side the corresponding prediction. Reasonable agreement is observed in the comparison of calculated and measured secondary flow. Also shown in Figure 4 (d) is the result at the S-duct exit by Harloff et. al. [6] using the Baldwin-Lomax model. By comparison of (c) and (d), the low pressure zone predicted the k-ϵ model is closer to the measured one. As can be seen from the above figures, the low total pressure zone starts to form in the firs bend due to the pressure differential between the outer and inner walls, and it gradually grows in size from the first bend to the exit of the second bend because the vortices formed in th first bend move to the outer wall of the second bend. This will be further illustrated by the secondary flow vectors in the following figures.

Figures 5 (a) and (b) show the secondary flow vectors on the exit plane by the k-ϵ (CH) mode and the Baldwin-Lomax (BL) model and their comparison with measurement. Two pairs of counter rotating vortices are clearly present on the exit plane, which are also observed in the experiment. I is found that the first pair of vortices are formed in the first bend and later they are pushed toward the outer wall of the second bend. They persist in the second bend because of the favourable pressur gradient and weak centrifugal force near the outer wall region. On the other hand, the fluid nea the high speed core is driven towards the wall at an azimuthal angle of slightly less than 90 degree and diverges into upward and downward streams. Thus a second pair of vortices are formed near th inner wall of the second bend as a result of higher pressure near the outer wall and lower pressur close to the inner. The influence of centrifugal force is negligible in the area because it is very clos to the wall and the fluid has lower velocity. In the experiment, it is observed that oil streak lines ar driven towards the lower surface in the first bend. In the second bend, the streak lines diverge near 9 degree angle. These observations agree with the computed results. It is found that although the stat pressure variations along the pipe wall at three azimuthal angles 0, 90 and 180 degrees agree we with the corresponding measurements, the static pressure contours on the transverse planes general are not in good agreement with the measurements. Further study in this respect is required. A observed in the experiment, the calculated results do not show any separation.

Compared with experimental measurement, the BL model predicts a too weak vortex near the outer wall while the vortex near the inner wall is too strong. On the other hand, the secondary flow patterns predicted by the CH k-ϵ model have close agreement with the measurement. The sizes and strength of the vortices near the outer and inner walls are all similar to their measured counterparts. Figure 5 (c) illustrates a comparison of computed and measured wall static pressures at three azimuthal angles. The agreement is reasonable. However, it is noticed that the quality of prediction deteriorates over the second half of the duct. This may be due to the influence of the stream line curvature on turbulence which the current model is unable to account for.

7 CONCLUDING REMARKS

A efficient multiblock/multigrid Navier-Stokes solver has been developed and validated in this study. The third-order upwind TVD scheme is superior to the Jameson's original centre-difference scheme with artificial viscosity, especially in the computation of supersonic flow. But the former is more computationally intensive than the latter. Preliminary results for compressible flow in a S-duct show that the FNS code developed is capable of capturing the pertinent flow characteristics and the convergence rate is relatively fast by the use of various convergence enhancement techniques. The use of the Chien's low Reynolds number k-ϵ turbulence model leads to better prediction of secondary flow patterns in the S-duct, compared with the zero-equation model. Application of the model to simulate flow in diffusing S-ducts with separations is being undertaken to further validate the program and model.

References

[1] TOWNE, C.E. and ANDERSON, B.H. "Numerical simulation of flows in curved diffusers with cross-sectional transitioning using a three-dimensional viscous analysis". AIAA Paper 81-0033, AIAA, 1981.

[2] TOWNE, C.E. and FLITCROFT, J.E. "Analysis of intake ducts using a three-dimensional viscous marching procedure". In *First World Congress on Computational Mechanics*, Austin, TX, Sept. 1986.

[3] VAKILI, A.D., et al. "Comparison of experimental and computational compressible flow in an S-duct". AIAA Paper 83-1739, AIAA, July, 1984.

[4] MALECHI, R.E. and LORD, W.K. "Parabolized Navier-Stokes analysis of circular-to-rectangular transition duct flows". SAE Paper 88-1480, SAE, 1988.

[5] JENKINS, R.C. and LOEFFLER, A.L. "Modeling of subsonic flow through a compact offset inlet diffuser". *AIAA Journal*, 29:401–408, March, 1991.

[6] HARLOFF, G.J., et al. "Navier-Stokes analysis of three-dimensional S-ducts". *Journal of Aircraft*, 30:526–533, July-Aug, 1993.

[7] JAMESON, A., SCHMIDT, W., and TURKEL, E. "Numerical solutions of the Euler equations by finite volume methods using Runge-Kutta time-stepping schemes". AIAA Paper 81-1259, AIAA, 1981.

[8] MARTINELLI, L. *"Calculations of viscous flows with a multigrid method"*. PhD thesis, Department of Mechanical and Aerospace Engineering, Princeton University, 1987.

[9] RADESPIEL, R. "A cell-vertex multigrid method for the Navier-Stokes equations". Technical Memorandum 101557, NASA, January 1989.

[10] VATSA, V.N. and WEDAN, B.W. "Development of a multigrid code for 3-D Navier-Stokes equations and its application to a grid-refinement study". *Computers & Fluids*, 18(4):391–403, 1990.

[11] BALDWIN, B. and LOMAX, H. "Thin layer approximation and algebraic model for separated turbulent flows". AIAA paper AIAA-78-257, 1978.

[12] CHIEN, K. "Prediction of channel and boundary layer flows with a low Reynolds number turbulence model". *AIAA Journal*, 20(1):33–38, 1982.

[13] LAM, C.K. and BREMHORST, K.A. "A modified form of the k-ϵ model for predicting wall turbulence". *ASME Journal of Fluids Engineering*, 103, 1981.

[14] PATEL, V.C., RODI, W., and SCHEUERER, G. " Turbulence models for near-wall flows and low Reynolds numbers:A Review". *AIAA Journal*, 23(2), 1985.

[15] CHAKRAVARTHY, S. "Development of upwind schemes for the Euler equations". NASA contract report 4043, NASA, January 1987.

[16] ROE, P.L. "Approximate Riemman solvers, parameter vectors and difference schemes". *J. Comput. Phys.*, 43:357–372, 1981.

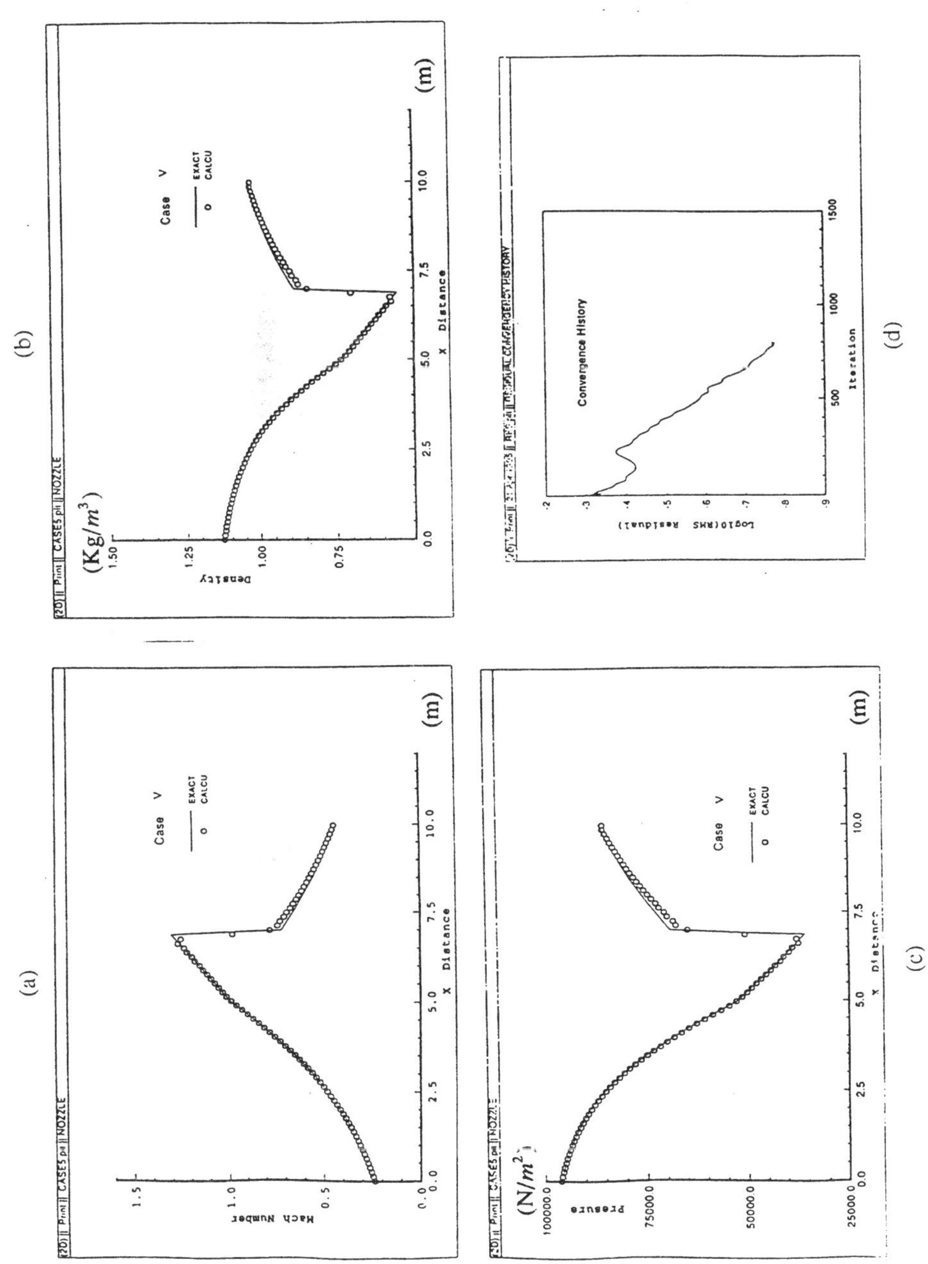

Figure 1: Transonic flow in a nozzle.

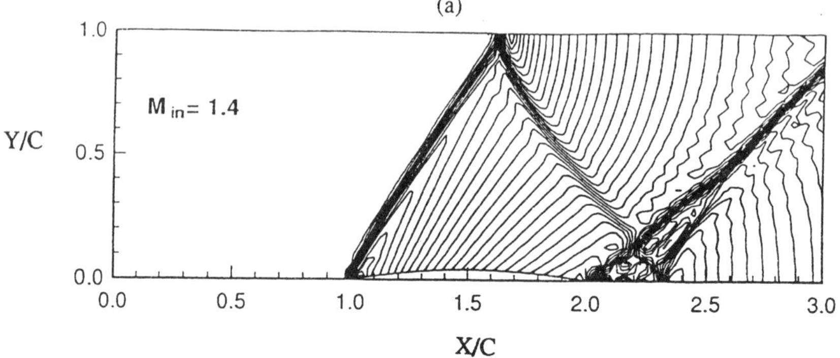

Iso - pressure contours (Jameson's artificial viscosity)

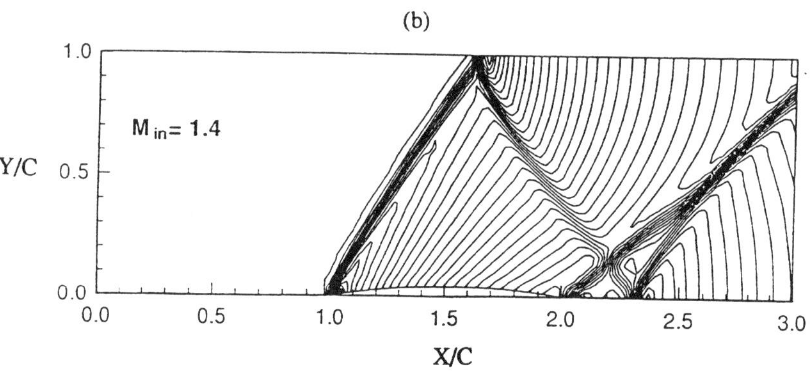

Iso - pressure contours (Third order TVD scheme)

Figure 2: Supersonic flow in a channel (C is the length of the arc bump).

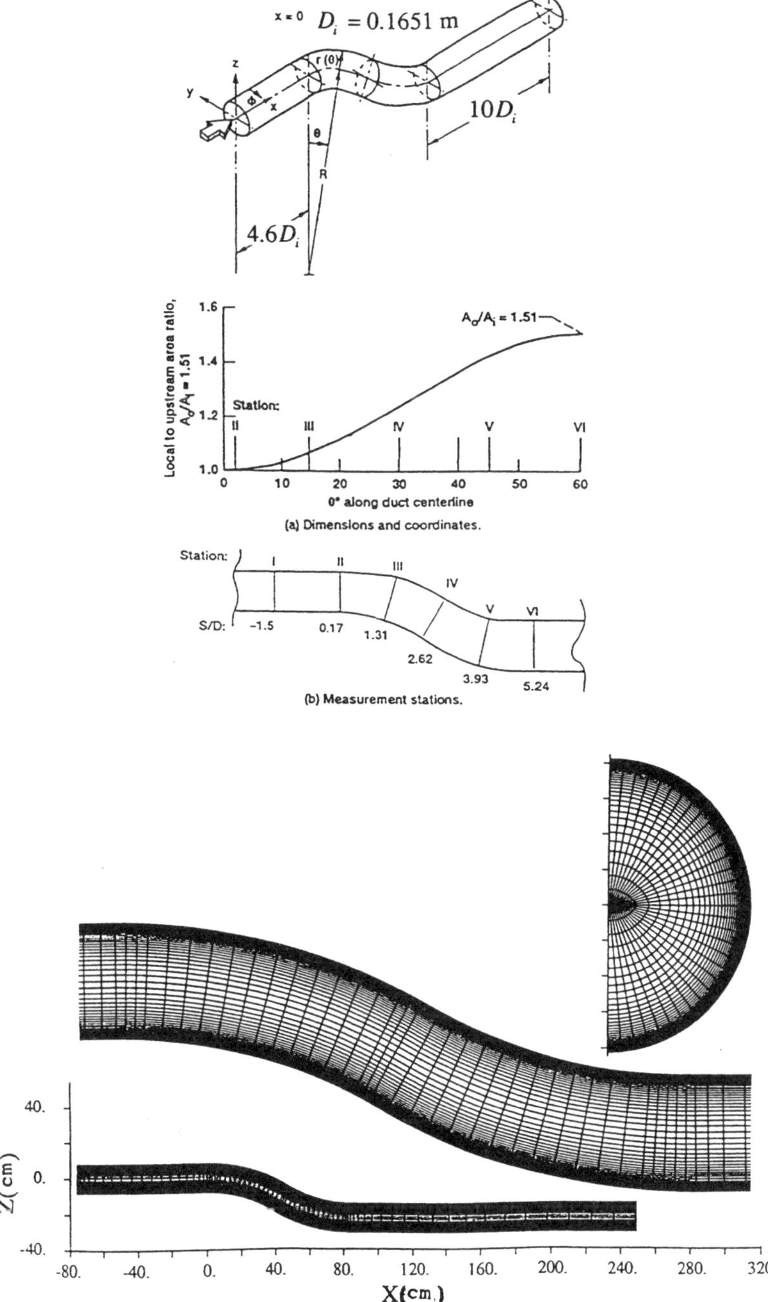

Figure 3: Geometry and computational grid of the non-diffusing duct.

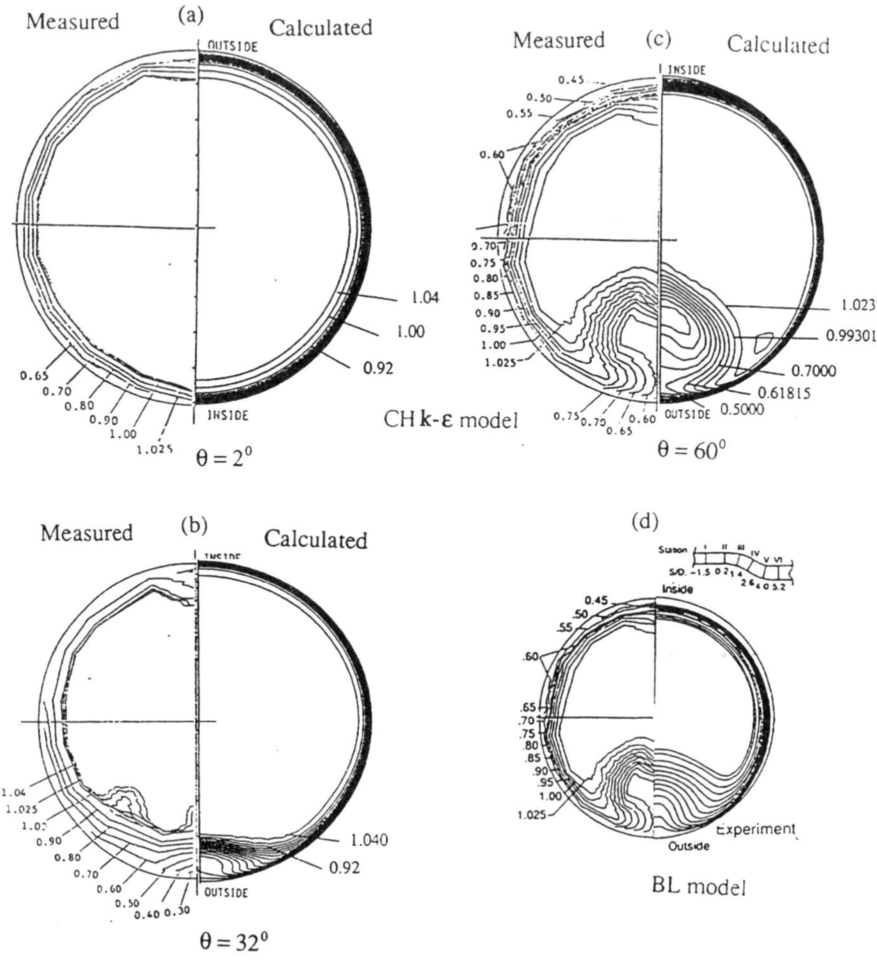

Figure 4: Comparison of total pressure recovery at various cross-sections.

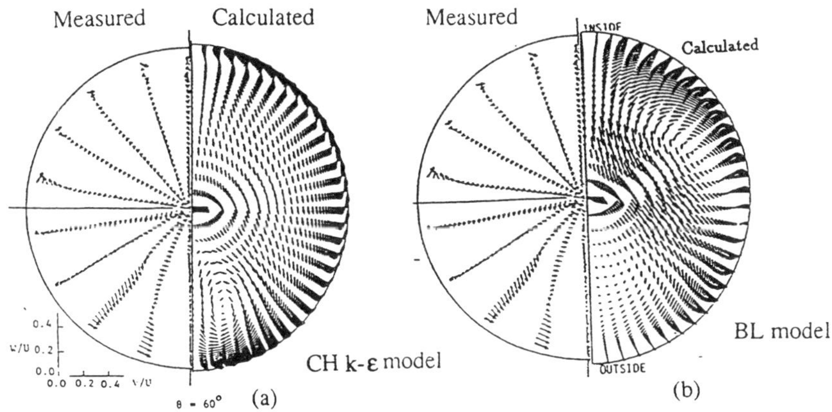

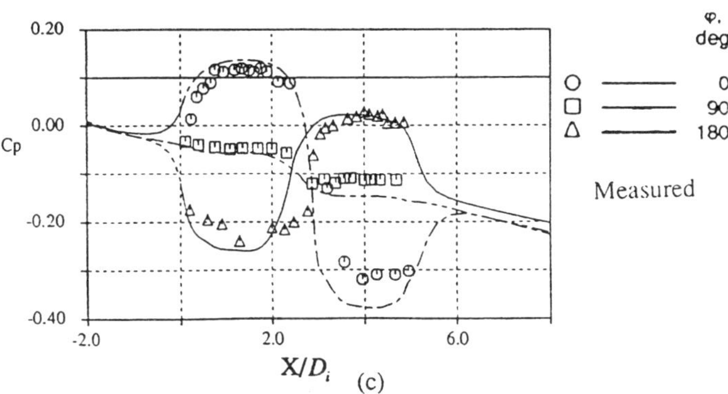

Figure 5: Secondary flow at exit and static pressure on solid wall.

C499/060/96

CFD simulation of combustion in a model gas turbine combustor

S ALIZADEH BSc, MSc, PhD
Technical University of Brno, Czech Republic
A ASKARI, R BENODEKAR, A GHOBADIAN, and **R SANATIAN** BSc, MSc, PhD
Computational Dynamics Limited, London, UK

SYNOPSIS

Calculations are compared with measurements for the reacting flow field of a model gas turbine combustor of the can type. The predictions were performed with an industrial CFD code embodying an unstructured, body-fitted mesh with embedded refinement capabilities. These features were used to faithfully represent the geometry of the combustor and achieve highest grid resolution in regions of steep flow field gradients. Different combinations of turbulence and combustion models were tested. In all cases the swirl-induced mixing in the primary zone was not accurately predicted leading to qualitative and quantitative differences with experiments. Encouragingly however, in the primary zone of the combustor results from the combination of an RNG version of the k-ε turbulence model and a Presumed-PDF combustion model used in conjunction with a laminar flamelet library appear to be of better quality than those previously reported.

NOTATION

AFR	air/fuel ratio
p	compressor delivery pressure
$P(\xi)$	probability density function for mixture fraction
T	combustor inlet temperature
ϕ	thermochemical scalar (temperature, species concentrations)
ρ	density
ξ	mixture fraction

1 INTRODUCTION

Since the pioneering days of the gas turbine the combustor has been the component which has required most attention in design and development. Due to the highly turbulent nature of the flow field inside the combustor and the additional complications of combustion, design remains generally a 'black art' guided predominantly by experimental methods.

However experiments are inherently slow and also very costly, especially at engine operation conditions. These drawbacks and the growing need to understand the complex flow field phenomena involved, has led to the development of fundamentally-based analytical approaches to assist the design process. CFD forms a major category of such procedures and is a key component in "right-first-time" strategies in advanced combustor development. Detailed combustor flow field predictions have proved effective for traditional concerns like exit temperature traverse quality and liner wall temperatures. However in the light of the challenging emission restrictions on gas turbine combustors, the use of predictive methods needs to be extended to the production of pollutants, which places higher accuracy demands in relation to temperature and species predictions.

This paper discusses the CFD simulations of a can-type model aero gas turbine combustor based on the Rolls Royce Tay design. Experiments under reacting flow conditions are reported by Tse (1) and Bicen, Tse and Whitelaw (2), providing thermochemical scalar field data, including temperature and major chemical species like fuel, CO and H_2, inside the combustor, as well as pollutant emissions. Previous reacting flow simulations of this combustor concentrated on the latter, in particular nitric oxide and particulate soot pollutant production (cf refs (3) and (4). The present study aimed to investigate the effect of various combinations of turbulence and combustion models on the main thermochemical scalar fields using a contemporary computer code and a various combinations of turbulence and combustion models, some recently developed. The objective was to determine whether improved results could be obtained. The predictions reported in the literature generally over predict the maximum temperature and the carbon monoxide levels (Alizadeh, 5).

The predictions were performed with the industrial CFD code STAR-CD, embodying unstructured body-fitted coordinates and embedded mesh refinement capabilities. These features were used to obtain a high degree of geometrical fidelity and achieve highest resolution of the grid in regions of sharp flow gradients.

2 PHYSICS MODELS

Different combinations of the standard and the RNG variants of the k-ε turbulence models and the Eddy-Break-Up (EBU) and Presumed-PDF (PPDF) combustion models were used and assessed in comparisons with the experimental data.

The RNG method is essentially a scale-elimination technique which can be applied to Navier-stokes and scalar transport equations. The RNG k-ε model is claimed by its originators to be more fundamentally-based and hence more accurate than the standard model, particularly for complex flow fields exhibiting strong streamline curvature (6). Although subsequent experience with the new model has not always supported this expectation, it is nonetheless of interest to apply it to the gas turbine combustor, where the weaknesses of the standard k-ε model for the prevailing high streamline curvature and large rates of strain are well recognised.

Combustion in this generation of gas turbine combustors in general, and the model combustor under consideration in particular, is essentially non-premixed. With the assumption of fast chemistry, the combustion process can be conveniently modelled via the distribution of the conserved scalar mixture fraction ξ, where $\xi = 1/(1+AFR)$, see Bilger (7).

As a conserved property, ξ is free from the direct influence of the highly complex chemistry of combustion. Hence its statistics, usually characterised in terms of its Favre -averaged mean and variance respectively, can be computed readily, thereby mapping the turbulent mixing field. The simplification of fast chemistry allows state relationships to be written for any thermochemical scalar $\phi = \phi(\xi)$, where ϕ may be temperature, density, species concentration etc. This is allowed if the time scales associated with the chemical reactions are much smaller than the those of turbulent mixing. The nature of the relationships depends on the assumptions made about the progress of reactions at the molecular level. Under such conditions the mean thermochemical fields can be related to the statistics of the conserved scalar ξ. For example the mean density $(\bar{\rho})$ can be obtained from the following relationship:

$$\frac{1}{\bar{\rho}} = \int_0 (1/\rho)\, p(\xi)\, d\xi \quad (1)$$

Here $p(\xi)$ is the probability density function for mixture fraction, assumed to take the form of a beta function (Jones and Whitelaw, 8), whose detailed shape is defined by its first and second moments ξ and ξ''^2 respectively, both of which can be determined from their own differential conservation equations (Spalding, 9). This approach has come to be known as the presumed-pdf method. The method is particularly attractive compared to simpler combustion representations like the EBU model as it allows incorporation of detailed chemistry information whilst only introducing two additional transport equations. The Favre pdf $\tilde{p}(\xi)$ is equal to

$$\rho p(\xi)/\bar{\rho}$$

The state relationships $\phi(\xi)$ used in this work were obtained from the assumption of either 'very fast' chemistry (although still fast enough not to include the slow minor species), in which the reactions go instantaneously to chemical equilibrium or 'finite rate' chemistry, assumed to locally occur in the laminar diffusion flame mode. The main assumption for validity of laminar flamelet concept is that burning occurs in moving laminar flamelets which are embedded in the flow. These laminar flames exist in a turbulent flow only if length scales of the turbulent flow field are sufficiently large in comparison with laminar flame thickness (Williams 10). The laminar flame structure is described as a function of mixture fraction and an instantaneous dissipation rate. Hence the flamelet library is generally two dimensional. The present predictions were obtained for a constant value of dissipation rate. Temperature and the reaction intermediates (e.g. CO, OH, O, H) can considerably vary between these two state assumptions and hence can also significantly influence emissions predictions. Equilibrium data can readily be obtained from chemical equilibrium programs like that of Gordon and McBride (11), used in the present study. The non-equilibrium flamelet data were obtained as by Alizadeh (4), from one-dimensional counter-flow diffusion flame calculations with a code developed by Warnatz and co-workers (12). In both cases the results were stored as libraries for use in the combustor simulations.

3 EXPERIMENTAL DETAILS AND MODELLING APPROACH

A schematic of the model can combustor is presented in Figure [1]. It has a 45° swirler (measured swirl number 0.6), six 10mm primary zone and six 20mm dilution zone air injection holes arranged at half pitch relative to one another and a circular to a rectangular transition exit nozzle. Gaseous propane (C_3H_8) at room temperature was injected through ten 1.7mm holes equally spaced circumferentially on the surface of a 90° cone positioned in the middle of the swirler on the can axis (Figures 1 and 2). The air mass flow was set to give an AFR of 57, corresponding to take-off conditions. For cooling purposes Transply material was used for the liner wall.

Although a number of experiments were performed at different operating conditions, the most extensive measurements were obtained at atmospheric conditions of 300K, 1bar: hence the present calculations were performed for this case.

The simulation was performed for the complete geometry. The outline of the air injection holes, swirler and fuel injector can be seen on the computational grid, displayed in Figure 2. Embedded mesh refinement was used near the swirler, primary and dilution zone holes where sharp gradients in the field variables were expected. A second mesh with further refinement in the primary zone of the combustor was also used, as a check on mesh sensitivity. The total numbers of computational cells for the 'coarse' and the 'fine' meshes were about 94000 and 184000 respectively. The calculations were performed using a TVD-type self-filtered centered spatial differencing scheme, nominally second-order. The refinements were based on reduction of higher order derivatives (Muzaferija 13).

The liner walls were modelled as non-porous. The effect of the Transply cooling flow was accounted for through the air flow splits in accordance with the recommendations of Bicen, McGuirk and Palma (14), leading to apportioning between the swirler, primary and dilution holes of 17%, 33% and 50% respectively. Zero-gradient boundary conditions were applied at the exit plane of the nozzle and all walls were assumed adiabatic. employing log-law wall functions. A turbulence intensity of 5% and a turbulent length scale of 1% of aperture size were specified at all inlets.

4 RESULTS AND DISCUSSION

Of the various turbulence and combustion model options tested, which also included the well-known Eddy Dissipation form of EBU model (Magnussen 15), the best results were obtained from the combination of the RNG turbulence model and the PPDF combustion model, used in conjunction with the laminar flamelet library. These results will therefore be reported in greatest detail. The contrasting behaviours of the other models will be briefly described, where of interest.

Figures 3 through 6 show plots of the velocity and thermochemical scalar fields on a section passing through the burner axis and a pair of primary zone injection holes. The last three figures also contain plots of the measurements.

In general the different combustion models did not have a significant impact on the flow structure, but a change in the turbulence model did, especially in the primary zone, as is evident in Figure 3. The recirculation zone in this region is seen to be markedly stronger when using the RNG variant. In previous simulations of this flow field (refs 4 and 5) based on the standard k-ε turbulence model, a lack of upstream penetration of the primary zone recirculation was reported. In this respect the stronger recirculation obtained here is an improvement.

However, the fuel concentration in the core region downstream of the fuel injector remains leaner than the measurements, a problem which existed, albeit to a greater degree, in the previous predictions. Although the errors in mixture fraction are only about 0.05 at maximum, in lean mixtures the effect on temperature and concentrations of species like CO and H_2 can be dramatic, due to the steepness of the $\phi(\xi)$ functions in these circumstances.

The comparison of the predicted temperature field with the experimental data (Figure 4) shows that the overall pattern and the regions of maximum temperatures in the primary zone and in the wake of the primary jets are reasonably well predicted. The EBU model grossly over-predicts the peak values to be about 2400°K. Due to allowance for dissociation, the PPDF-equilibrium chemistry combination predicts a drop of maximum temperature to about 2300°K. As can be seen in the predicted plot, the introduction of non-equilibrium chemistry causes the peak values drop still further to around 1950K, more in line with the experiments. These peak temperatures and their location are very important to the designer in relation to liner wall cooling requirements, temperature pattern factor and pollutant production.

The plots of CO and H_2 (Figs 5 and 6) show that the predicted patterns are broadly representative of the data in most, but not all, regions. The peak values are well predicted (4.3% CO and 2.8% H_2 for computations compared to 6% CO and 3% H_2 for experiments). The value of using the laminar flamelet libraries becomes particularly apparent when it is noted that the assumption of equilibrium combustion yields maximum CO values of about 21%.

As anticipated earlier, the greatest discrepancies between the thermochemical fields and the data occur in the lean core area downstream of the fuel injector. Although the temperature field close to the injector is better reproduced than in previous predictions for this combustor, further downstream in the core region serious discrepancies of over 600°K can be observed. In the same regions, the predicted levels of CO and H_2 are negligible, whilst the experiments show concentrations of about one third of maximum values.

Mesh refinement did not have any significant influence in further reducing the discrepancies, indicating that their origins still lie in the turbulence modelling (and probably also to some extent in the combustion modelling as well) rather than in numerical discretisation issues. This is consistent with previous findings.

5 CONCLUSIONS

This study has confirmed some of the current weaknesses of CFD in application to combustor flow fields. Nevertheless, it has also shown that with appropriate choice of contemporary physics models CFD can be a valuable tool in the gas turbine industry. It provides a qualitative appreciation and at times a quantitative impression of the velocity and thermochemical fields, helping engineers to tackle the numerous problems encountered in the design and development processes.

In particular use of flamelet data in conjunction with presumed probability density function improves the temperature and species concentration predictions. Use of two dimensional flamelet data based on local dissipation rate and joint probability density function may further improve the predictions and will be tried in the future.

The RNG variant of k-ε predicts a lower turbulent viscosity in the recirculation zone than the standard k-ε model and hence a stronger recirculation region. This trend of events improved predictions, hence turbulence models such as Chen's variant of k-ε and a fuel differential Reynolds stress model may improve the predictions even further.

REFERENCES

(1) TSE, D.G.N. Flow and combustion characteristics of model annular and can type combustors, PhD Thesis, Imperial College of Science and Technology, Department of Mechanical Engineering, 1988.

(2) BICEN, A.F., TSE, D.G.N. and WHITELAW, J.H. Combustion characteristics of a model can type combustor, Combustion and Flame, 1990, vol.80, pp.111-125.

(3) ALIZADEH, S. and MOSS, J.B. Flowfield prediction of NO_X and smoke production in aircraft engines, AGARD CP-536, Fuels and Combustion Technology for Advanced Aircraft Engines, 1992.

(4) ALIZADEH, S. Flowfield prediction of NOx and smoke production in aircraft engines, PhD Thesis, Cranfield Institute of Technology, School of Mechanical Engineering, 1988.

(5) ALIZADEH, S. Flow field prediction of NO_X and smoke production in aircraft engines, Cranfield Institute of Technology, 1994.

(6) YAKHOT, V., and ORSZAG, S.A. "Renormalization group analysis of turbulence-I: Basic theory", J. Scientific Computing, 1, pp. 1-51. 1986.

(7) BILGER, R.W. Reaction rates in diffusion flames, Combustion and Flame, 30, pp.277-284, 1977.

(8) JONES, W.P., and WHITELAW, J.H. "Calculation Methods for Reacting Flows in practical systems, [Presented at the Fluid Engineering conference. Boulder, Colorado], Morel,T. (ED), The American Society of Mechanical Engineers. 1981.

(9) SPALDING, D.B., Concentration fluctuations in a round turbulent free jet, Chem. Eng. Sci. 26: 95

(10) WILLIAMS, F.A. Turbulent mixing in non-reactive and reactive flows. (Ed. S.N.B. Murphy), P.189 Plenum Press. 1975.

(11) GORDON, S. and McBRIDE, B. Computer program for calculation of complex chemical equilibrium compositions, rocket performance, incident and reflected shocks, and chapman-jouget detonations, NASA SP-273, 1971.

(12) WARNATZ, J. Numerical methods in laminar flame propagation, 1982, Friedr. Vieweg & Sohn, Braunschweig/Wiesbaden.

(13) MUZAFERIJA, S. Adaptive finite volume method for flow prediction using unstructured meshes and multigrid approach, PhD Thesis-University of London,1994.

(14) BICEN, A.F, McGUIRK, J.J. and PALMA, J.M.L.M. Modelling gas turbine combustor flowfields in isothermal flow experiments, Proc. Instn. Mech. Engrs., 1989, vol.203, pp.113-122.

(15) MAGNUSSEN, B.F., and HJERTAGER, B.W. "On the structure of turbulence and a generalised eddy dissipation concept for chemical reaction in turbulent flow", 19th AIAA Aerospace Meeting, St Louis, USA. 1981.

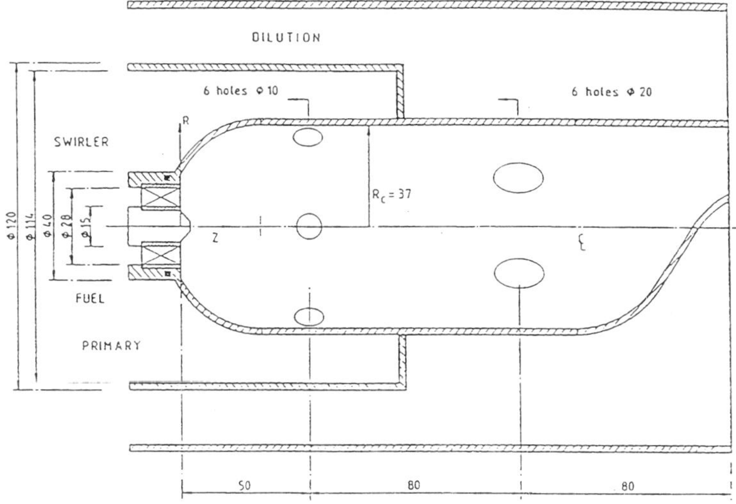

Fig 1 Schematic of the Model Can Combustor experimental rig

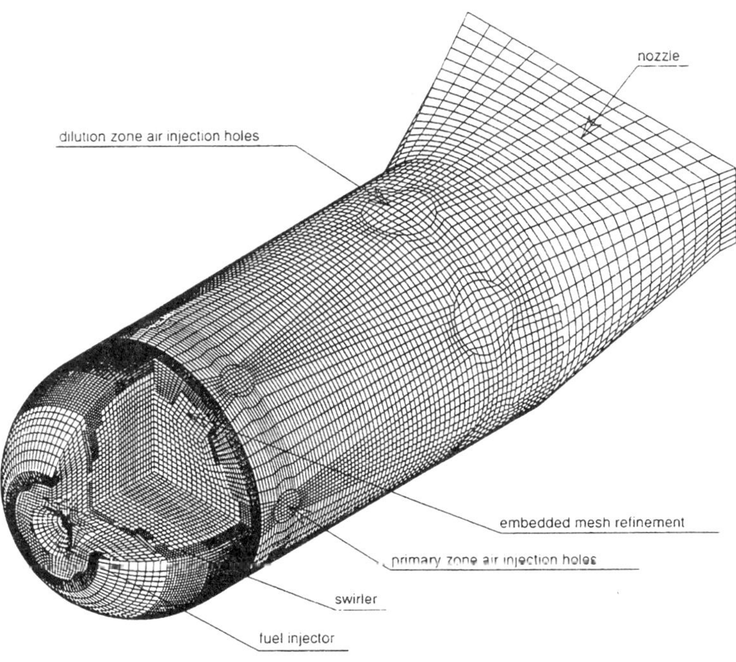

Fig 2 Calculation mesh for Model Can Combustor

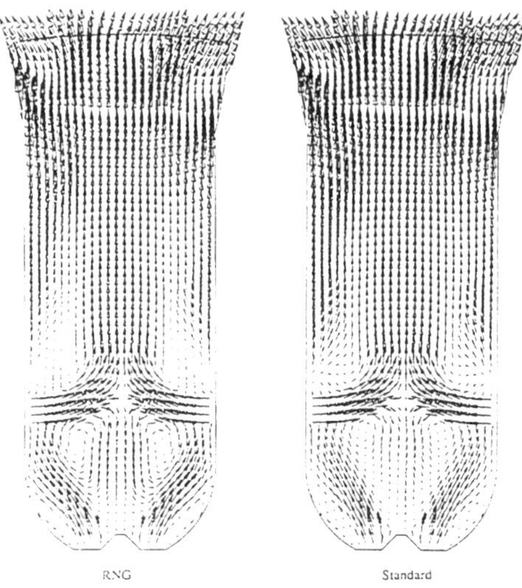

Fig 3 Comparisons of flow fields predictions using standard k-e and RNG turbulence models in conjunction with PPDF (Equilibrium) combustion model

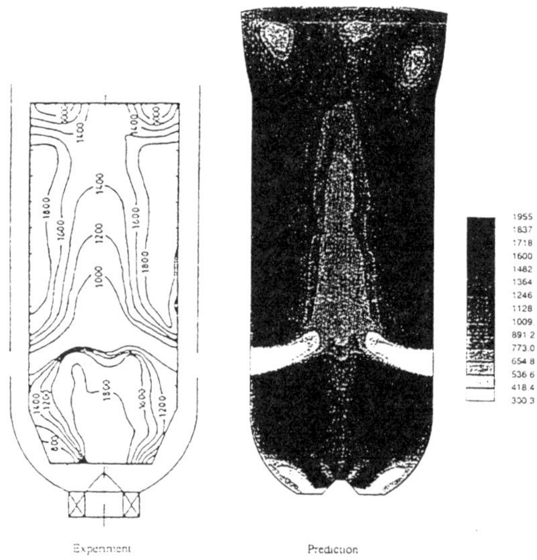

Fig 4 Comparisons of predicted and measured temperature fields

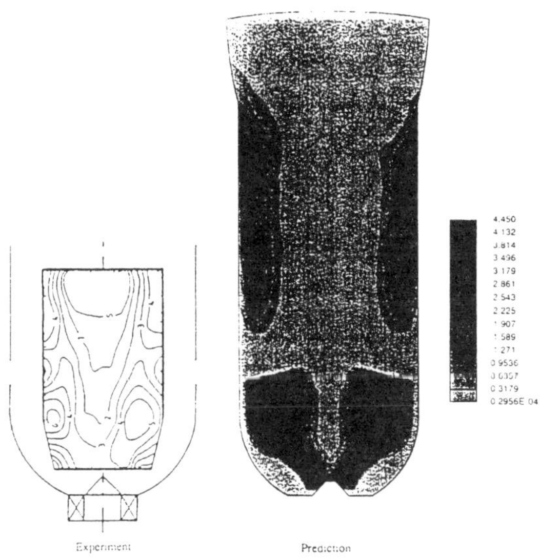

Fig 5 Comparisons of predicted and measured CO concentration fields

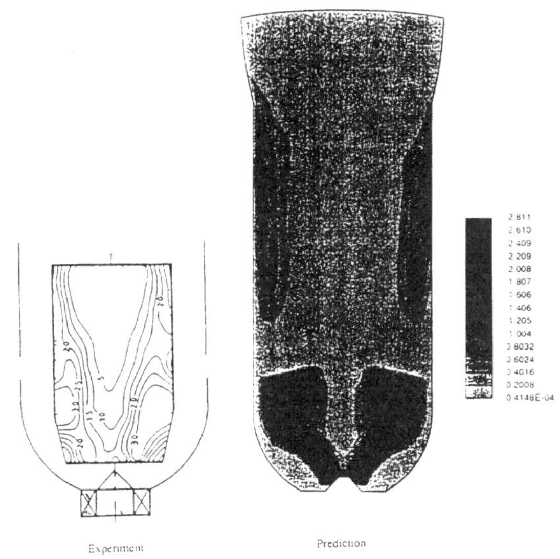

Fig 6 Comparisons of predicted and measured H_2 concentration fields

C499/002/96

Flame image processing in a four-valve spark-ignition engine

C ARCOUMANIS PhD, DSc(Eng), MSAE, MASME, FIMechE
Department of Mechanical Engineering, Imperial College of Science, Technology and Medicine, London, UK
C-S BAE BSc, MSc, PhD, DIC, MSAE
Department of Aerospace Engineering, Chungnan National University, Korea

Synopsis

The flame development in a four-valve, single-cylinder, spark-ignition optical engine was analysed under lean mixture conditions by direct flame imaging as well as by planar imaging of the Mie-scattered light from titanium dioxide particles introduced into the inlet port; illumination along a thin light sheet was provided by a pulsed Cu-vapour laser synchronised to the engine and to an image-intensified CCD camera. Software enhancement of these images allowed comparison of the cycle-resolved enflamed area and the displacement of its centre calculated from the two imaging methods, for two tumble flow fields generated by non-sleeved and sleeved intake ports.

1. INTRODUCTION

Combustion in a spark-ignition engine is a complicated process involving a short duration spark discharge, development of a flame kernel which grows under the influence of the local flowfield into a highly-wrinkled fast-burning turbulent flame, and propagation of this flame across the combustion chamber until it reaches the chamber walls where it extinguishes. During the flame kernel development, part of the electrical energy deposited into the mixture is lost by heat conduction to the spark plug electrodes and it is the local mixture composition and flame stretch which govern the kernel growth rate. The initial strong effect of the mean flow is succeeded by the flow turbulence which gradually transforms the small flame kernel into a turbulent flame as the propagating flame front experiences an increasing spectrum of length scales (1). Evidence of this is the increasingly wrinkled and distorted thin reaction zone which leads to faster burning rates until the flame reaches the wall where successively smaller scales of turbulence exert an influence on the flame and the burning process slows down (2).

Lean-burn combustion represents one of the most promising concepts for improving fuel economy and reducing exhaust emissions including carbon dioxide. Unfortunately, lean burning is associated with increased cycle-to-cycle pressure and torque output variations, due to the reduced flame initiation and propagation rates, and high engine-out hydrocarbon emissions. It is the lack of understanding of how flames slow down and eventually extinguish, whether during flame kernel development or near walls, which limits the ability of engine designers to resolve the problem of cycle-to-cycle combustion variations under lean mixture or high dilution engine operation (3).

The most practical approach for improving engine stability under lean mixture conditions is to increase burning rates through enhanced mean flow and turbulence of the mixture. At present, the formation of a tumbling vortex along the mid-plane through the intake and exhaust valves in four-valve, pentroof combustion chambers represents the most effective method for enhancing turbulence levels at the time of ignition. This is achieved through distortion of the tumble by the moving piston in the second half of the compression stroke, which releases its turbulent kinetic energy at a time when it is too late for turbulence to decay. The various scales of turbulence successively wrinkle the flame, causing local stretching or straining of the reaction sheet which is usually quantified by the product of the Lewis number, Le, and the Karlovitz stretch factor, K; Le is the ratio of the thermal to the molecular diffusivity while K is the product of the turbulent strain rate and the transit time of the flow through the laminar flame. The higher the value of the product KLe, the stronger is the impact of flame stretching on the laminar burning velocity which, at high enough values, can lead to flame quenching (4). Under lean mixture conditions where the laminar flame speed is lower than for stoichiometric mixtures, the value of Le exceeds unity and, if combined with the high K values associated with the high turbulence levels resulting from strong tumble and/or high engine speeds, can lead to misfires near the lean-burn limit and, thus, high exhaust hydrocarbon emissions.

Visualisation has been traditionally the most effective method for analysing the flame behaviour in reciprocating engines as a function of the flow and mixture distribution at the time of ignition (5). In addition to direct imaging of the self-illuminated propagating flame (6), schlieren photography (7) and shadowgraphy (8), planar imaging has been recently used (9-12) to examine details of the flame structure with an improved spatial and temporal resolution.

In the present study, the flame development in a four-valve, single-cylinder, optical petrol engine was investigated through a pair of small quartz windows in the cylinder liner and a large window in the piston crown by means of direct flame imaging along the line of sight and Mie-scattering. In the latter case, the pulsed light from a copper-vapour laser was scattered along a plane by titanium dioxide particles introduced into the port and imaged onto an image-intensified CCD camera. Software processing of these images allowed calculation of the enflamed area and the displacement of its centre as a function of crankangle from the time of ignition. Comparison of the results obtained with the two flame imaging methods was performed under lean mixture (A/F =18) conditions for two flowfields generated by sleeved and non-sleeved intake ports (13), corresponding to high and low mean flow and turbulence levels, respectively.

2. EXPERIMENTAL SYSTEM

2.1 Research Engine

Figure 1 shows the configuration of the four-valve optical research engine which is based on a Hydra crankcase and incorporates the cylinder head of a production Jaguar engine. The geometric and operating characteristics of this engine are summarised in Table 1. The entry section of the inlet manifold and port have a direction of 26° to the bottom surface of the cylinder head and approximately 47° to the stem of the intake valves. Previous studies (13, 14) with the same head showed that this inlet port geometry imposes directional characteristics to the valve exit flow, thus generating a tumbling motion (barrel swirl) in the engine cylinder. Figure 1 also shows the modifications which have allowed access to the combustion chamber for flame visualisation as well as laser Doppler velocimetry (LDV) measurements through windows in the liner and in the piston. The side window module was installed between the gasket face of the cylinder head and the engine liner and contained two parallel quartz windows of 15mm height and 20mm thickness which could be rotated in two directions. More complete optical access into the combustion chamber was provided through a hollow elongated piston incorporating a large flat quartz window.

Table 1 Geometric/operating characteristics of four-valve research engine

Bore	83 mm
Stroke	92 mm
Mean Piston Speed, V_p 2,000 rpm	6.134 m/s
Compression ratio	10.5
Connecting rod	154 mm
Intake Valves : diameter	33.5 mm
seat angle	45°
open at	6° BTDC
close at	224° ATDC
Max. lift	9.13 mm

One of the belt-driven overhead camshafts was modified to accommodate an optical shaft encoder for the synchronisation of the LDV measurements and flame images with crankangle, offering a temporal resolution of 0.36°CA. The twin inlet ducts were fitted in some of the tests with sleeves to partially block their lower parts and enhance the induction-generated tumbling vortex strength. This proved to be a very convenient method for enhancing the mean flow and turbulence at the time of ignition (Fig. 3) without a significant adverse effect on the volumetric efficiency of the engine.

The cylinder head incorporated a pentroof combustion chamber with an almost centrally-located spark plug and was equipped with a port-fuel injection system for the firing tests. The engine operation was under the control of the engine test bed so that the coolant water and oil temperature were maintained at 80±3°C, and the engine rotational speed remained constant within ±3rpm. A commercial engine management system (LUCAS 15CU), which comprised an electronic control unit, various sensors and a transistorised coil ignition system, was modified to control the ignition timing and duration of fuel injection. The intake air/fuel ratio was estimated by a corona discharge air flowmeter and the fuel consumption by gravimetric means. The temperature of the fuel prior to its injection into the port was around 30°C and the pressure in the fuel line was regulated at 3±0.001 bar. The firing tests in the engine were performed at 2,000 rpm for a lean mixture (A/F =18) using 95 ULG fuel at a partial load of 2.1 bar IMEP.

Two inlet port geometries were examined (for details see Fig. 1), one with purpose-built sleeves fitted in the intake ports and the other without sleeves, representing high and low tumble configurations, respectively.

2.2. Combustion Analysis

The fuel injection duration and throttle position were adjusted for each test condition so that the engine output remained constant at 2.1 bar IMEP. The cylinder pressure was measured with a piezoelectric pressure transducer (Kistler 6121) and a charge amplifier. The output of the charge amplifier was interfaced to a microcomputer via an interface card (Dostek 1400 DMA) which allowed simultaneous acquisition of both the pressure and the crankangle information provided by the camshaft-mounted optical shaft encoder. For each of the test conditions, cylinder pressure data were collected from over 350 cycles. The timing uncertainty comprised that of the trigger position, engine speed fluctuations and ignition timing. The uncertainty in the trigger position was around 0.36°CA which is the resolution of the shaft encoder. Considering the engine speed variability (±3rpm), the timing uncertainty relative to TDC of compression was estimated to be 0.5°CA similarly to the uncertainty in the spark timing. Therefore, the overall timing uncertainty relative to the crankangle of ignition was not more than 1°CA.

Combustion angles were derived from the in-cylinder pressure history according to the conventional heat release analysis (15). Although the cylinder pressure responds to both changes in the instantaneous volume of the combustion chamber and the heat transfer to the chamber walls, a heat release analysis can provide useful information about the combustion development. By considering the combustion chamber as an open system and assuming that the ideal gas law is applicable to the charge mixture, the chemical energy release rate or the gross heat release rate can be estimated according to the first law of thermodynamics. The cumulative heat release variation as a function of crankangle, integrated from either of the heat release rates and normalized by its maximum value, is considered equivalent to the burned mass fraction. A typical calculation shows, for example, that for central ignition and typical engine operating conditions, only about 10% of the charge has burned when the flame radius is half the cylinder radius due to the 4:1 density ratio between the unburned and burned gas. The heat release analysis can be used conveniently to characterise different stages of the combustion process according to their duration in crankangle degrees which are defined as :

i) Flame development angle q_d - the crankangle interval between the spark discharge and the time when 5% of the fuel chemical energy is released.

ii) Rapid burning angle q_b - the crankangle internal between the end of the flame development stage and the end of the flame propagation process (energy-release fraction of 5-95%)
iii) Overall burning angle q_o - the sum of the crankangle intervals corresponding to the flame development and the rapid burning angle (energy-release fraction of 95%).

In addition to these typical combustion angles, the crankangle at which 2% of the energy is released was estimated. Prior to the heat release analysis, the measured instantaneous cylinder pressure was smoothed to reduce any effects of noise. The combustion angles were evaluated from the in-cylinder pressure trace of individual cycles which were then averaged to obtain cycle-averaged information.

2.3 Visualisation Techniques

Direct imaging - The propagating flame was first visualised directly by means of an image intensified CCD video camera (Proxitronic HF1) at lean mixture conditions (A/F =18) and 2,000 rpm through the quartz piston window and mirror. This was done for each sequence of engine cycles with the same relative location in each cycle by gating the image intensifier at a range of crankangles. The electronic shutter was triggered externally by the shaft encoder signal at a particular crankangle after ignition for about 35 successive cycles. The intensified detector was able to detect the blue-coloured flames generated during part-load operation with lean mixtures which are difficult to observe otherwise due to their low luminosity (16). The spectral response of the photocathode elements ranges between 300 and 800nm with a flat peak region at about 480nm.

The flame images were subsequently analysed with a software package capable of calculating the enflamed area, the gravitational centre of this area and the distance between the spark plug and the flame centre. This distance is very important during the early stages of flame development since a small flame can be more easily convected by the mean flow and is more susceptible to quenching.

Two-dimensional imaging with laser Mie-Scattering - The set-up of cycle-resolved 2-D flame visualisation is shown in Fig. 2 and consists of a Cu-vapour laser as the light source and the image-intensified CCD video camera mentioned previously as the detection device. The mixture was seeded with TiO_2 particles introduced into the port. A thin planar sheet of laser light was passed through the liner windows and the Mie-scattered light from the particles in the unburned mixture which could be easily separated from the dark burned gas regions, allowed the flame to appear as an interface between the bright and dark areas.

The Cu-vapour laser was operated at a frequency of 8.3kHz which is a quarter of the frequency of the pulse train generated by the shaft encoder at an engine speed of 2000rpm. Pulse duration was about 20ns and the maximum single pulse energy at 511nm was ~2.75mJ. The laser light sheet of 40mm width and ~0.5mm thickness was generated by a spherical/cylindrical lens system; the sheet entered the combustion chamber through the liner windows, at a plane 3mm below the bottom surface of the cylinder head and 2mm above the piston at TDC. The scattered light from the seed particles was passed first through a band-pass filter before being focussed onto the CCD camera.

Titanium dioxide particles of submicron size were used for seeding. The seeding rate of the cyclone generating these particles was regulated by the pressure of the compressed air convecting the particles downstream. TiO_2 has been widely used as seeding particles in laser Mie-scattering studies; it can also be formed by adding gaseous $TiCl_4$ to the air which then reacts with water to form TiO_2 particles and HCl. It has been reported however, that extensive wear and corrosion may occur due to the presence of HCl.

The flame propagation images from the lean petrol/air mixture obtained by the CCD camera were stored in a S-VHS video tape and transferred into the computer by a Matrox IP-8 grabber card. These images were subsequently analysed with an image processing software to obtain information about the flame structure under the different tumble flow conditions generated by sleeved and non-sleeved intake ports. The software package allowed enhancement of the flame images and calculation of the enflamed area and its displacement from the spark plug.

3. RESULTS AND DISCUSSION

The velocity field in the cylinder of the four-valve optical engine was characterised by LDV and detailed results were reported in (13). Figure 3 presents the mean and rms tumble velocities in the vicinity of the spark at an engine speed of 2000rpm. The four-valve cylinder head generates an organized tumbling motion which was evident in the symmetry plane from the early stages of induction. During compression, the turbulence produced by the increasing shear of the tumble velocity component was redistributed into the other two velocity components and resulted in volume-averaged turbulence levels which were more than 60% higher than conventional two-valve induction systems. Since inlet ports with inserted sleeves were found to give a stronger tumbling vortex (14), similar purpose-built sleeves were inserted into the present four-valve head to examine the degree of flow enhancement that they can induce; for details about the sleeve geometry see Fig. 1. In Fig. 3, bold lines represent ensemble-averaged mean and rms velocities for the sleeved inlet ports and thin lines represent those for the non-sleeved ports; both are normalised by the mean piston speed Vp, where Vp=6.13m/s at 2,000rpm. After 310°CA, the tumble velocities in both cases start decreasing. With the sleeved-ports, by 330°CA which is a typical ignition angle for lean engine operation, turbulence has decayed to ~1.0Vp while the mean velocity remains relatively high, of the order of 2Vp. It is interesting to note that in absolute terms the mean velocity magnitude at this location is much higher than the 3-5m/s recommended by Pischinger and Heywood (17) for stable lean-burn operation. The basis for their recommendation was that the local velocity magnitude at the spark plug should be large enough to convect the flame kernel away from the electrodes but not so large as to quench the flame. Figure 3 also shows that in the case of the non-sleeved ports the tumble mean velocity during the period between 310°CA and 330°CA decays to 1Vp and the turbulence to 0.8Vp which is 10-20% lower than that with the sleeved ports at the time of ignition.

The higher mean and rms velocities present in the combustion chamber in the case of the stronger tumble resulted in significant enhancement of the burning rates. Figure 4(a) shows the pressure traces for the sleeved and non-sleeved inlet port configurations while Fig. 4(b) presents the combustion angles evaluated from the heat release analysis based on the in-cylinder pressure traces; these are cycle-averaged values over more than 350 cycles. Figure 4(b) confirms that the flame development angle for the high tumble configuration generated by the sleeved-ports is, as expected, shorter than that for the low tumble configuration by about 5°CA. On the other hand, the rapid burning angle which represents an average period of flame propagation throughout the combustion chamber is only slightly longer with the low tumble configuration. This difference in the overall burning angle indicates that the slow burning rates of the low tumble configuration are mostly due to the longer flame development period and, therefore, early ignition angle. The low-tumble configuration resulted in 7.4°CA longer overall combustion angle than the high tumble sleeved-port configuration, with 5.1°CA out of the 7.4°CA attributed to differences in the flame development angle.

In order to identify and define the flame contour in the images obtained with direct flame imaging and 2-D laser Mie-scattering, the intensity data of each image generated by the image-intensified CCD video camera require several data processing steps (Figure 5). The basic element of a digitized image is the pixel or picture element and an image is composed of pixels distributed in a rectangular array. Each pixel in the image can have a value between 0 (black) and 255 (white), that is 256 different grey levels are defined as pixel values. Figure 6(a) shows a raw image before processing obtained by direct imaging which contains all the image data including noise. The x and y axes represent the image array and the z axis characterises the pixel value.

The first step in processing the image was to look for the region of the combustion chamber which excludes background noise outside the 45deg mirror in the extended piston (Fig. 5(b)). A threshold technique was applied to identify the flame contour which defines the flame region (see Fig. 5 (c)), thus allowing the calculation of the enflamed area and the gravitational centre of the flame. It is clear that the identification of the flame front is accurate if the edges of the flame are sharp as shown in Fig. 5(d). Although in most cases, the flame front was optically sharp, some parts of it exhibited a gradual intensity variation from the core of the kernel towards the unburned gas, especially in the case of very early flame images. Nevertheless, changing the threshold level did not seem to have an appreciable effect on the flame region. The Mie-scattered image data had to be inverted during the above processing steps because the Mie- scattered light from the particles in the unburned mixture was much brighter than that of the dark burned regions. Figure 6 shows examples of processed images from both direct flame imaging and 2-dimensional visualisation with laser Mie-scattering, together with the corresponding observation areas. The two inlet valves are positioned on the left-hand side of these images and the exhaust valves on the right-hand side. The spark plug position is marked by a cross, while the calculated gravitional centre is represented by a dot.

Figure 7 shows typical images of the developing flame for the sleeved inlet ports configuration based on the flame natural luminosity in the absence of external illumination. The obtained flame images provided evidence of highly turbulent and asymmetric flames with the initial flame kernel convected away from the spark plug and stretched towards the direction of the exhaust valves, due to the prevailing tumbling motion. The strong tumbling flow past the spark plug convects the early kernel away from the gas, allowing it to grow more rapidly and preventing it from being cooled and slowed down through contact with the electrodes.

Figure 8 shows typical images of the developing flame obtained with 2-D visualisation of the laser Mie-scattered light in the case of the sleeved inlet ports and at the same engine condition to that of Fig. 7. These planar, indirect flame images obtained over a very short period (20ns) along a thin laser sheet are capable of revealing more details about the geometry of the flame front than the direct flame images integrated over a much longer period (50-100μs) along the line of sight. As it is clear from Figs. 7 and 8, the cyclic variation of these early flame images is significant with variations more pronounced in the direction and the extent of flame convection, as reflected in the shape and size of the flame images.

In order to derive more quantitative information, the flame images obtained with both sleeved and non-sleeved inlet ports were processed over at least 65 engine cycles to allow calculation of the projected enflamed area, the gravitational central point of this area and the distance between the spark plug and the centre of the flame.

Figure 9 shows the calculated enflamed area versus crankangle for both imaging methods. The flame areas were computed by adding the number of pixels that have intensity values above a predetermined threshold level, to allow first identification of the flame front, and then multiplying them by a weighting factor derived from calibration. There is, obviously, a degree of subjectivity in the definition of the edges of the flame during its early development which can lead to errors in the calculated flame areas; the overall uncertainty in the calculation of the flame area is estimated to be about 10%. Figure 9(a) shows the flame area of the Mie-scattered flame images for both sleeved and non-sleeved inlet port configurations. It is clear that the enflamed area of the sleeved inlet configuration is always larger than that produced by the non-sleeved inlet ports which demonstrates the advantages of sleeves as a simple method to improve the lean burn performance of the engine, by increasing the charge burning rate through the combined effect of higher mean and turbulent velocities. Figure 9(b) compares the area of the flame images obtained with direct imaging and 2-D Mie-scattering which shows that the calculated flame area from the Mie-scattered image data is smaller than that obtained from direct flame illumination by 3-5 times; this is justified from the different method of observation in the two cases. The Mie-scattering technique applies to a thin slice of the flame in the combustion chamber imaged over a very short time (20ns), thus allowing identification of the highly turbulent flame structures present which include islands of reactants and products associated with turbulent mixing and air entrainment. On the other hand, the projected images of the 3-D flame over the line of sight, exposed over the relatively long period of 50-100μs, are the result of optical integration of the flame illuminated areas at different planes. In addition, there is a difference in the observation areas between the two methods (see Fig. 6); the area of the Mie-scattered images is just 40mm wide compared to the 58mm diameter of the piston window, thus restricting the observation region of the flame in this case. All the above imply that conclusions based on direct comparison of the two imaging methods should be treated with caution but that each method provides certain type of information which can be useful to engine designers and computer modellers alike.

Closer examination of the rate of change of the enflamed area estimated from the two flame imaging methods reveals that direct flame imaging gives a faster rate of area change of ~22 mm^2/°CA over the crankangle period 16-26°CA after ignition compared to only ~14mm^2/°CA as estimated from the 2 -D Mie-scattered images. This seems reasonable when considering the three-dimensionality of engine flames which are more closely represented by the integrated information along the line of sight provided by direct flame imaging than the limited information along a single plane cutting through the flame.

Figure 10(a) shows the comparison of the flame centre displacement obtained with the sleeved and non-sleeved inlet port configurations. The flame centres of the sleeved configuration are more displaced towards the exhaust valve side compared to those of the non-sleeved case. This underlines the influence of the higher mean velocities generated by the stronger tumbling motion in the case of the modified intake ports which convect the small flame away from the spark plug. This stretching and convection of the flame by the mean and turbulent flow in the case of the sleeved inlet ports gives rise to a highly wrinkled and convoluted flame about 10mm downstream, corresponding to a larger flame surface area between the burned and unburned gases and resulting in faster burning rates.

Although the high tumble flow obtained with the sleeved ports has led to faster burning rates, reduced cyclic variations and extension of the lean operating limit (13), it should be borne in mind that this does not necessarily mean that the tested configuration represents the optimum inlet port geometry and tumble strength for this particular engine. It has been found that a high tumble configuration may result in lower power output at wide open throttle due to the reduced volumetric efficiency of the induction system and can lead to a weak knock, indicating the need for higher octane number fuel. It seems that a compromise is necessary in terms of the optimum tumble strength which allows stable operation at both part-throttle and wide-open-throttle conditions for the case of lean mixtures.

Figure 10(b) shows the comparison of the flame displacement deduced from the direct flame images and the 2-D Mie-scattering images at various crankangles after ignition. It is clear from these results that although both methods indicate a similar order of flame displacement away from the spark plug, the exact direction of displacement varies. Nevertheless, both methods confirm that the flame is convected towards the exhaust valve side and that at a distance of about 10mm from the spark plug the turbulent flame has grew to a size which is less susceptible to convection by the mean flow, as evidenced by the relatively small displacement observed in the period between 16-28°CA after ignition.

Overall, it can be concluded that both flame imaging techniques are capable of identifying the influence of the local flow near the spark plug on early flame development and estimating the flame displacement distance but they differ in terms of the quantitative information they provide about the enflamed area due to their very different observation method. Direct flame imaging seems to provide a better indication of the rate of area change with time by including 3-D effects along the line of sight while 2-D Mie scattering is more capable of resolving the flame structure along a plane and providing validation data for CFD modelling. However, both techniques are unable to characterise the three-dimensional structure of the turbulent flames present in engine cylinders which only multiple laser sheets (18) can resolve.

4. CONCLUSIONS

Two-dimensional visualisation with laser Mie-scattering and direct flame imaging was undertaken to examine the flame propagation under tumbling flow, lean mixture conditions in a single-cylinder engine equipped with a four-valve pentroof combustion chamber for two tumbling vortex strengths generated by sleeved and non-sleeved inlet ports. In particular :

1. Laser Mie-scattering was used to visualise the 2-D structure of the developing flame;. the deduced flame images were processed and compared with direct flame images in terms of the flame structure, enflamed area and displacement of the flame centre away from the spark plug. The 2-D Mie-scattering technique revealed a more irregular boundary between the burned and unburned gas regions than is apparent with direct flame imaging which, on the other hand, allowed a better estimate to be obtained of the size and direction of the flame.

2. Comparison of the flame development between the two tumble flow configurations using the two flame imaging techniques revealed that the high tumble configuration induced by the sleeved inlet ports resulted in larger projected flame area and stronger convection of the flame towards the exhaust valves during the early stages of flame development. This convection and stretching of the flame by the higher mean velocities near the spark plug at the time of ignition resulted in 6% faster burning rate due mainly to the faster flame development time.

5. ACKNOWLEDGEMENTS

The authors would like to acknowledge the financial support provided for this project by the Science and Engineering Research Council of the United Kingdom (Grant GR/H46879) and the technical support offered by Mr S. Barraclough of Jaguar Cars Ltd. They would also like to thank T. Haas and S. Ali for their contributions to the processing of the flame images.

6. REFERENCES

(1) BRACCO, F.V., "Modeling and Diagnostics of Combustion in Spark-Ignition Engines", Proc. Intl. Symp. on Diagnostics and Modeling of Combustion in Reciprocating Engines, COMODIA 85, Tokyo, Japan, pp. 1-13, September 4-6, 1985.

(2) MERDJANI, S. and SHEPPARD, C.G.W., "Gasoline Engine Cycle Simulation Using the Leeds Turbulent Burning Velocity Correlations", SAE Paper 932640, 1993.

(3) HEYWOOD, J.B., "Combustion and its Modeling in Spark-Ignition Engines", Proc. Intl. Symp. on Diagnostics and Modelling of Combustion in Reciprocating Engines, COMODIA 94, Tokyo, Japan, pp. 1-15, July 11-14, 1994.

(4) BRADLEY, D., LAU, A.K. and LAWES, M., "Flame Stretch Rate as a Determinant of Turbulent Burning Velocity", Phil. Trans. R. Soc., London, A338, pp.359-387, 1992.
(5) BERETTA, G.P., RASHIDI, M. and KECK, J.C., "Turbulent Flame Propagation and Combustion in Spark Ignition Engines", Combustion and Flame, Vol. 52, pp. 217-245, 1983.
(6) BATES, S.C., "Further Insights into SI Four-Stroke Combustion Using Flame Imaging", Comb. Flame, Vol. 85, pp. 331-352, 1991.
(7) BARITAUD, T.A., "High Speed Schlieren Visualisation of Flame Initiation in a Lean Operating SI Engine", SAE Paper 872152, 1987.
(8) WITZE, P.O. and VILCHIS, F.R., "Stroboscopic Laser Shadowgraph of the Effect of Swirl on Homogeneous Combustion in Spark-Ignited Engine", SAE Trans. 90, pp. 979-992, SAE Paper 810226, 1981.
(9) BARITAUD, T.A. and GREEN, R.M., "A 2-D Flame Visualisation Technique Applied to the I.C. Engine", SAE Paper 860025, 1986.
(10) ZUR LOYE, A.O. and BRACCO, F.V., "Two-Dimensional Visualisation of Premixed-Charge Flame Structure in an IC Engine", SAE Paper 870454, 1987.
(11) ZIEGLER, G.F.W., ZETTLITZ, A., MEINHARDT, P., HERWEG, R., MALY, R. and PFISTER, W., "Cycle-Resolved Two-Dimensional Flame Visualisation in a Spark-Ignition Engine", SAE Paper 881634, 1988.
(12) MANTZARAS, J., FELTON, P.G. and BRACCO, F.V., "Three-Dimensional Visualisation of Premixed-Charge Engine Flames : Islands of Reactants and Products; Fractal Dimensions and Homogeneity", SAE Paper 881635, in SP-759, 1988.
(13) ARCOUMANIS, C., BAE, C-S. and HU, Z., "Flow and Combustion in a Four-Valve Spark-Ignition Optical Engine", SAE Paper 940475, 1994.
(14) BAE, C-S., "Flow and Flame Interaction in Spark-Ignited Premixed Mixtures", PhD Thesis, Imperial College, University of London, 1993.
(15) CHUN, K.M. and HEYWOOD, J.B., "Estimating Heat-Release and Mass-of-Mixture Burned from the Spark-Ignition Engine Pressure Data", Comb. Sci. Tech., Vol. 54, pp. 133-144, 1987.
(16) NAKAMURA, A., ISHII, K. and SASAKI, T., "Application of Image Converter Camera to Measure Flame Propagation in a SI Engine", SAE Paper 890322, 1989.
(17) PISCHINGER, S. and HEYWOOD, J.B., "How Heat Losses to the Spark Plug Electrodes Affect Flame Kernel Development in an SI Engine", SAE Paper 900021, 1990.
(18) HICKS, R.A., LAWES, M., SHEPPARD, C.G.W. and WHITAKER, B.J., "Multiple Laser Sheet Imaging Investigation of Turbulent Flame Structure in a Spark-Ignition Engine", SAE Paper 941992, 1994.

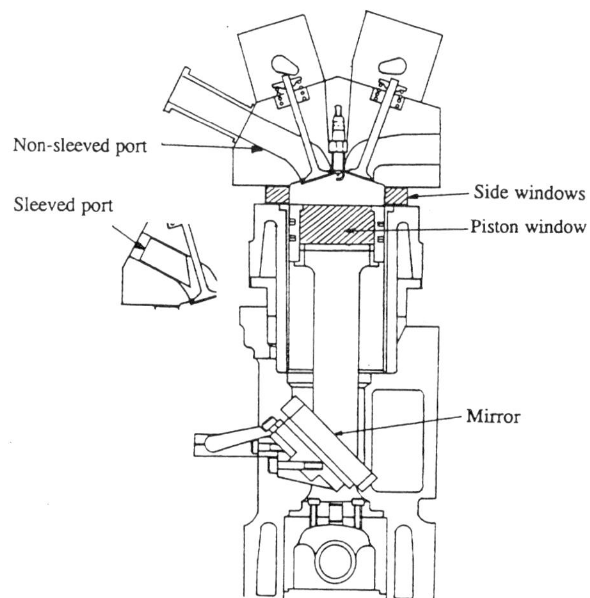

Figure 1 Jaguar four-valve engine with the details of the optical windows

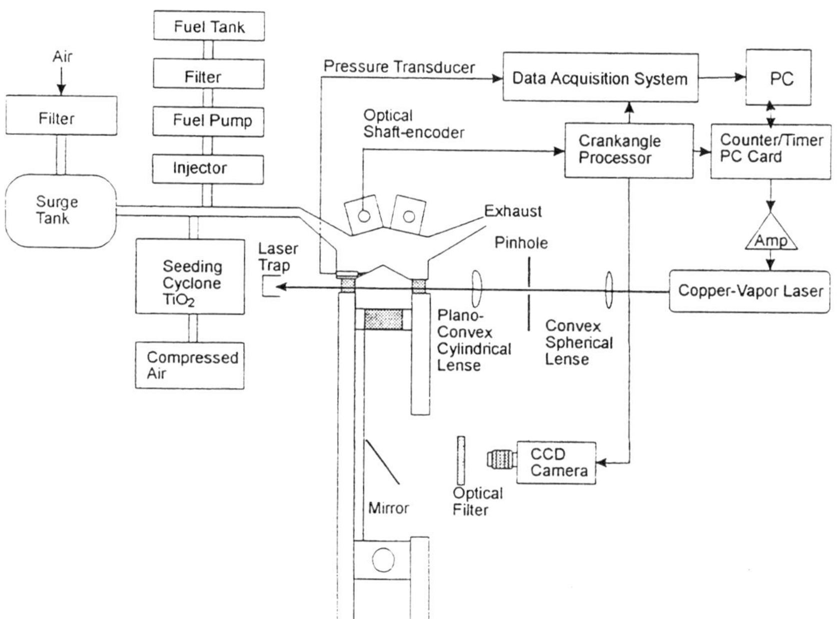

Figure 2 Schematic of the experimental set-up for two-dimensional visualisation with laser Mie scattering

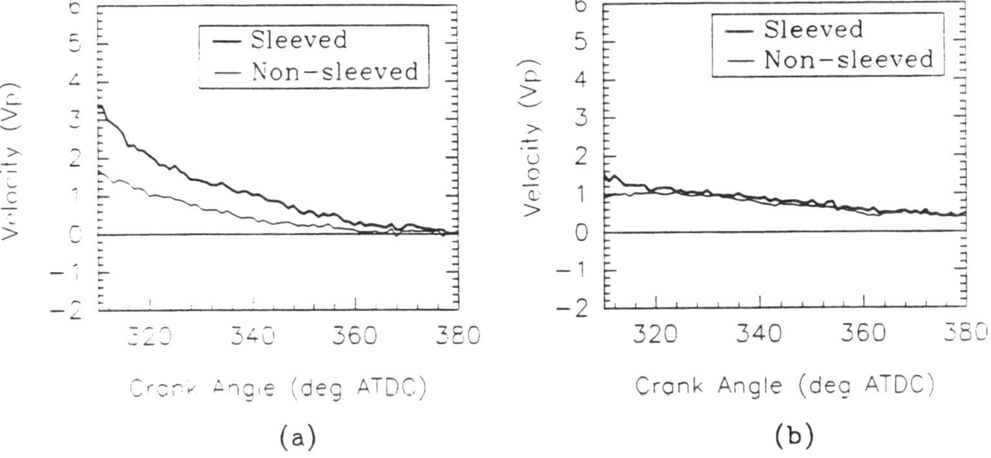

Figure 3 Ensemble-averaged velocities at the spark gap for the sleeved and non-sleeved port configuration at 2000rpm
(a) mean velocities (b) rms velocities

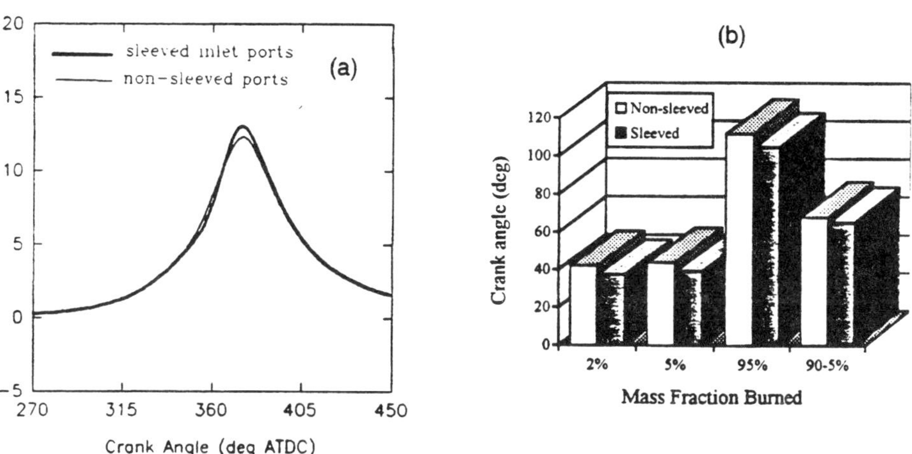

Fig. 4(a) Pressure trace for the sleeved and non-sleeved inlet port configurations
4(b) Mass fraction burned for the sleeved and non-sleeved inlet port configurations (2000rpm, 2.1 bar IMEP, MBT ignition, 95 ULG fuel)

(a) before processing

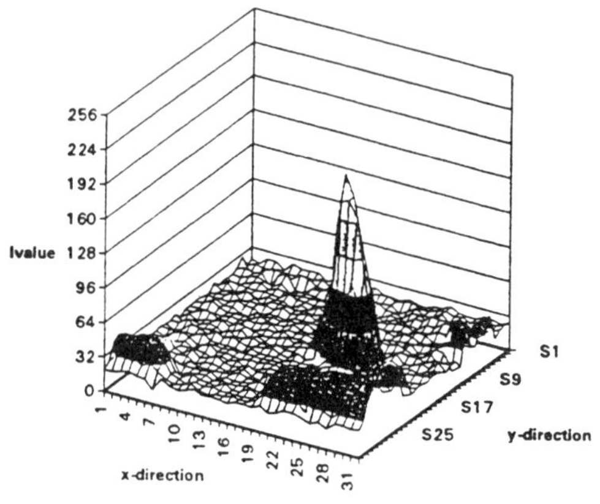

(b) smallest circle around the centre of the combustion chamber that comprises the flame

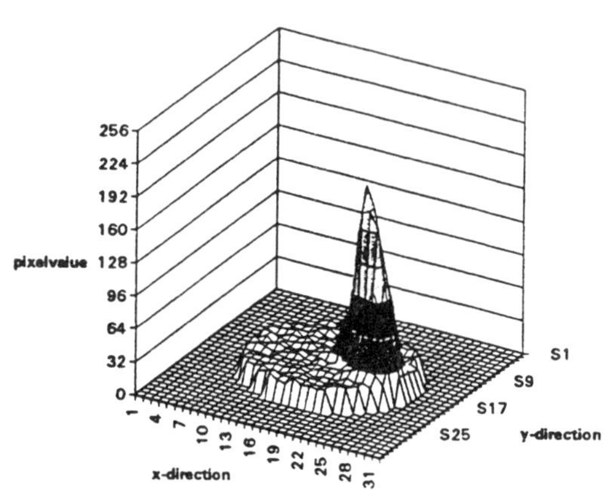

(c) after thresholding the image

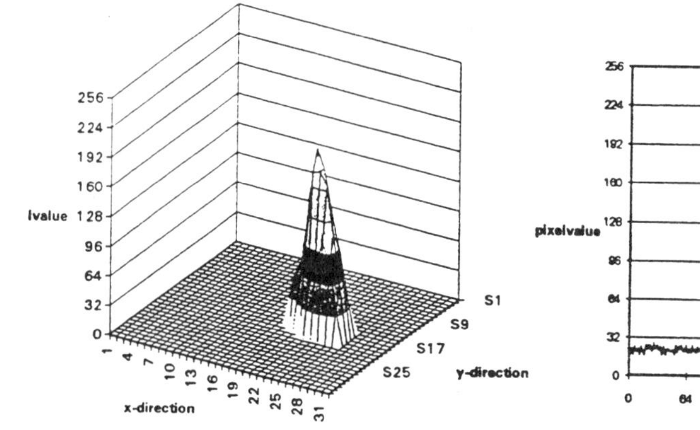

(d) 2-D slice of the image in y-direction through th flame front (x = 330)

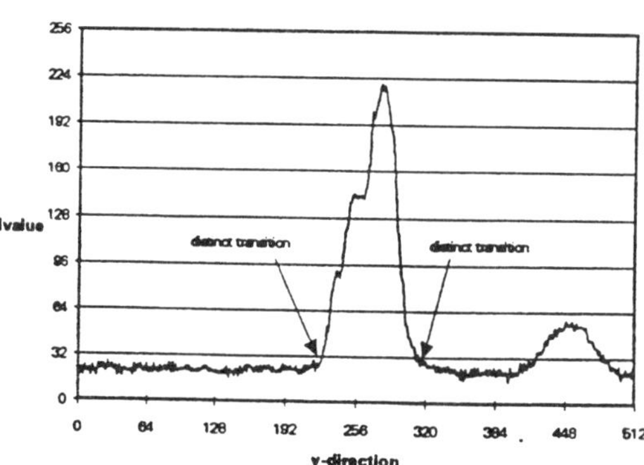

Figure 5. Processing steps of a typical flame image array

Before processing *After processing*

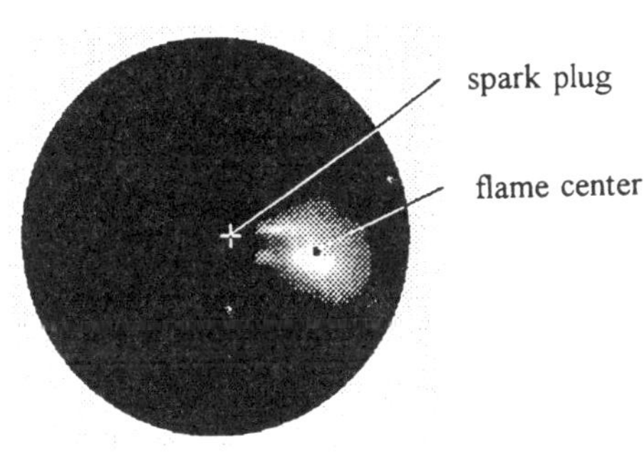

(a) direct imaging

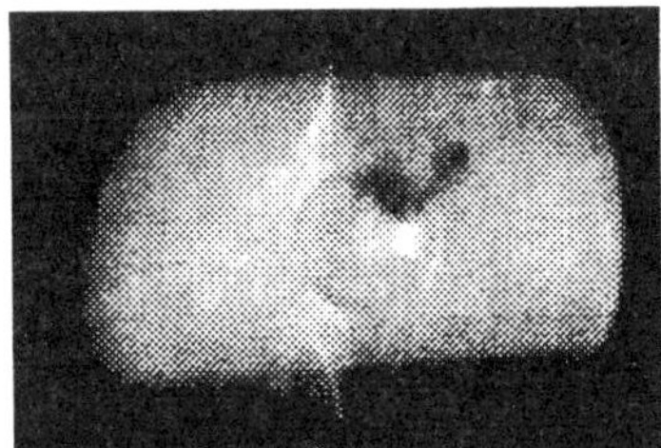

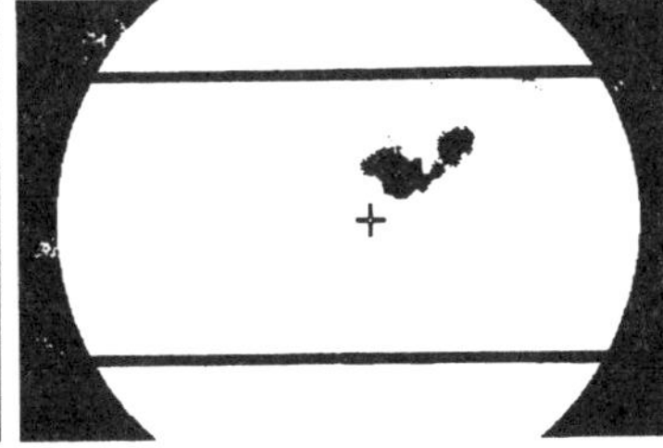

(b) 2-D imaging with laser Mie scattering

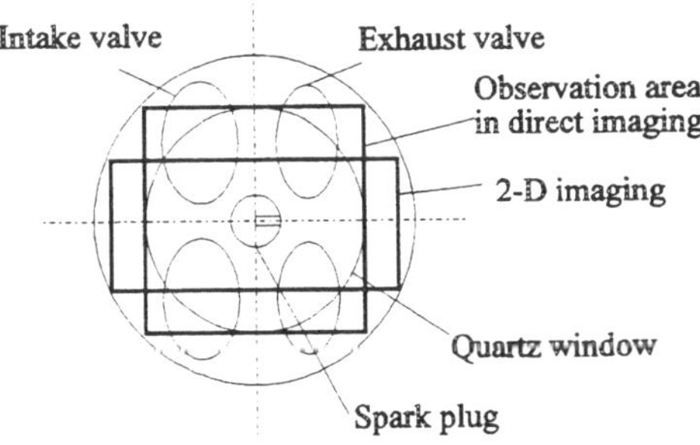

Figure 6. Examples of images before and after processing
(a) direct imaging (b) 2-D imaging with laser Mie scattering

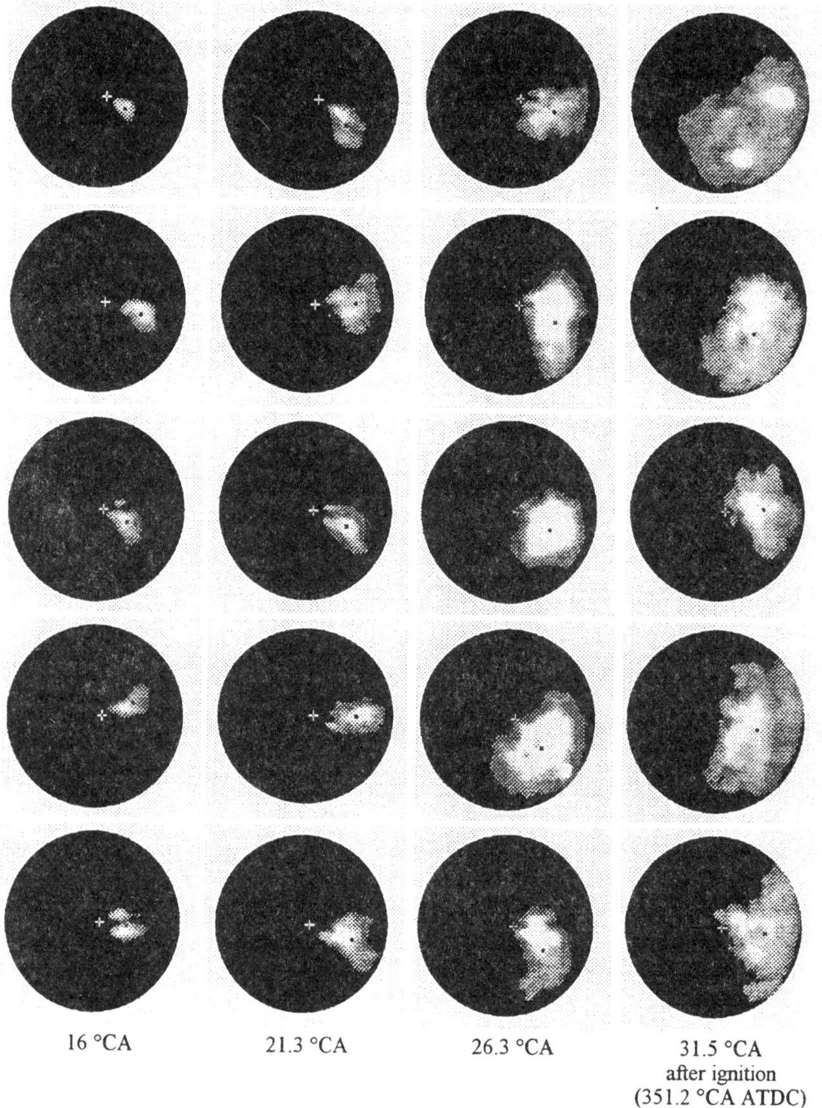

Figure 7. Images of the developing flame obtained with direct imaging with sleeved inlet ports under part-throttle condition (A/F=18.0, 2000rpm, 2.1bar IMEP)

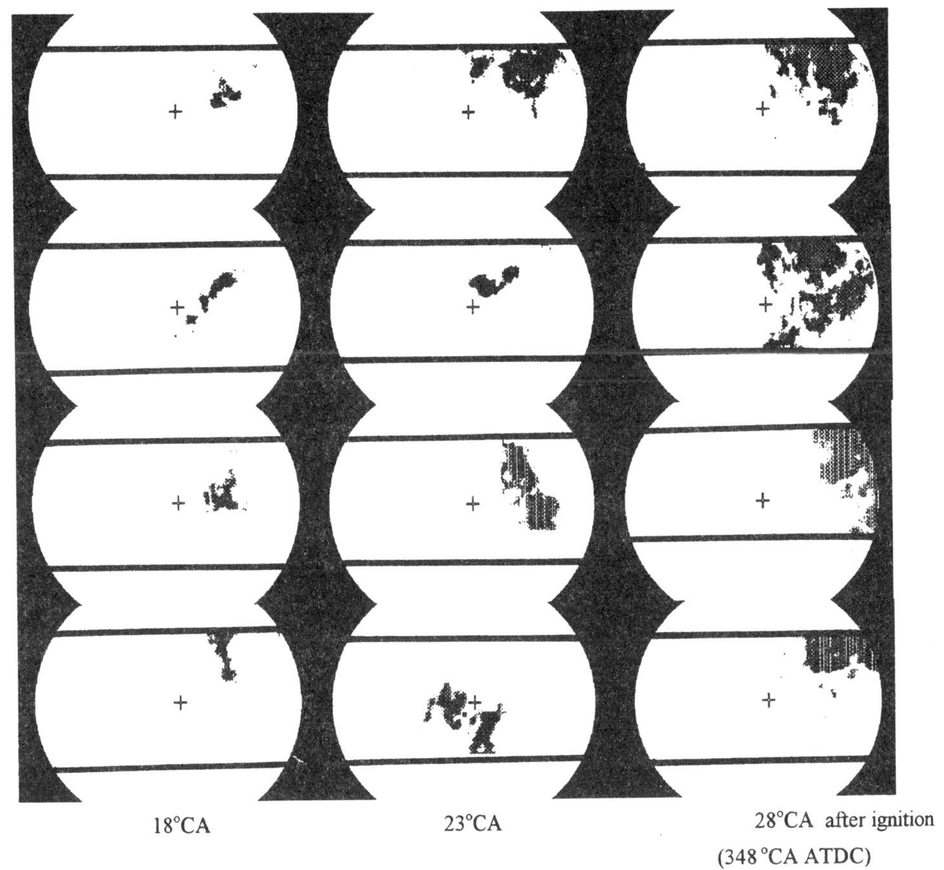

Figure 8. Images of the developing flame obtained with two-dimensional visualisation of laser Mie scattering with sleeved inlet ports under part-throttle condition (A/F=18.0, 2000rpm, 2.1bar IMEP)

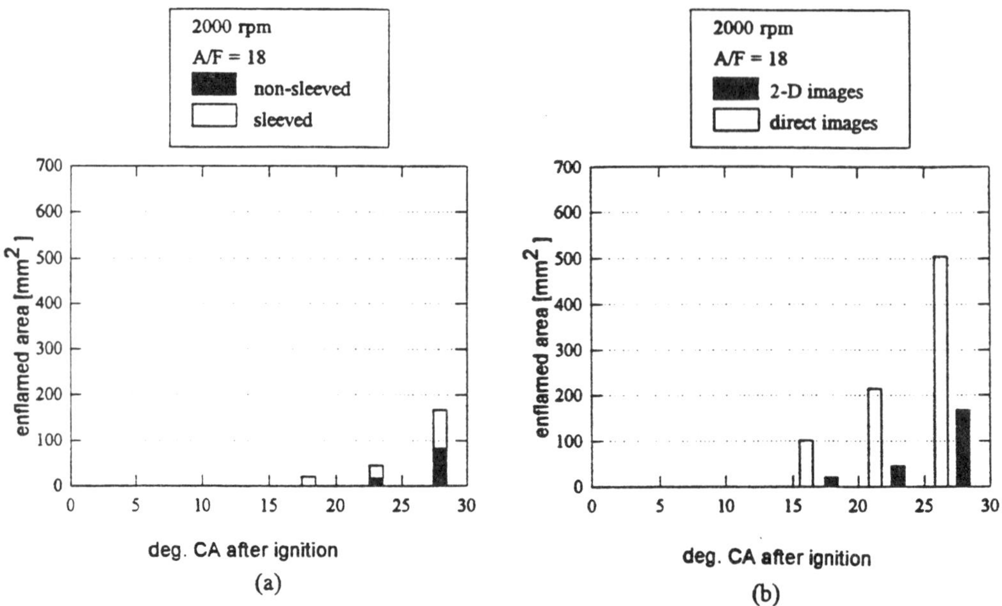

Fig. 9 Calculated enflamed areas from flame images
(a) comparison between sleeved and non-sleeved configurations with 2-D imaging
(b) comparison between direct flame imaging and Mie-scattering for the sleeved-ports configuration

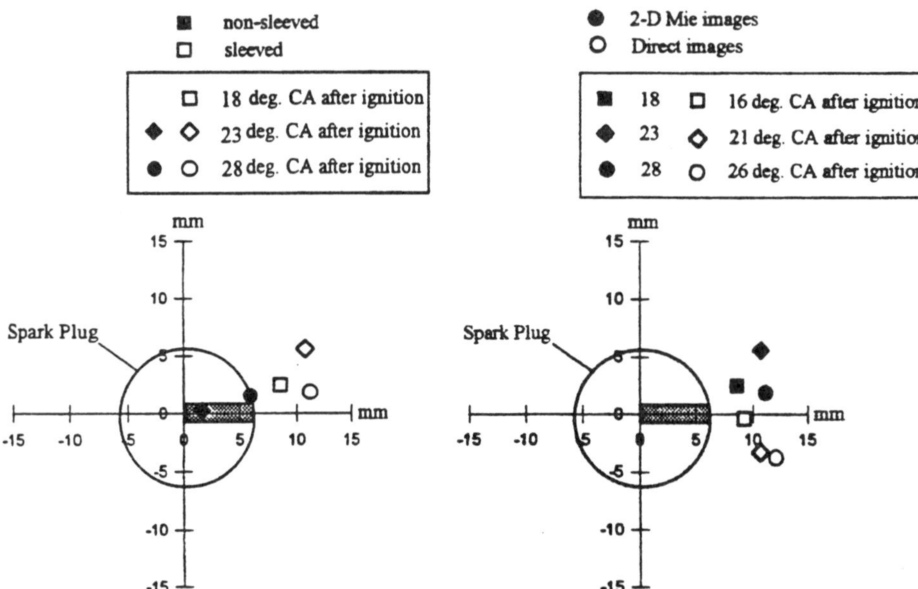

Fig. 10 Calculated flame displacement
(a) comparison between sleeved and non-sleeved configurations with 2-D imaging
(b) comparison between direct flame imaging and Mie-scattering for the sleeved-ports configuration

C499/013/96

Computers give added value to combustion photographs: the evaluation of temperatures, 'soot', and gas velocities

D E WINTERBONE FEng, MSAE, FIMechE, **D A YATES** BSc, PhD, MIMechE, **E CLOUGH, K K RAO, P GOMES,** and **J-H SUN**
Department of Mechanical Engineering, UMIST, UK

Abstract

This paper reports work on a high-speed direct-injection diesel engine, and the discussion focuses on the contribution of computers to procuring and analysing the experimental data. Computers have filled two separate roles in the work described. The first has been the control of the experiment itself and the acquisition of the experimental data. In particular the use of the computer has enabled the illumination of the fuel-injection and early combustion phases to be controlled to allow the fuel jets to be seen without interfering with the views of the combustion process. The second has been the analysis of the photographic records obtained from the experiment. The advent of the CCD camera and the digital computer has made it possible to perform this operation very quickly, and the image files thus obtained are then immediately available for further computer processing and analysis. The method of obtaining quantitative information from combustion images is described, and examples are given of the calculated temperature, soot density, velocity and vorticity fields in the combustion chamber.

Introduction

In a programme of work carried out over the last five years many films of combustion in a direct-injection diesel engine have been obtained by photographing upwards through the transparent-bowled piston. The apparatus used for the experiments has already been described in a number of papers[1-6], which also contain surveys of previous work and fully develop the theory used for this paper. The quartz window in the bowl of the piston was 46.4 mm in diameter, while the cylinder bore was 93.67 mm, making 24.5% of the combustion chamber visible. However, due to the high compression ratio of this engine the piston bowl accounted for approximately 42.6% of the clearance volume at top dead centre, and this is the region in which the most important processes take place.

The important processes in diesel combustion can be divided into two parts, the first being the injection and fuel preparation phase, and the second being the burning phase. Illumination is required to make the fuel jets visible in the first phase, while the second phase is largely self illuminating (except for the initial part of the pre-mixed period). This is a major difference between combustion in petrol and diesel engines: in the former the premixed near-stoichiometric flame gives very little light, and to render it visible requires either the use of fuel additives or else an image intensifier with the photographic device. After combustion has commenced in a diesel engine the fuel breaks down by pyrolysis and produces carbon particles which radiate with a spectrum characteristic of their temperature: this is the red or yellow cloud seen on combustion photographs.

Cine-photography was chosen in preference to direct recording by a CCD (charge coupled device) camera because of its vastly superior speed and image resolution: the day when CCD catches up, at least in terms of speed, is eagerly awaited, but until then the complication, expense and inconvenience of film have to be tolerated. The flame images used in this project have been obtained using a 16 mm high-speed camera (Hadlands Photec) with a maximum framing rate of 10,000 frames per second. By converting the camera to a half-framing device the rate could be increased to 20,000 frames per second while retaining ample resolution for the purpose. For some initial experiments colour reversal film was used, and this was adequate for the evaluation of the velocity field. However the much broader exposure latitude available from colour negative film (seven stops, as against one) dictated its use for temperature measurement. Positive images can be made for projection of the films (because the negative images are difficult to interpret) but the analysis was performed on the negative film. The images were "grabbed" into a PC using a monochrome CCD camera, using filters when it was necessary to obtain colour information for temperature evaluation. Arcoumanis *et al*[7] describe similar experimental results obtained using a high speed CCD system, but the spatial resolution was about an order of magnitude poorer than that obtained using film.

Test bed control and data acquisition

A schematic diagram of the computer control and data acquisition system is shown in figure 1; this shows how the computer plays a major role in sequencing the various devices used to obtain the data from the engine. A significant source of difficulty in obtaining diesel combustion photographs is the propensity for soot and unburned fuel to collect on the viewing windows. The problem is particularly serious in small high-speed engines where impingement of the fuel jet on the piston may occur. It is therefore necessary to limit the number of firing cycles before the camera is activated to that required to establish broad cycle to cycle consistency. In this investigation combustion was initiated towards the end of the acceleration phase of the camera, this degree of control being made possible by a solenoid operated electronically controlled fuel injector. Consistency of combustion was generally established after one or two cycles.

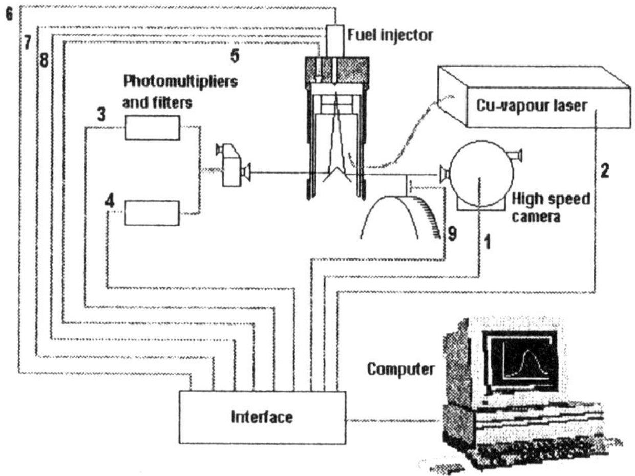

Fig 1: Schematic diagram of computer control and data acquisition system

A typical test sequence starts with the engine being driven for several minutes at its operating speed using the dynamometer with an intake of preheated air to warm up the piston and combustion chamber. On the initiation of the test sequence the computer takes control and starts the film camera. The film has to be accelerated to a speed of about 76 m/s before photographing commences, and a very high rate of acceleration is required because the film is only 400 ft (122 metres) long - breakages are not uncommon. When the film reaches its full speed, a process which takes 1.8 seconds, the computer switches on the fuel injector, and 7 or 8 firing cycles are recorded before the film runs out. Synchronisation of the laser, when in use, is also carried out by the computer. While these control functions are being performed the computer also starts to acquire data from the cylinder and fuel injection equipment; this data acquisition is synchronised with crank angle by an encoder mounted on the end of the crankshaft. The computer therefore fulfils the essential function of orchestrating the whole experiment, which is completed within three seconds of initiation. The results presented here could not have been obtained without it.

Legend for fig 1

1. Camera synchronising pulses
2. Laser actuating pulses
3. Photomultiplier 1 signal
4. Photomultiplier 2 signal
5. Cylinder pressure signal
6. Fuel injector actuating signal
7. Fuel line pressure signal
8. Needle lift signal
9. Crankshaft encoder signal

Laser illumination

A copper vapour laser, chosen for its high illumination intensity, short pulse duration (necessary to freeze the image) and high repetition rate, was used to illuminate the pre-combustion phase of the process. Early in the project the images were obtained by flashing for every frame of the photograph, as depicted by series 1 in fig 2. This meant that every frame of the film was flooded with green laser light, and while this did not have a major deleterious effect on the visual quality of the images it did interfere with subsequent analysis. It made it difficult to locate the onset of ignition, and impossible to use the spectrometric method to evaluate the temperature. A number of different strategies were considered, labelled series 2 to 4, to overcome this problem: these all required the use of a computer to achieve the desired triggering sequence. In series 2 the laser was not triggered, and only the self-illuminated part of the film was visible. This meant that two films were required to obtain a sequence of photographs showing both injection and combustion: such are the irregularities of the combustion process that there is no guarantee of the two films being compatible. Series 3 attempted to overcome this problem by illuminating the injection phase and not triggering the laser for the combustion phase. This was successful, but meant that there were no images showing the interaction of injection and combustion in the early stages of burning. Series 4 overcame this problem by illuminating alternate frames after combustion had commenced. Photographs taken using the series 4 illumination are shown in figure 3. On one frame the fuel jets can be seen, on the next the first small patches of self-illuminating combustion are visible. These cannot be seen on laser-illuminated frames. While the principle of skipping pulses seems simple enough, its realisation was made unexpectedly difficult by the propensity of the laser to return to its resting mode after a skipped pulse and ignore the next trigger signal. This technique is reported in more detail by Winterbone[6].

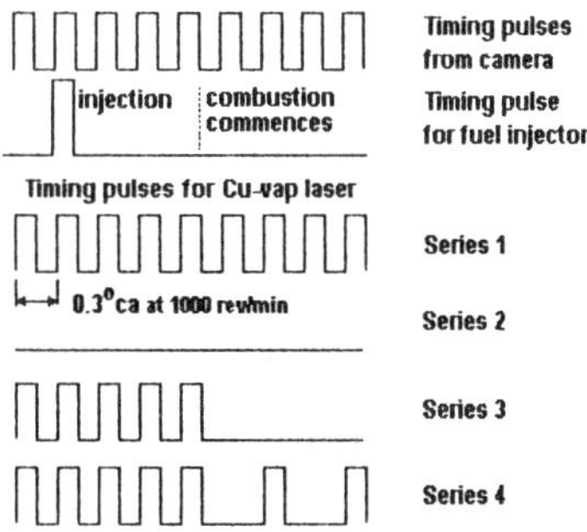

Fig 2: Timing pulses for Cu-vapour laser

Frame grabbing

The images on the high speed film are only useful for qualitative assessment of the combustion process, unless laborious and time consuming densitometer measurements and manual tracing techniques are adopted. They were therefore converted to digital images by a frame grabber incorporated into a PC. The frame-grabbing system is illustrated in figure 4. A light source is positioned on one side of the film. On the other side a standard monochrome CCD camera collects the light through a filter. The image from the camera is relayed to the frame grabber board in the computer which converts and stores it as a 512 x 512 pixel monochrome image with 256 shades of grey. The PC used was a 386DX (Opus PC7) and the frame grabber board and software were supplied by Integral Vision. Each image file (.image format) occupies 250 kbyte. This inexpensive system has proved quite satisfactory for the purpose. The rectangular format of the CCD camera, with an aspect ratio of 4:3, is distorted into a square by the frame grabber. The circular combustion chamber thus appears as an ellipse. The images are restored to their correct shape by interpolation by the AVS software used in the analysis.

The digitised images are used for two purposes. The first is two-colour analysis to obtain soot temperatures and density estimates, and the second is a cellular cross-correlation technique to estimate gas velocities in the combustion chamber. Both techniques require a pair of images to be compared.

The two images required by the two-colour method are obtained by grabbing each frame of the film twice, once through a filter with a wavelength of 581 nm and then through a filter with a wavelength of 631 nm. The wavelengths were chosen to give good discrimination between the light transmitted by the dyes in the colour film, which, of course, constitute the real filters being used for the two-colour process. Narrow band filters were used for convenience because, quite coincidentally, they were the same filters used for the direct single-spot two-colour calibration procedure. No problems with image registration arise because the film remains stationary while the two images are grabbed.

The two images required for cellular correlation are derived from successive frames of the film, since the process is one of feature tracking. However, the technique requires that the two images are accurately registered with each other, and this cannot be done mechanically because the frames are not accurately located on the film. Instead, it has to be done after the images are grabbed.

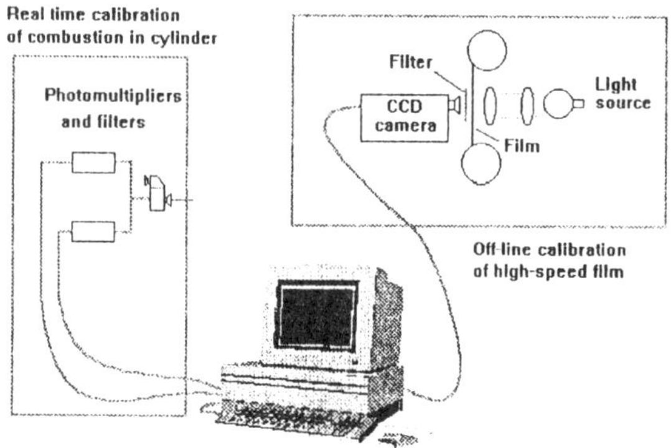

Fig 4: Method of calibrating and grabbing high-speed films

Computer processing of image files

A substantial part of the image processing was carried out using the Applications Visualisation Software (AVS), supplied by Advanced Visual Systems, installed on a Hewlett-Packard workstation; this is a comprehensive package for image manipulation and display. It can import numerical data fields with up to four dimensions in a variety of formats, including the .image files created by the frame grabber. It can display data graphically or pictorially. Amongst operations which can be performed on 2-D images are edge detection and convolution, both of which facilities have been used in this work. It can also compare two data fields, and perform arithmetical or logical operations between them. A very useful facility, also used in this investigation, is that which enables the user to program in modules for special tasks not catered for in the AVS library.

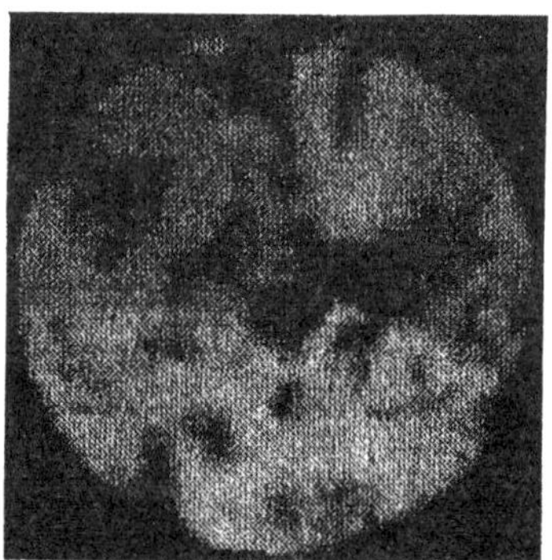

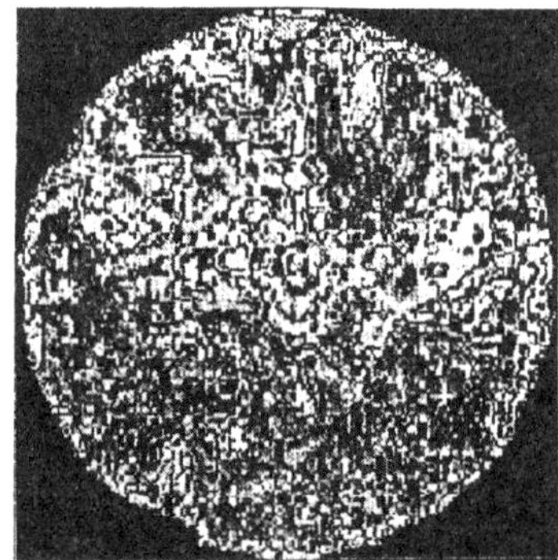

Fig 5: Comparison of image with 256 grey levels and false colour image

The images contained in this paper have been transformed to a format compatible with Microsoft packages, to make them more readily accessible by a wide range of users. This has the disadvantage of sacrificing resolution in some cases as the processed images are converted from vector form (as used for the velocity fields) to bit map form, but it can give greater flexibility in the manipulation of data on relatively simple computers. For example, the pseudo colouring of grey images reveals far more detail to the eye than was originally apparent, since the eye is relatively insensitive to subtle variations of grey level but readily distinguishes changes of colour. The images in figure 5 were transformed from 256 shades of grey to the false colour images shown by manually recolouring the CCD image using Microsoft Paintbrush.

Two colour method for evaluating temperatures and soot

This well established technique was first described by Hottel and Broughton[8] in 1932, who applied it to open flames. Since then it has been used by various workers to study diesel combustion: see for instance, refs 9 to 11. The explanation will not be repeated here, except to describe the method of calibration used, since it differs from that described elsewhere, and the computer processing of the data.

Figure 1 shows how one mirror was used to reflect an image of the combustion zone into the high-speed camera. At the same time another mirror reflected light from the combustion zone through a lens on to a split fibre optic. This light, emanating from an area only 2 mm in diameter, was divided into two channels and passed via filters to photo multipliers. Thus calibration curves of image density versus brightness temperature could be derived for the different dye layers of the film, since the temperature at the point surveyed by the fibre optic varied continually with the changing temperature of the gas passing that point. The fibre optic and photo multipliers were themselves calibrated by a tungsten ribbon standard radiation source.

As described above, to analyse a frame of the film a red image and a green image were grabbed by a CCD camera. These two images were then transferred to the AVS package installed on an HP workstation. This software was then used to

(i) mirror the image, to make it consistent with the convention adopted for the vector plots;
(ii) crop the image, to remove the overlap from adjacent frames;
(iii) correct the aspect ratio by interpolation to restore the circular shape of the combustion chamber;
(iv) set all pixels outside the circle defining the combustion chamber to zero;
(v) write the pixel values to an ASCII file for transfer to the software specially written to derive the soot temperature and density fields.

Unimap software was preferred to AVS for the presentation of the results, a sample of which is displayed in figure 6.

Velocity vectors

Once again the process begins with the acquisition of two images, this time from successive frames of the film, and transfer of the files to the AVS system, which was used this time for the pre-processing, processing and post-processing stages of the analysis.

Pre-processing.

This stage included cropping and interpolation, as also carried out for the two colour analysis, and registration. Correct registration of the two images could not be ensured at the time when they were grabbed because jitter in the camera and frame-grabber displaces the images slightly from frame to frame. To add to the difficulties it was not possible to use fixed features in the combustion chamber for alignment because the valves and injector were frequently concealed by soot and flame, and the rim of the chamber is lost in reflections from the inner walls of the piston. The solution was to overlay the two images and "twitch" one until maximum cross-correlation of the viewing area was obtained.

Processing.

The processing of the two images was carried out by a cellular cross-correlation algorithm written as a user-generated FORTRAN module and incorporated into the AVS program. The theory of the method is described by Winterbone *et al*[4] and Sun *et al*[12], and was developed from a technique proposed by Shioji *et al*[13].

Post-processing.

The output from the image processing operation was a vector map representing the 2-dimensional velocity distribution in the combustion chamber. Since the maps were generated by tracking features which generally change shape slightly from frame to frame, and are not always clearly differentiated, spurious vectors inevitably creep in. These were eliminated, firstly, by setting a lower limit to the cross correlation coefficient and, secondly, by removing vectors whose magnitude and direction differed markedly from those of their neighbours. This process left a number of holes in the vector plot. These were filled in by interpolation based on the surrounding vectors, using a Hardy multi-quadric approach. Finally a smoothing operation was carried out on the vector field by convolution with an axi-symmetric Gaussian kernel, the justification for this procedure being that the vector field still contained distortions arising from such factors as temperature changes and combustion initiation and extinction. All of these post-processing operations were carried out by the addition of user-generated modules to the software package. A typical vector velocity plot before and after data refinement is shown in figure 7. This work is described by Sun *et al*[12].

A facility was included in the software for finding the div, grad and curl of a vector field. This enabled the vorticity (grad) to be calculated; the provision of a facility to calculate the divergence of the vectors would also enable the out of plane component of the flow to be assessed.

Conclusions

From their beginnings as calculating devices and test bed controllers, computers have found innumerable and constantly increasing applications in engine technology. It is probably true to say that most of the internal combustion engine research work being carried out today would be impossible to conduct without the aid of digital computers. The work described in this paper demonstrates three facets of their usefulness.

i. The control of a test schedule consisting of a very rapid sequence of events. This involved the accurate sequencing of the high-speed camera, fuel injection and the laser illumination.
ii. The acquisition of data relating to the combustion process during the test.
iii. The analysis of the results. This stage included the transfer of images from the high-speed film to digital storage, and the processing of the digitally stored images to obtain temperature and velocity maps.

References

1. **Rao K K, Winterbone D E and Clough E** Laser illuminated photographic studies of the spray and combustion phenomena in a small high speed di diesel engine. SAE Paper 922203, San Francisco, 1992.
2. **Rao K K, Winterbone D E and Clough E.** Combustion and emission studies in high speed di diesel engines. IMechE Conference Proceedings C448/070, 1992.
3. **Winterbone D E, Yates D A,Clough E, Rao K K, Gomes P, Sun J-H** Combustion in High-Speed Direct Injection Diesel engines - a Comprehensive Study. Proc.Instn. Mech.Engrs. Vol.208, 1994.
4. **Winterbone D E, Sun J-H and Yates D A.** A study of diesel flame movement by using the cross-correlation method. SAE Paper 930979, Detroit, 1993.
5. **Winterbone D E, Yates D A,Clough E, Rao K K, Gomes P, Sun J-H** Quantitative Analysis of Combustion in High-speed Direct Injection Diesel Engines Commodia 95, Yokohama, Japan
6. **Winterbone D E** Advances in photographic studies of sprays and diesel combustion
7. **Arcoumanis C, Bae C, Nagwaney A, and Whitelaw J H.** Effect of EGR on combustion development in a 1.9L DI diesel optical engine. SAE 950850, Detroit, 1995.
8. **Hottel H C, and Broughton F P** Determination of true temperature and total radiation from luminous gas flames Industrial and Engineering Chemistry, Vol 4, No 2, pp 166-175, April 1932
9. **Uyehara, O A, Myers P S, Watson K M, and Wilson L A** Flame temperature measurements in internal combustion engines Trans ASME, pp 17-30, Jan 1946
10. **Matsui Y, Kamimoto T, and Matsuoka S** A study on the time and space resolved measurement of flame temperature and soot concentration in a di diesel engine by the two-colour method"SAE paper 790491, Detroit, 1979
11. **Matsui Y, Kamimoto T, and Matsuoka S** Formation and oxidation processes of soot particulates in a di diesel engine - an experimental study via the two-colour method SAE paper 820464, Detroit, 1982

12. **Sun J-H, Yates D A, and Winterbone D E.** Measurement of the flow field in a diesel engine combustion chamber after combustion by cross-correlation of high-speed photographs. *Submitted to J Expt Fluids*
13. **Shioji M, Kimoto T, Okamoto M and Ikegami M.** An analysis of diesel flame by picture processing. *Jap Soc Mech Engrs,* 1989, Ser II, 32 (3)

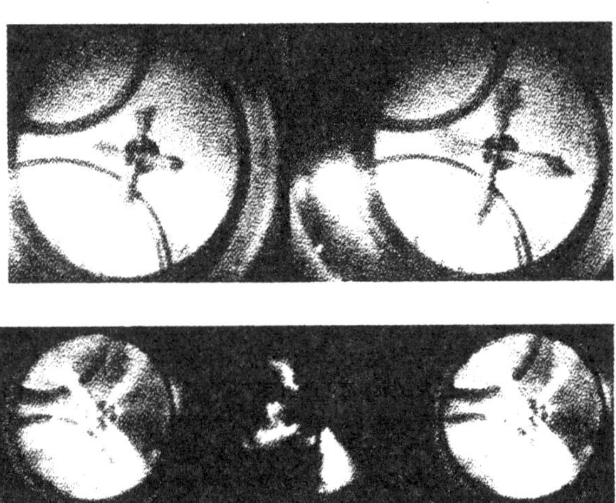

Fig.3 Successive frames of high speed photographs with and without laser illumination

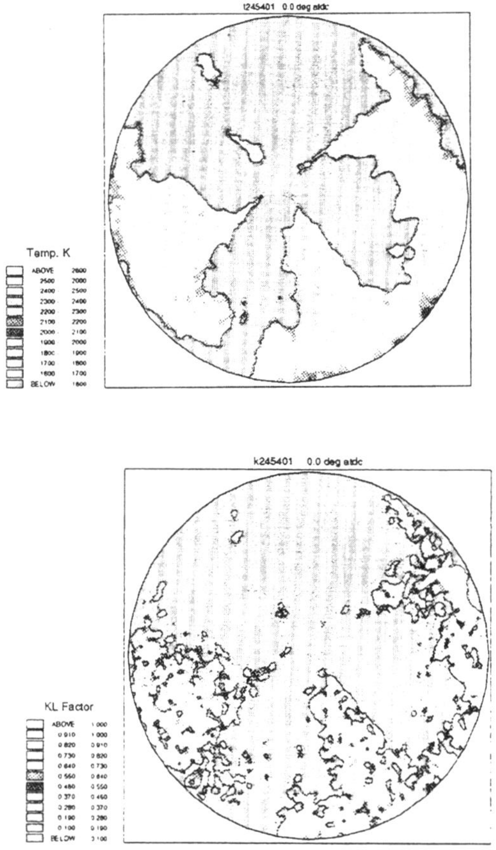

Fig.6. Combustion chamber temperatures and KL factors by two-colour method

C499/030/96

A two-camera CCD imaging system for in-cylinder spray imaging and analysis

D G SALTERS BEng, PhD, A R GREIG BSc, MSc, PhD, MIMechE, MIMarE, and P WILLIAMS BEng, PhD, AMIMechE
Department of Mechanical Engineering, University College, London, UK

SYNOPSIS

A two-camera high resolution CCD based imaging system has been developed to study fuel spray structure within a direct-injection engine. The two cameras can be asychronously triggered to activate an in-built CCD electronic shutter process. A short duration light source is used to freeze droplet motion. Image processing techniques are used to determine droplet size distributions and the comparison of the temporally phase-differenced images from the two cameras enables the determination of droplet path and velocity. The digital video output from the cameras is recorded by a high speed digital acquisition system which can store up to 20 consecutive images.

1 INTRODUCTION

Advances in computer technology are enabling computer based imaging and image analysis to become increasingly useful in understanding fuel spray behaviour in the IC engine field. The rapid capture of video pictures and almost instant display of the images by personal computer based hardware enables one to obtain a rapid qualitative understanding of spray behaviour from 'raw' and enhanced images, whilst image processing routines provide quantitative results such as drop size distributions, velocity and path data. The rapid progress in hardware, particularly the improving resolution of the video imager is starting to redress the balance with film, which has traditionally had significantly higher resolution than video, but has been time consuming to process and analyse. Also, the increasing ability to control the video acquisition hardware and imager itself is now allowing novel imaging systems to be built. This paper discusses the development of a video imaging and analysis system that has been designed to study fuel sprays within an optically accessed direct-injection research engine. The system utilises two asynchronously triggered and electronically shuttered CCD (charge coupled device) cameras to provide both spatial and temporal spray information. Prior to a description of the system the basic functional aspects of CCD technology and the corresponding problems of compatibility with high speed spray analysis are discussed.

2 THE OPERATION OF A CCD VIDEO IMAGER

The first commercial charge coupled device appears to have been developed in the 1970's in the United States. The CCD is allied to a branch of semiconductor electronics similar to that of a transistor, however, the unique properties of the CCD have enabled its application to areas such as video camera image sensors and its early use as computer memory.

2.1 The CCD element

A CCD imager consists of a two-dimensional array of photosensitive elements which are combined to form an imaging device (1). An individual CCD element is shown in Figure 1. Light incident on each of the silicon semiconductor elements, results in the generation of electrons due to a photoelectric effect. These electrons are trapped in potential wells produced by small electrodes. Electrons produced by the incident radiation on the silicon accumulate in this storage region and consequently the magnitude of the electron charge formed is a function of the illuminating radiation. Hence, the unit is analogous to a radiation driven capacitor. As each element develops a charge proportional to its illuminating intensity, one thus has a spatially digitised electric analogue of the original image.

2.2 Charge coupling and video output

In order to provide a useful imaging device a means is needed to retrieve this "electron charge" image. This is accomplished using **charge coupling** (hence the term charge coupled device). The process of charge coupling involves the transfer of all the electric charge stored within the semiconductor storage element to a similar, adjacent storage element by external manipulation of voltage. A periodic wave form called a "clock" voltage is applied to the electrodes of a CCD array and result in the electron charge being moved due to the continuous lateral displacement of the "local potential wells" in which the electrons have accumulated. Once the charge is moved from the CCD it is converted to a signal that is compatible with a standard video signal. The European and American television standards define an "interlaced raster scanned technique" that eliminates perceptible flicker when the image is displayed. Interlacing takes advantage of the human eye's characteristic persistence of vision to eliminate flicker. The horizontal lines which make up a frame are divided into two fields, that is, the even field and the odd field, as shown in Figure 2. Interlaced scanning results in the display of one field at a time. The European standard, CCIR, has a frame rate of 25 Hz, and hence, a field rate of 50 Hz. One field consists of 313 lines, whilst the other has 312 lines, these are interlaced to give a total of 625 television lines, of which 576 are active, the remainder are for other signals (i.e. test and teletext signals). Hence, the television line rate is 15625 Hz/line or 63.5μs per line. The American standard, RS-170, operates at 30 Hz/frame with 525 total

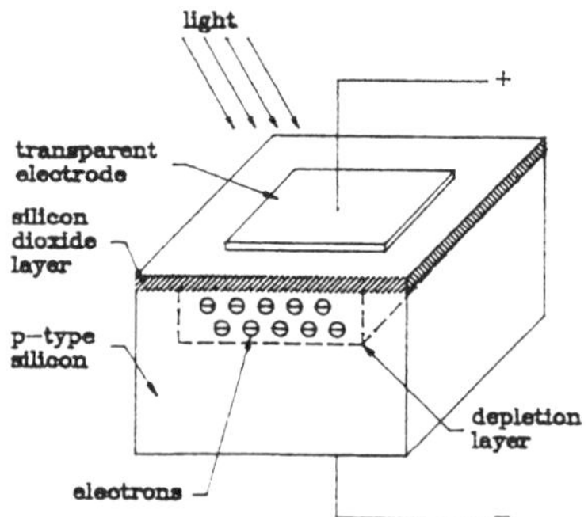

Figure 1 The basic element of a CCD

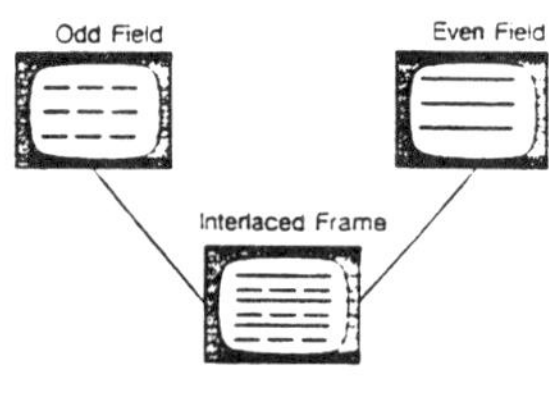

Figure 2 An interlaced display

lines and 486 active lines. Synchronisation information is carried with the video signal . One synchronisation pulse occurs at the beginning of each TV line, this is referred to as the Horizontal Sync., Blank or Drive (HD), a group of pulses also occur at the beginning of each field and are referred to as the Vertical Sync., or Drive (VD). Hence, for CCIR, the Horizontal Drive synchronisation marks occur at a period of 63.5μs and the Vertical Drives, occur at a period of 20ms.

2.3 Imager array architecture

Two types of CCD imager array architecture are used, that is, **frame transfer** and **inter-line transfer** (2). The functional disparity between the two is mainly the way in which the movement of the charge is organised. The frame transfer CCD imager consists of an upper arrangement of light sensitive sensors, whilst the lower half forms the storage/readout area which is opaquely masked to exclude light. This is shown in Figure 3. During the vertical blanking the charge corresponding to the picture information is clocked out rapidly down into the frame storage array, which is emptied line by line during line (or horizontal) blanking, into the horizontal (read out) shift register. The contents of this are then clocked to the output terminal during the active picture line. The sensors are still responsive to light whilst acting as shift registers during the vertical blanking period, which can result in the transferred signal being contaminated with extraneous light information, and blurring of the picture. This effect is known as frame shift smear.

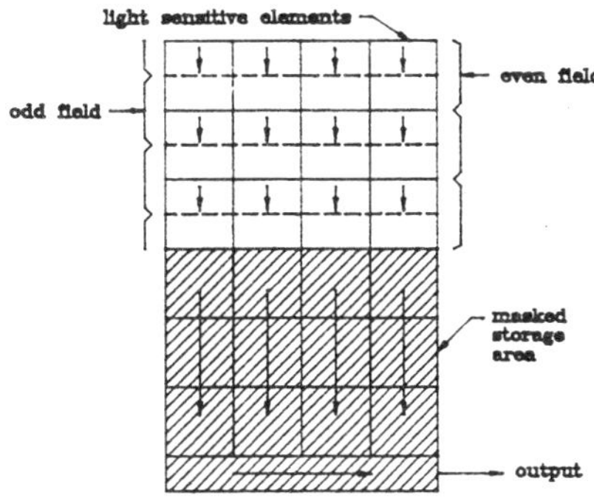

Figure 3 A frame transfer device

The inter-line transfer imager, shown in Figure 4, consists of interleaved columns of sensors and vertical shift register, feeding into a horizontal (read out) shift register. All the CCD shift registers are masked from incident light. During field blanking, the charge from the photo sensor representing the picture information are moved sideways into the vertical shift registers and then clocked downwards, line by line, during the line blanking period, into the horizontal register, the contents of which are read out to form the video signal. Figure 4 illustrates the detail of inter-line transfer, specifically the sideways movement of the charge by the transfer gate and the interlaced operation of the device.

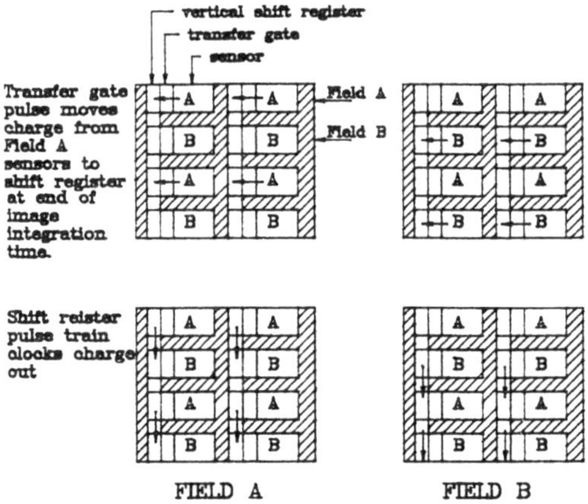

Figure 4 Detailed operation of interlaced scanning within an interline transfer device.

Performance comparisons between the two types of CCD imaging device centre on pixel resolution and the control of the CCD. The frame transfer device should provide higher resolution, as this type of device has more active imaging silicon than the inter-line transfer CCD. However, a frame transfer device requires channel-stops on the CCD wafer to stop charge migration and so does not provide 100% coverage. A further consideration is that of CCD control, and particularly frame shift smear. As noted above, the frame transfer device is susceptible to "contamination" by extraneous light when the charge, forming the exposed field, is being shifted into the storage area. Using transfer gates, the inter-line device can shift the charge rapidly (in parallel) through the gates into "light-safe" vertical shift register. Control of this facility provides the ability to produce short integration times and a pseudo-shuttering facility (this is discussed in more detail in the following sections).

2.4 CCD imager scanning and electronic shuttering mechanisms

In trying to understand the operational characteristics of various high resolution machine vision type CCD cameras, a number of serious implications were noted regarding image degradation when such cameras were to be used with a short duration pulsed light source. The first problem regards the way a standard high resolution monochrome CCD camera accumulates and processes the image to provide a television/video compatible interlaced image. Figure 5 illustrates the timing diagram of how the CCD is scanned and the output is

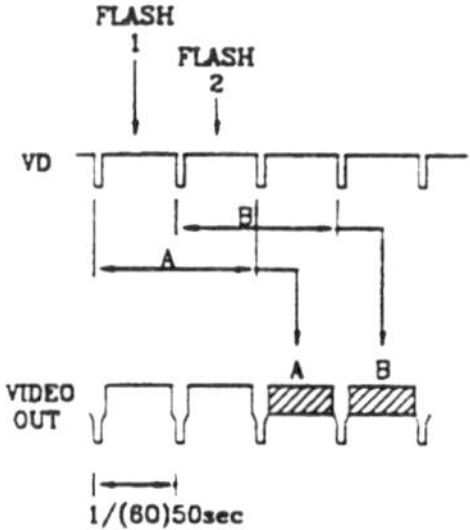

Figure 5 Integration timing of a standard CCD

displayed in an interlace format (3)(4). The pixel elements on the CCD chip are physically scanned as two fields (i.e. the even lines and then the odd lines), the scanning of these fields is overlapped with respect to time and the two fields are output one after the other in an interlaced format. With reference to figure 5, it can be seen that the time at which a light source is pulsed is critical to whether all, or only half, the pixels are exposed. If the flash occurs at position FLASH 2 on figure 5, then all the pixels are scanned and a full resolution image is realised. However, if the light pulse occurs at FLASH 1 then as a result of the scanning arrangement half of the image information is "lost" for that particular frame.. Hence, if one were to record such an event with a video recorder or triggered frame grabber then a significant proportion of the frames would only contain half the pixel information (one would see a flickering image, this has been observed and validated by the authors). Hence, in order to use a short duration pulsed light source with a CCD camera, one must either synchronise the pulse to the CCD scanning cycle, or provide an alternative method of camera synchronisation.

A number of high resolution machine vision type CCD cameras now posses the ability to asynchronously electronically control or trigger the CCD so that it only integrates for a short time (i.e. the camera only accumulates an image for say 0.1ms as opposed to the normal integration time of 20ms for a 50Hz field rate). The ability to asychronously electronically shutter the camera and to accurately synchronise a shutter to a light pulse through electronic control was thought to be extremely useful for high speed spray visualisation. The mechanism of controlling the CCD involves dumping the charge accumulated outside the shutter period, usually moving the charge off the CCD and dumping it until the shutter is needed, at which point the charge is kept and processed. Alternatively, the CCD chip is constructed with a substrate drain underneath, this allows charge to be dumped from the CCD until the shutter period is signalled, at which time the charge and corresponding image is accumulated. The advantage of this system is that the substrate drain acts on all pixels simultaneously and has the potential for short shutter periods. However, until recently, as a result of the compromises required in the design of CCD chip architecture to provide interlaced output, inspection of the camera specifications show that when the asynchronous shutter period is used, only one field is scanned. Also, most cameras can only be reset at the end of the horizontal line scan (i.e. 63.5μs), so that they do not disturb the video recorder/frame grabber phase lock-loop circuitry which synchronised the recording with HD timing. Hence, when the camera is reset there is an unknown random delay, or jitter, of 0 - 63.5μs, which could prove to be problematic for high speed spray analysis work.

The above discussion is intended to give a broad outline of the operation of a CCD and details the problems encountered during the design of the system that is outlined in the following paragraphs.

3 A TWO CAMERA ELECTRONICALLY SHUTTERED CCD IMAGING SYSTEM

A two-camera electronically shuttered CCD based imaging system has been developed to investigate fuel spray behaviour within an optically accessed direct injection engine. Figure 6 illustrates a schematic of the system.

3.1 The camera system

The system consists of two factory modified progressive scanning Pulnix TM-9700, full frame shutter, asynchronous reset, high resolution monochrome CCD cameras. The cameras use an interline transfer 2/3 inch monochrome CCD with a resolution of 768 (H) x 484 (V). The

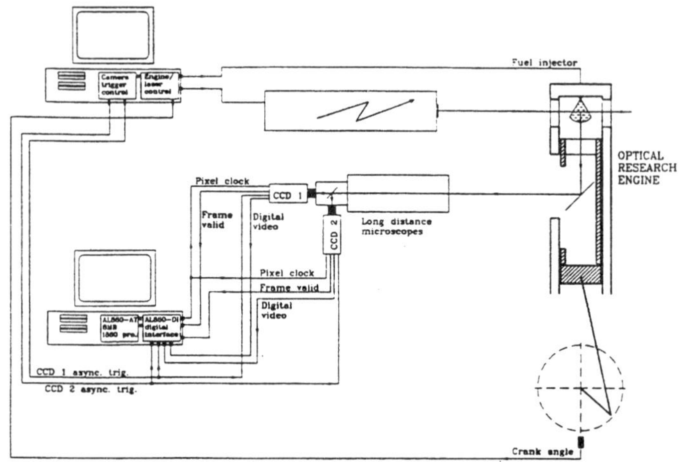

Figure 6 A schematic of the two-camera electronically shuttered CCD imaging system

cameras use a substrate drain type shutter process, the maximum shutter "speed" is 1/16000s or 63μs. A fundamental asset of the cameras which makes them ideal for this application is that they use state-of-the-art progressive scanning CCD architecture, that is, all the pixels are scanned during the shutter process, as opposed to interlaced/interleaved scanning. Hence, the full resolution of the camera is obtained during shuttering. The image data from the two cameras is captured by a triggered high speed Alacron digital data acquisition system. For general spray characterisation work both cameras are focused through a long distance microscope lens, via a beam splitter, to observe the same point in the engine. A short duration pulsed light source is used to illuminate and freeze the motion of the spray/droplets. The light source used is either a copper vapour laser with a duration of 30ns or an argon spark flash unit with a duration of 500ns. The sequence of events regarding the timing of the camera shutter period and the laser pulses are shown in figure 7. The shutter period for camera 1 is activated, within this 63 μs period the first light pulse is made. Sometime after the end of camera 1's shutter period camera 2's shutter is activated and correspondingly during this 63μs period the second light pulse is made. This sequence is repeated once every engine cycle The timing and triggering of the camera shutter periods are usually taken from an engine timing mark (i.e. top dead centre), the engine is run at speeds between 200-2000 r/min (the cameras can operate at up to 30 frames per second and can thus accommodate the maximum engine speed of the four-stroke research engine). The specification of the cameras and the use of digital control enables the shortest duration between the two light pulse to be reduced to 7μs This is achieved by making the first light pulse at the end of camera 1 shutter period and the second light pulse at the start of the camera 2 shutter (with an allowance for the frame transfer and reset periods). Previous work has indicated that such high temporal resolution is required to ensure that droplets do not move off-frame for the subsequent picture when high magnification in-cylinder pictures are taken.. In order to achieve the necessary communication between cameras and to provide accurate shutter timing, the authors have worked with the manufacturer to include a number of camera modifications. The main one being that the shutter can be activated within 100ns of a trigger signal. This is contrary to the standard camera where the camera "waits" to the next HD timing signal (i.e. jitter of up to 63.5μs).

3.2 Data transfer and acquisition

The cameras offer digital output (EIA-422), of the progressive scanned video data, this image data is captured by a high speed Alacron, Intel I860 processor based, digital data acquisition system. As the cameras are reset within 100ns without latching to the horizontal line timing, conventional VCR/frame grabbers would be unable to capture the image data due to the

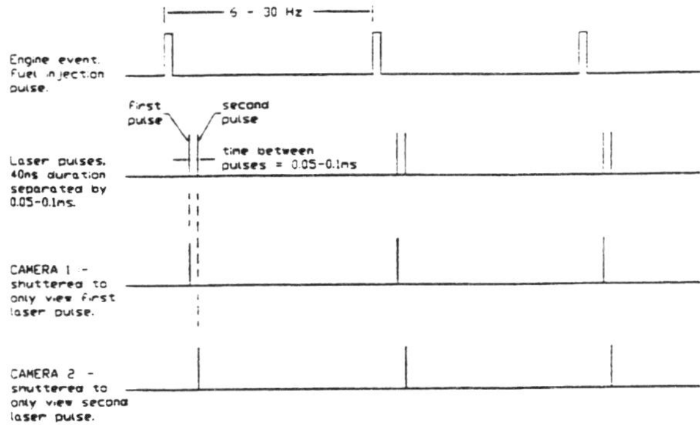

Figure 7 Light pulse and camera exposure timing

sudden resetting of the analogue video timing signals. The use of the camera's digital video output and a digital data acquisition system overcomes this problem due to the fully asynchronous triggered operation of the Alacron system. As well as ensuring that the complete imaging/capture system is fully asynchronous, the digital signal is less susceptible to "noise" which could affect and degrade an analogue video signal in an engine test-bed environment. The Alacron 64 bit I860 based digital acquisition system consists of three elements, an I860 programmable stand-alone motherboard which manages data-acquisition and processing and contains the acquisition memory, a digital interface which buffers the incoming data, and a high resolution display card. The system is resident on a PC and can communicate on the standard AT bus. The system uses a 64 bit digital interface buffer which can accommodate input data rates of 160MB/sec, each camera has a data output rate of 14 MB/sec. Modification of the two cameras by the manufacturer has allowed the cameras to output the image data as a parallel 16 bit stream (i.e. one 8 bit stream from each camera). The two cameras are electronically linked together in a MASTER-SLAVE arrangement, where the first camera holds its image data until the second camera has finished its shutter period, upon which the two cameras simultaneously output their image data. The two cameras use the same pixel clock to synchronise their data output, this clock runs at 14.13Mhz and is also used to clock the data into the Alacron digital interface. As the data is clocked into the digital interface it is transferred directly to the I860 system memory. Presently the system has 16 Mbytes of memory which will store up to approximately 20 images at the present camera resolution. As noted the system utilises a high resolution graphics card and monitor to display the captured images. Noesis Visilog image processing software resident on the system PC was used for image analysis.

4 PRELIMINARY RESULTS

A number of preliminary results obtained during the recent commissioning of the system are presented in the following pages in order to illustrate some of the capabilities of the system. Figure 8 illustrates the penetration of a spray from a pintle-type port fuel injector. The time interval between the two images was 100μs. Once the system is set-up it was relatively straightforward to determine the penetration rate and cone angle of the spray, also, tests could be rapidly repeated to quantify any cyclic variations present. Figure 9 illustrate the spray further downstream from the nozzle where it was well atomised. Analysis of images using the present optical system have indicated that the smallest droplet size that can be imaged is approximately 20μm, this compares well with previous high speed photography using a similar optical arrangement (6). Figure 10 illustrates the overlaying of the two consecutive images from camera 1 and camera 2 after basic image processing involving image enhancement and thresholding. With reference to figure 10, the darker image is from camera 1 whilst the lighter image is from camera 2, the time between the two images is 7μs. Using this data, one can rapidly deduce droplet velocities and path information, also, the direction of travel of each droplet is implicitly known, which is often difficult to determine by other techniques. As figure 10 consists of two overlaid images and is not itself a multi-exposed snail-trail type image, it is also possible to deduce droplet sizes from each original image.. Figure 11 illustrates the droplet size distributions. The imaging system and processing routines are calibrated using a graticule with known drop sizes. The graticule also enables the physical alignment of the cameras (typically to within 100μm), followed by final alignment using software to within one pixel.

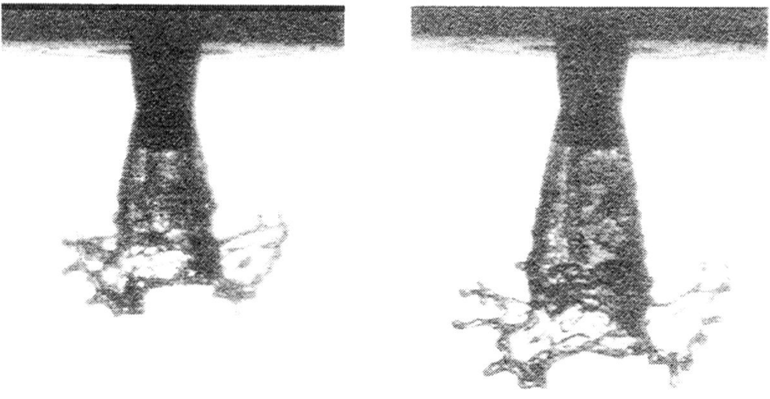

Figure 8 Penetration of a pintle injector spray, Δt=100μs.

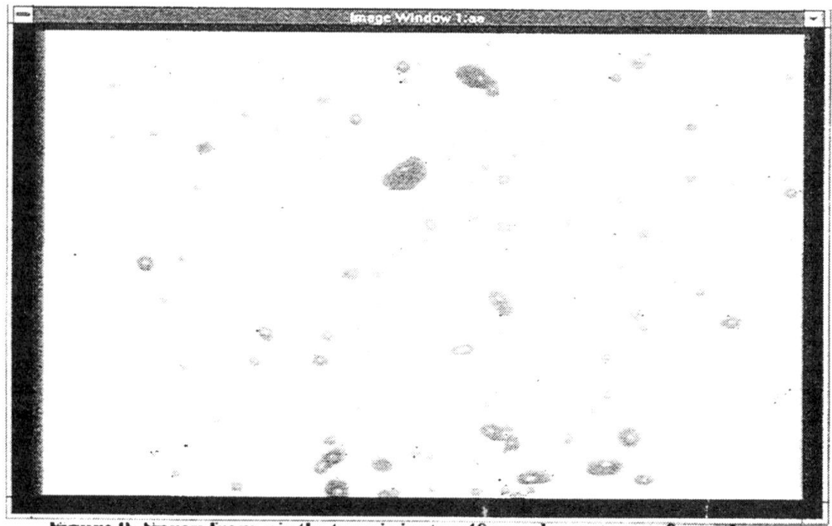

Figure 9 Spray from pintle-type injector 40mm downstream from the nozzle (actual image area = 9x7mm)

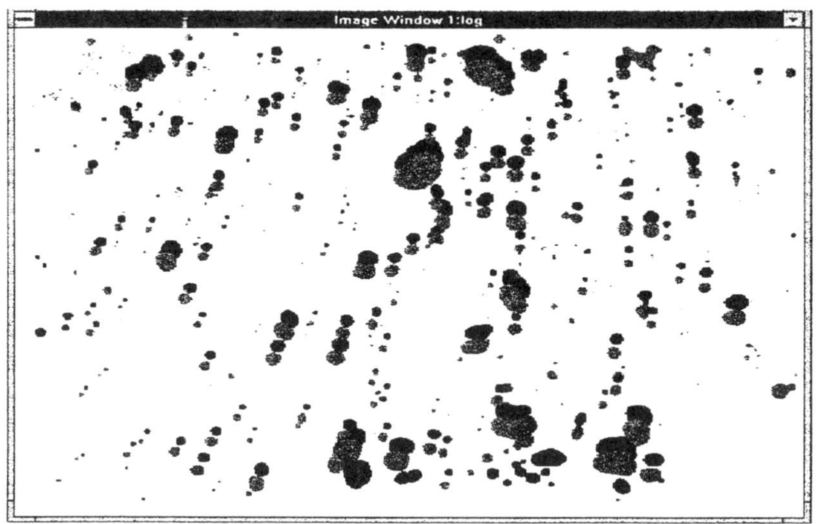

Figure 10 Overlaying of processed image from camera 1 (black image), and camera 2 (grey image) , the time between the two images is 7μs.

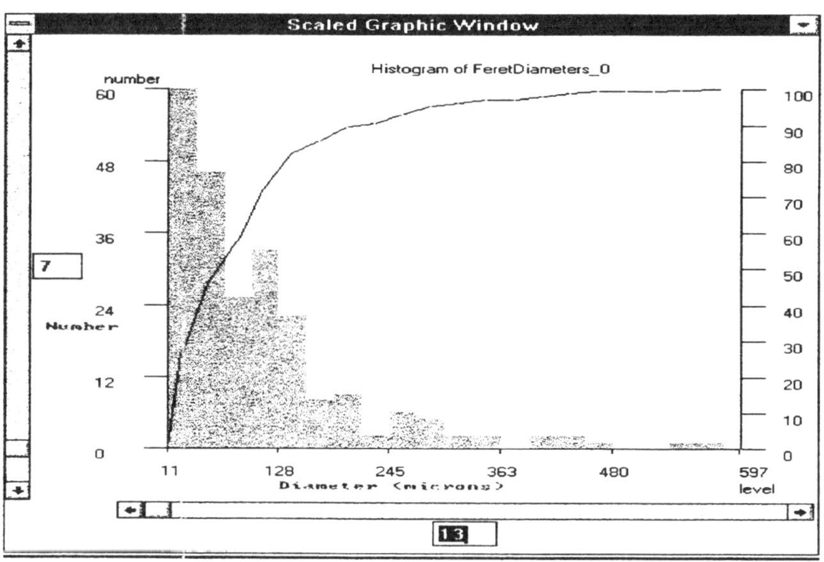

Figure 11 Droplet size distribution

Figure 12 illustrates an image of a fuel spray within the cylinder of the optically accessed engine, the image is taken though the quartz window of an extended piston. The spray can be seen in the centre of the image emerging from the injector 300μs after the start of injection.

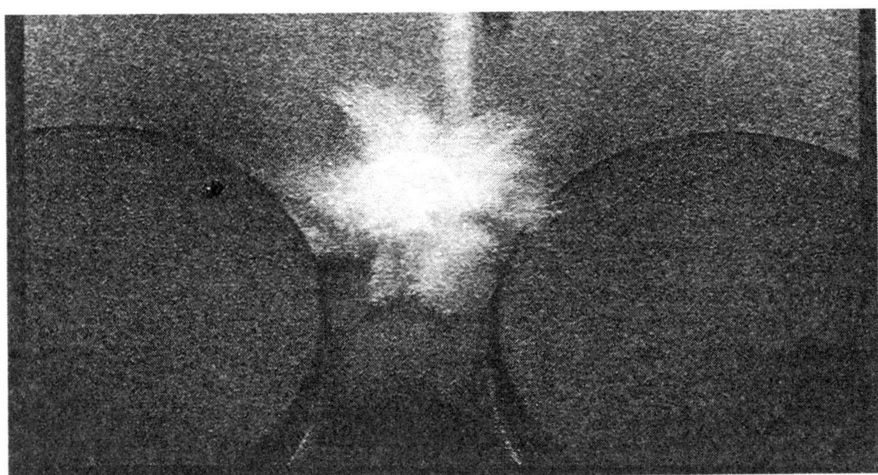

Figure 12 Spray emerging from an injector within the optically accessed engine

5 CONCLUSION

Utilisation of state-of-the-art CCD imager technology and high performance computational and data acquisition hardware has enabled the development of a computer based imaging system to investigate the fuel spray behaviour within an optically accessed engine. The system has provided an extremely cost effective method of rapidly obtaining qualitative image data and through image processing, quantitative spray data, such as droplet size, velocity and path.

ACKNOWLEDGEMENTS

The authors would like to acknowledge the financial support of the EPSRC. The assistance is acknowledged of Pulnix (USA) who modified the CCD cameras and Pete Swinburn at Data Cell who provided software integration for the Alacron system.

REFERENCES

(1) BARBE D.F., Charge Coupled Devices, McGraw Hill, 1980.

(2) EEV., CCD imaging. EEV Ltd. Chelmsford, Essex., 1994.

(3) PULNIX, Pulnix machine vision CCD camera specifications, Pulnix Ltd, Basingstoke 1994.

(4) SONY, Sony machine vision CCD camera specifications, Sony Ltd, Basingstoke, 1994.

(5) HITACHI, Hitachi machine vision CCD camera specifications, Hitachi Densai, London, 1994.

(6) FRY M., and NIGHTINGALE C., High speed photography and image analysis techniques applied to study droplet motion within the porting and cylinder of a 4-valve SI engine, SAE 952525.

C499/022/96

An integrated system for the design of radial compressors and turbines

P M CAME BA, CEng, FIMechE, **E SWAIN** BSc, PhD, CEng, MPhil, MIMechE, **C J ROBINSON** BSc, PhD, CEng, MRAeS, MEng, and **M S RUTHERFORD** BA
PCA Engineers, UK

Synopsis - A design system for radial-flow turbomachinery is described. The system comprises modules for preliminary design, off-design performance prediction, impeller three-dimensional geometry modelling, 2D and 3D aerodynamic analysis, and stress and vibration analysis. The system is the result of sustained development over a long period and the impact of today's computers on its implementation and user-interface is emphasised. The system is in routine use in industry in a variety of applications including gas turbines, turbochargers, and industrial machines.

NOTATION

A	flow area (m^2)
C	absolute velocity (m/s)
C_x	axial component of axial velocity (m/s)
m	mass flow (kg/s)
N_f	number of full vanes
N_s	Specific speed
o	throat width (m)
P	stagnation pressure (Pa)
p	power input factor
Q	non-dimensional flow function
r	radius (m)
T	stagnation temperature (K)
t_n	normal vane thickness (m)
W	relative velocity (m/s)
VR	relative velocity ratio
U	blade speed
u,v	vane geometry patch parameters
β	vane angle (deg)
ΔH	Change in stagnation enthalpy

Subscripts

h,hub	at hub
s,shr	at shroud
th	at throat
1,7	compressor calc, stations, see Fig.3
0,6	turbine calculating stations, see Fig 4
u	Euler

1 INTRODUCTION

Radial-flow turbomachines are ubiquitous. Their inherent ability to achieve a high pressure change with good efficiency from a single stage, given even a limited design capability, have rendered them attractive to engineers in diverse industries for many years. Radial machines are pre-eminent in small gas turbines.

The design of the key components, notably the impeller, was in former times achieved with modest design knowledge skillfully applied through semi-manual calculations and draughting expertise, supported over the years by successes and failures on the test-bed or in the application.

There was no prospect of calculating the internal flow so great reliance was placed on the inlet and exit velocity triangles. Good performance was often obtained in this way without the use of computers for a number of reasons. For example, the radial turbine, in common with most expanding machines, is a forgiving device in which unintentional adverse pressure gradients tend to be relatively undamaging. This is due to the overall high expansion which will encourage temporarily separated boundary layers to re-attach. In addition, the slip factor - in both compressors and turbines - could be calculated with adequate accuracy from the early published methods (1). In the case of the compressor this enabled the correct work input to be obtained with reasonable certainty, irrespective of the impeller internal flow. Therefore, for the modest pressure ratio requirements of the time of 4:1 or less, sufficiently good efficiencies were often obtained.

As computers became available to industry during the 1960's, 2D axisymmetric calculations whose theory and numerical solution methods had preceded the means of solving them, could now be brought to bear. At last the designer could begin to think about the impeller internal flow and apply existing knowledge of, for instance, the effects that govern boundary layer development, albeit within the confines of a very approximate model (2). During the 70's new applications for both compressors and turbines posed new performance targets for the designers and so encouraged the application and development of the new techniques.

Centrifugal compressor technology benefited immensely from an influx of research funding for the study of gas turbines for high-performance military helicopters and this resulted in much published work, eg (3) to (5). Turbine technology received encouragement from other aerospace applications, notable air-turbine starters, APU's, and turbopumps for liquid fuel rockets (6).

Now the performance targets were becoming more severe: for example pressure ratios as high as 7:1 or even 10:1 were being considered by the US military programmes. The empirical knowledge that a compressor impeller seemed to deliver an exit flow which could be represented by two distinct flow types - jet and wake - encouraged further computer-based methods to be evolved. The need to generate grids on which the flow equations could be solved encouraged the vane geometry to be transferred from the drawing board to computer-based geometry models, (7,8).

Although methods linking 2D hub-to-shroud with 2D blade-to-blade calculations were pursued for axial machines, they saw relatively little application in their radial counterparts and it was not until the mid-80's that the first computer codes for solution of the 3D Reynolds-averaged Navier-Stokes equations were published (9, 10). Initially the run-times were too long for routine design use but nevertheless the potential offered was recognised and investigated at an early stage. In the last five years computer power per price has risen to the point where 3D methods can be used as an integral part of the design process.

Within this backdrop of history the system described in this paper has been developed. It is now entirely computer-based and utilises many of the advantages offered by modern computer hardware and software.

2 ELEMENTS OF THE DESIGN SYSTEM

The essential elements of a design system for radial turbomachines are:

1. Preliminary design
2. Off-design performance (map) prediction
3. Impeller geometry definition
4. Impeller aerodynamic analysis
5. Aerodynamic analysis/models of other aerodynamic components eg diffusers, nozzles etc
6. Mechanical analysis
7. Generation of impeller manufacturing data

While an integrated design system is necessary, it is advantageous to preserve the modularity implied above. Firstly, this allows newly available methods to be embraced and to be integrated into the system with minimum effort. Secondly, there is commonality in many of the modules in their applicability to both turbines and compressors. The 1D analyses (preliminary design and off-design prediction) of course require modules that are particular to the turbine or the compressor; the remaining modules are all equally applicable to either type of device.

Modern workstations and systems software offer an excellent choice of graphical displays and graphical interactive facilities. In order to achieve the minimum time for a design iteration, to reduce effort and manual activity on the part of the design engineer, and to provide clear and rapid hard-copy for reporting, the system employs these facilities to advantage.

Figure 1 shows a flow-chart of the combined compressor and turbine design system.

In the compressor system the module OONA is a preliminary design program which, given the compressor duty, will establish a "one-dimensional" or "mean-line" design. This is followed by a performance prediction module, CFPRED; using only "1D" geometric data, this program is able to predict off-design performance and provide the operating boundaries of the compressor in terms of choke and surge for the required range of pressure ratio and rotational speed.

At the start of the turbine design system, TOONA is the preliminary design program which determines a 1D design from an initial specification of the duty and constraints while the module RITA predicts off-design performance.

In both the compressor and the turbine systems the 1D programs provide the overall dimensions of the impeller. From this point the modules are substantially the same for both compressors and turbines and, in both cases, the next design step is to determine the detailed geometry of the impeller vane. Initially the 2D meridional geometry is generated and optimised by the GEMIMA module and is then transferred to the 3D interactive modeller, VANESSA. VANESSA enables the designer to create and modify an impeller vane geometry which can then be passed through interface modules (see Fig 1) for aerodynamic or mechanical analysis or for manufacturing purposes.

Much of the aerodynamic analysis is carried out with the axisymmetric inviscid method, KATRINA whose calculating grid is supplied by the interface KATLINA. The 3D viscous code BTOB3D is used for the final design refinements and for off-design analysis. BTOB3D's computational mesh is provided by the interface module VANTIME.

Mechanical analysis is performed by FENELLA whose finite element model is created by the interface PAMELLA. Finally, for machined components, manufacturing data for a 5-axis NC machine is generated using the module MATILDA. For cast components, appropriate geometric details are obtained direct from VANESSA.

3 PRELIMINARY DESIGN

It would be difficult to over-emphasise the continuing importance of preliminary design within the overall design process. While the contribution and significance of the 2D and 3D analyses is acknowledged, they cannot compensate for incorrect 1D or preliminary design. Possibly 50 percent of the engineer-time expended on a radial turbomachine design is rightly devoted to the preliminary design - either at the beginning of the design process, or later when it has been found necessary to return from some later stage to revise the initial assumptions or decisions. This balance may be contrasted with the division of computer time in which the 1D calculations account for less than 1 percent of the computer time used to achieve the final design.

Thus, in a good design system, the emphasis in the configuration for the 1D codes is on maximising convenience of use, thus making best use of engineer-time. The principal features will now be described.

3.1 Design point

For some applications the duty of a turbine or compressor may not require optimisation at a single design point since there may be equal emphasis on performance at several points on the characteristic. However, it is still preferable and convenient to begin the design at a nominal design point but the design calculation should make available information which enables the designer to maintain an awareness of the off-design performance requirements at as early a stage as possible. Figure 2 depicts the data flow for the compressor design point module OONA. The data for the code comprise:

1. The duty - pressure ratio, mass flow, rotational speed

2. Fixed geometric constraints - hub inlet diameter, for example, which is often determined by shaft requirements.

3. Geometric and aerothermal data which may be specified as best-guess values in the first instance but which are free to be varied as the optimisation proceeds - vane number for example.

The output comprises all the major dimensions and the key aerothermal parameters on which design judgements will be based. The latter include the specific speed (Ns), work factor ($\Delta H/U_5^2$), Mach numbers at inlet and exit, and so on. The calculation for each case takes less than a second on a modern machine and so the designer can afford many iterations, adjusting the free input parameters using an integrated data editor in accord with his own expert knowledge until an optimised design is achieved. When the preliminary design is complete, datafiles for the next codes in the sequence (CFPRED and GEMIMA) can be automatically generated.

A full description of the methods used is outside the scope of the present paper. The methods used are for the most part well-known, see for example (11). In summary the flow is calculated at the discrete stations as illustrated in Figure 3. The calculation of the impeller inlet conditions departs from the purely 1D or mean-line approach (Fig 3). Because of the importance of the shroud inlet conditions (eg relative Mach number, incidence), the flow and geometry are calculated at three radial stations, hub, root-mean-square, and shroud. The approximate calculation makes use of two simple assumptions as follows:

1. The variation of axial velocity C_x from hub to shroud is linear.

2. The variation in vane angle from hub to shroud obeys one of a selection of simple relationships, for example:

$$\tan\beta \propto r$$

One of the more critical initial design parameters is the estimated inducer throat area. This is needed to calculate the inducer choke flow and thereby enable the margin between design mass flow and choking flow to be determined. At any radial station the throat width can be approximately expressed by:

$$O = \frac{2\pi r}{N_j} \cos\beta - t_n$$

For prescribed values of vane thickness at hub and shroud, using the simple relationship for tan above, and assuming a linear thickness variation, an approximation for throat area can be derived (12).

From the computed values of impeller inlet stagnation temperature and pressure, and assuming no change in these parameters up to the throat, it is then possible to calculate a non-dimensional mass flow function at the throat which may be compared with the value of Q_{th} for choking.

The impeller tip conditions are calculated in the standard manner from the Euler equation, the isentropic efficiency, and continuity. Points to note are:

1. The slip factor is calculated from Wiesner's well-known formula.

2. The relative exit velocity is calculated using a relative diffusion velocity ratio VR where

$$W_5 = VR.W_{3,shr}$$

3. Parasitic work (the contribution to stagnation temperature rise due to windage and re-circulating flows) is acknowledged by means of a Power Input Factor

$$\Delta T_u = \frac{\Delta T}{p}$$

The turbine design code, TOONA, is arranged in a similar way to the compressor module. The input data fall into the same generic categories of duty, fixed geometric constraints, and initial guesses. In this case the duty is specified by expansion ratio, mass flow, rotational speed, and turbine inlet temperature. The geometric constraints include hub outlet diameter ratio (hub diameter/inlet diameter) and the rotor inlet vane angle of zero. The geometric and aerothermal data used to iterate towards an optimum design include U/C, exducer shroud diameter ratio (shroud diameter/inlet diameter), and vane number.

The calculation procedure can have a number of variations. For example, the inlet velocity triangle can be determined from minimum Mach number considerations or can be based upon a supplied inlet vector. Also, the exit velocity triangle can be determined using the assumption of axial absolute velocity at exit or by introducing specific values of exit swirl. In a similar manner to the preliminary compressor design the experience of the designer is of paramount importance, and a satisfactory design requires a balanced judgement using the various aerothermal output such as work factor ($\Delta H/U_4^2$), discharge velocity ratio (C_{m5}/U_4), relative velocity ratio (W5shr/W), degree of reaction, and specific speed. In a similar manner to the compressor code the non-dimensional mass flow functions at the nozzle and rotor throat sections are calculated and when compared with Q_m for choking provide additional useful guidance. The calculating stations (Fig 4) are arranged to fall in-line with the requirements of RITA, but in this preliminary design module only the core stations of rotor and nozzle inlet and outlet are used.

When the designer is satisfied with the preliminary design, datafiles can be automatically generated for the next codes in the system (RITA and GEMIMA).

3.2 Performance prediction

The performance prediction modules CFPRED and RITA perform the following functions:

1. Prediction of entire operating map at an early stage in the design process working with relatively simple data.
2. Assistance in the definition of a family or series of stages.
3. Prediction of the effects of geometric changes to an existing compressor or turbine design.

For functions 1 and 2, the codes are self-sufficient. Using internal correlations which have been developed through comparisons with test results, the code can provide reasonably accurate predictions for new designs. For an existing design for which test data is available it is of course preferable to match the prediction as well as possible with the test data before proceeding to examine the effect of design changes as in 3.

3.3 Compressor performance prediction

The several published methods of centrifugal compressor performance prediction can for convenience be considered in two categories. In the first category are those procedures which attempt to model the 1D flow within the impeller with the objective of obtaining an estimate of impeller stagnation pressure loss through, for example, boundary layer calculations. Probably the most determined and elaborate example of this approach is reported in (13). The second approach accepts the difficulty and possibly dubious return for effort of modelling the internal flow; instead a more empirical approach is followed, but by experience, a sufficiently accurate procedure can be obtained. Rodgers' method (14, 15) is the oldest and probably the most successful of this genre.

The procedures used in CFPRED, the performance prediction module in the present system, generally follow the approach of Rodgers, but embody a number of modifications and additions to cater for modern designs (16, 17). Additionally, considerable development through validation against test results from a diverse range of compressor types has enabled the drawbacks of the empirical approach to be minimised and thus a useful design and analysis tool has been developed. For a full account reference should be made to the earlier publications. However, a summary of the methods used in CFPRED is appropriate here:

Idealised efficiency curves for both impeller and diffuser are defined. Examples of impeller efficiency characteristics are shown in Figure 5. The impeller efficiency at any point on a constant $U/\sqrt{T}$ curve is a function of the specified peak efficiency and the ratio of the mass flow to the choking mass flow at that speed. Figure 5 indicates that the mass flow ratio corresponding to the peak efficiency (ie efficiency ratio = 1) varies with $U/\sqrt{T}$; the linear relationship is given in reference 16 where also it is shown that the locus of peak-efficiency mass flow corresponds closely to that associated with constant inducer incidence. Empirically-derived expressions are used to determine the efficiency at any value of $U/\sqrt{T}$ (16).

Figure 6 shows a comparison between the predictions of CFPRED and the experimental results for a compressor of moderate pressure ratio.

As the impeller characteristics are related to the inducer choke flow it is necessary to use the throat area previously calculated by OONA in the design point analysis. Alternatively, if the value of throat area is known more accurately, for example after using the VANESSA module, this can be supplied; otherwise a modified $\cos^{-1}$o/s rule can be used within CFPRED.

Inducer stall prediction uses standard two-dimensional cascade theory to relate inducer turning (deflection) and incidence to nominal deflection. The method adopted follows the approach of Rodgers (4), with a number of additions such as the ability to model backswept impellers (16). Stage surge is usually determined at moderate to high pressure ratios by diffuser stall. However, at lower pressure ratios the compressor more often operates in areas where the presence of inducer stall coincides with regions of lower stage stability. In these regions inducer stall can influence the stage surge behaviour and cause an uneven surge line often accompanied by audible and other effects. An example of this is shown in Figure 6 where the surge line is affected on the 30000 rev/min constant speed line.

The diffuser performance is modelled in a similar way to the impeller although in this case the diffuser efficiency is only a function of the ratio of the diffuser flow to the diffuser choke flow. Diffuser choke mass flow is determined in a similar way to the inducer by relating the throat area and stagnation conditions at diffuser inlet to the diffuser choke flow.

Invariably it is stage surge which is the most difficult performance phenomenon to model in a simple method (or even in more complex methods). The method used in CFPRED for vaned diffuser stages, although entirely empirical, has been refined over many years so that reliable surge predictions are possible for a wide variety of compressor stages. More recently work has been carried out to develop an equally reliable method for vaneless diffuser stages and this has shown promise (17). In the cases studied stage surge appears to correlate well with the larger of the two mass flows equivalent to constant values of either inducer incidence or mixed-out absolute flow angle from the impeller. Unfortunately, the actual values of these two parameters can vary slightly with compressor type, and so this makes a reliable method of surge prediction for an "unknown" vaneless-diffuser stage more difficult to achieve.

3.4 Turbine performance prediction

The methods used in RITA, the turbine off-design prediction module, are fully described in (18). An outline of the principle features will be given here.

The turbine is modelled as a number of discrete components - scroll or housing, nozzle, impeller, and exhaust diffuser (Fig 4). Turbine characteristics as constant $N/\sqrt{T}$ curves are generated by computing mass flow and efficiency at a range of points specified in terms of expansion ratio. The overall efficiency is a consequence of the losses occurring in each of the components. The loss models used generally follow the method published in (19) but were tuned to give improved agreement with a range of turbine test results.

The loss models may be summarised as:

- A viscous loss model to represent frictional and other viscous losses in which the loss is assumed to be proportional to the mean dynamic head within the component:

$$\Delta P_t = \tfrac{1}{2} C_1 C_2 (D_{in} + D_{out})$$

- A scroll shock loss model of the form:

$$\frac{P}{P'} = 1.0 + C_3 (M_t - 0.9)^n$$

for $M_t \geq 0.9$ and

$$\frac{P}{P'} = 1.0$$

for $M_t < 0.9$

- A nozzle inlet incidence loss model based on loss of stagnation pressure being a function of kinetic energy normal to the vane at inlet, ie

$$\Delta P = \tfrac{1}{2} \rho c_n^2$$

- A rotor incidence loss model of the same form as the nozzle inlet loss. A typical comparison of test and predictions is shown in Figure 7.

4 IMPELLER GEOMETRY

4.1 Analytic versus interpolative

Compressor and turbine impeller vanes are complex 3D geometric shapes and their definition has posed a challenge to turbomachinery engineers from the earliest generations of designs which pre-dated by many years the use of computers. Initially, of course, vanes were defined in manufacturing drawings, often by means of families of sections parallel to, and perpendicular to, the impeller axis. Their shapes were defined by 2D analytic functions whose discrete values were manually calculated. Between the sections the vane shape was implied rather than defined and had to be determined by interpolation. An obvious, but not the most advantageous, way of utilising the power of computers when they became commercially available was simply to automate the traditional definition methods, thus perpetuating the use of interpolative techniques and the attendant ambiguity of definition. At a time when aerodynamic and stress analysis techniques were themselves highly approximate, there was little disadvantage ensuing if the manufactured vane differed somewhat from the intended vane.

However, even as early as 1970 it was recognised that newly available analysis methods and the power of the latest computers could engender a new approach to vane definition. Smith and Merryweather (7) published the first known comprehensive system for representing impeller vanes by analytic patches. Casey (8) described a similar, but independently developed system. Analytic definition in place of interpolation eliminated ambiguity and encouraged the development of iterative techniques for manipulating the vane geometry to achieve design changes. Additional computer code to transpose the geometry into FE stressing models and aerodynamic models was a relatively straightforward matter and ensured accuracy of geometry definition for analysis purposes. (It is interesting to note here that designers of axial machines which employ blades of simpler, but still quite complex shape, remained dependent on interpolative methods some twenty years longer; the analytic methods now in use in large gas turbine companies tend to be derived from the proprietary products of computer software vendors rather than the developments of turbomachine engineers and, as such, have until recently been lacking in usability).

4.2 Principal features

The present method, VANESSA, has its roots in the system described in (7) and has for many years, see (11) for example, been used for the design of high performance gas turbine compressors. More recently it has been extended for use in radial flow turbines. Its salient features will be described:

The vane is modelled as three surfaces, the pressure surface, the suction surface, and an imaginary surface lying between the actual surfaces called the camber surface. Each surface comprises a number of connected patches, typically four or six. Figure 8 shows a typical compressor arrangement where four patches have been used; for a turbine more patches are usually required and Figure 9 gives an example. The type of patch chosen is a cubic-linear patch in which the patch may be thought of as an infinite number of cubic curves aligned in the streamwise direction or as an infinite number of straight lines in the hub-to-shroud or spanwise direction. This was deemed to provide sufficient flexibility in the streamwise direction to enable the required aerodynamic shape to be obtained and allowed the vane to be machined with a straight-sided conical cutter. (Later requirements suggested the need to include a curved-sided vane option and this is described below).

The equations which describe a cubic-linear patch may be arranged in terms of the parameters u and v as follows:

$$x = a_x u^3 + b_x u^2 + c_x u + d_x$$

$$y = a_y u^3 + b_y u^2 + c_y u + d_y$$

$$z = a_z u^3 + b_z u^2 + c_z u + d_z$$

which express the coordinates in terms of u along curves such as AB and MN where u varies from 0 to AM to 1 on BN, and:

$$x = e_x v + f_x$$

$$y = e_y v + f_y$$

$$z = e_z v + f_z$$

which expresses the coordinates in terms of v along lines such as AM and BN where v varies from 0 on AB to 1 on MN.

The boundary conditions at the patch corners are chosen such that continuity of slope and curvature is guaranteed.

Turbine wheels are frequently cast rather than machined and so the advantage of a straight-sided vane disappears. Furthermore, lower vane stresses can be achieved if the vane has curved tapering sides (frequently known as Eiffel Tower sections). Vanes of this type are modelled by superimposing upon the cubic-linear patch a thickness distribution which varies as follows:

$$t_r = t_s + \left(\frac{r_s - r_r}{r_s - r_h}\right)^n (t_h - t_s)$$

An example of a vane with curved sides is shown in Figure 10.

VANESSA creates vane models as described above. The additional features of VANESSA are:

- Graphics display of virtually any geometric parameter (Figs 10 and 11 for example).
- Interactive-graphical modifications of the vane shape, including vane thickness changes, scaling of the entire vane, and direction of rotation reversal
- Automatic history trace of all changes
- Calculation of throat area
- Interfaces with aerodynamic and stressing programs

The interfacing programs are shown in Figure 1. On occasions special interfaces have been required, for example to transfer a vane geometry from another definition system into VANESSA. The combination of the analytic definition system and the modularity of the set-up allows this to be done without difficulty.

4.3 Meridional pre-processor

It is frequently required to generate new or slightly modified hub and shroud profiles. The simple meridional geometry module GEMIMA, used in conjunction with VANESSA, greatly facilitates this task. The module can generate full meridional geometry including patch boundaries directly from both OONA and TOONA input, if necessary without additional manual input, with the aid of default values. The hub and shroud profiles are produced from combinations of straight lines and circular arcs manipulated interactively. The module may be thought of as a pre-processor to VANESSA and allows the user readily to construct new geometries or modify existing ones.

5 IMPELLER AERODYNAMIC ANALYSIS

5.1 2D inviscid analysis

In spite of the widespread application of 3D Navier-Stokes solvers in turbomachinery analysis, 2D inviscid methods are still used extensively in radial-flow machines. There are a number of underlying reasons:

1. The run-time comparison of, say one minute as compared to 3 hours.

2. The calibration of inviscid methods

3. The sparseness of the output which takes little time to assimilate

Additionally, the 3D methods may now be used to check the 2D approach and, in some aspects, there is good corroboration. Thus the 2D inviscid code still tends to be the main design tool.

The method due to Katsanis (20) is the basis of the code used in the present system. The Euler equations are arranged in the well-known streamline curvature formulation and are solved iteratively on a relatively coarse mesh. The geometric data defining the vane and passage shape are provided by an interface (KATLINA) which reads the analytic patch definition from VANESSA.

5.2 3D viscous analysis

One of the earliest papers describing the use of 3D inviscid and 3D viscous codes (9, 10) for radial-flow machines was by Casey et al (21). They demonstrated the potential of the new codes for accurate prediction of the flow details in centrifugal impellers and for providing insight into the nature of impeller flow.

The BTOB3D code of Dawes (10) has been integrated into the present design system. The Reynolds-averaged Navier-Stokes equations are solved on a computational mesh comprising around 30,000 points which includes an inlet section and exit section. The geometric data for the program is prepared by a special interface (VANTIME) which enables the user to optimise the streamwise mesh interactively. Figure 12 shows VANTIME's streamwise mesh for a typical turbine wheel.

Figure 13 shows BTOB3D predictions of Mach number contours on a surface near the impeller shroud at a low inlet Mach number operation condition.

The authors have accumulated four years experience in the use of BTOB3D. Its usefulness can be summarised as follows:

- The code is able to predict choking flow quite well and to predict with very good accuracy the effect of small changes in, for example, vane thickness on choking flow.
- For compressor impellers, Euler work input is predicted quite well.
- The code has given significant guidance in the process of refinement of impeller geometry to achieve higher efficiency.
- Comparisons of BTOB3D with the 2D Katsanis calculation have shown that the simpler analysis is capable of predicting the effects of moderate to gross geometric changes with adequate accuracy.
- At the start of the four years experience typical computer run-times of around 15 to 20 hours restricted the use of the code to end-of-design refinements or urgent diagnosis. Now with run-times around 1/5 to 1/10 of these, the code can be used in a routine manner as a design tool.

6 IMPELLER MECHANICAL ANALYSIS

As gas turbine duties put more demand on the turbomachinery in terms of pressure ratio, blade speed and gas temperatures, so greater care and accuracy must be exercised in the mechanical design. Radial turbine and compressor impellers would pose a serious FE model generation problem due to their complex shape and would make rapid turn-round of trial vane shape designs difficult to achieve. In the present system, the VANESSA vane model is processed by the FE pre-processor PAMELLA to create a 3D FE model of the vane. PAMELLA performs the geometric manipulations, and allows the user to closely control the automatic generation of the FE model of the vane. The FE analysis code used in the system is FENELLA (22, 23), which has been developed as a general purpose code, but with a number of additional features of particular benefit when analysing turbomachinery components. The system enables the generation of a 3D impeller model to be achieved in less than half a day, with subsequent iterations in much less time. The procedure followed is:

1. Generate an axisymmetric FE model of the impeller using FENELLA
2. Achieve an optimum axisymmetric design through, for example, geometric changes to the rear face to reduce bore stresses and tip deflections
3. Using the interface PAMELLA, to map directly the three-dimensional vane patch geometry modelled in VANESSA onto the axisymmetric FE model to achieve a 3D FE model.

4. Carry out 3D stress and vibration analysis using FENELLA (Fig 14).

7 CONCLUDING REMARKS

A system for the design and analysis of radial-flow turbomachinery has been described. The system incorporates every aspect of the design process and its modular structure has enabled state-of-art technology to be included as it becomes available. The fluent interfacing between modules allows turn-round of an entire iteration (starting with revision of the 1D design) to be achieved in under a day. Over several decades of development for compressor applications, consistently good performance has been achieved from impellers designed with the system. Turbine experience is smaller but it is envisaged that the application of the same design philosophy and tools will enable equally good results to be attained.

8 REFERENCES

(1) STANITZ, J D. Some theoretical aerodynamic investigations of impellers in radial and mixed-flow centrifugal compressors, Trans ASME 74 (1952) 473.

(2) DALLENBACH, F. The aerodynamic design of centrifugal and mixed flow compressors,SAE Tech Prog Series Vol 3, 1961.

(3) MORRIS, R E and KENNY, D P. High pressure ratio centrifugal compressors for small engines "Advanced centrifugal compressors", ASME publication 1971.

(4) MCANALLY, W J. 10:1 pressure ratio single-stage centrifugal compressor program, USAAMRDL-TR-74-15, 1974.

(5) SCHORR, P G, WELLIVER, A D and WINSLOW, L J. Design and development of small, high pressure ratio, single-stage centrifugal compressors "Advanced centrifugal compressors", ASME publication 1971.

(6) RODGERS, C. Small high pressure ratio turbines, von Karman Institute Lecture Series 1987-07, 1987.

(7) SMITH, D J L and MERRYWEATHER, H. Representation of the geometry of centrifugal impeller vanes by analytic surfaces, International Journal for Numerical Methods in Engineering, Vol 7 1973.

(8) CASEY, M V. A computational geometry for the blades and internal flow channels of centrifugal compressors, ASME Journal Engineering Power, Vol 105,1983.

(9) DENTON, J D. The use of a distributed body-force to simulate viscous effects in 3D flow calculations, ASME Paper 86-GT-144, 1986.

(10) DAWES, W N. Development of a 3D Navier-Stokes solver for application to all types of turbomachinery, ASME Paper 88-GT-70, 1988.

(11) CAME, P M. The development, application and experimental evaluation of a design procedure for centrifugal compressors, Proc IMechE 1978 Vol 192 No 5.

(12) CAME, P M, SWAIN, E, BACKHOUSE, R J and WOODS, I H. A computational design system for centrifugal compressors, VDI Berichte Nr 1185, 1995.

(13) HERBERT, M V. A method of performance prediction for centrifugal compressors, ARC R&M 3843, 1978.

(14) RODGERS, C. Typical performance characteristics of gas turbine radial compressors, Journal Engineering Power, Trans ASME Vol 86; 1964.

(15) RODGERS, C. Influence of impeller and diffuser characteristics and matching on radial compressor performance, SAE Tech Prog Series, Vol 3, 1961.

(16) SWAIN, E. A simple method for predicting centrifugal compressor performance characteristics, IMechE Paper C405/040, Turbochargers and Turbocharging Conference 1990.

(17) SWAIN, E and CONNOR, W A. The use of simple methods to predict the performance of industrial centrifugal compressors, VDI-GET Conference Thermische stromungmaschinen Turbokompressoren im industriellen einsatz II, VDI Berichte 947, VDI Verlag, Hannover 1992.

(18) CONNOR, W A and FLAXINGTON, D A. A one dimensional performance prediction method for radial inflow turbines, IMechE Conference on Turbochargers and Turbocharging, 1994.

(19) FUTRAL, S M and WASSERBAUER, C. Off-design performance prediction with experimental verification for a radial inflow turbine, NASA TN D-2621, February 1965.

(20) KATSANIS, T. Use of arbitrary quasi-orthogonais for calculating flow distribution in the meriodional plane of a turbomachine, NASA Technical Note NASA TN D-2546, 1964.

(21) CASEY, M V, DALBERT, P and ROTH P. The use of 3D viscous flow calculations in the design and analysis of industrial centrifugal compressors. ASME Paper 90-GT-2, 1990.

(22) FENELLA user manual rev.1, PCA Engineers.

(23) DAWES, W N and WOODS, I W. A simulation of the forced response of a centrifugal impeller by coupling a finite element code to an unsteady aerodynamic code, ASME, IGTI conference Birmingham 1996.

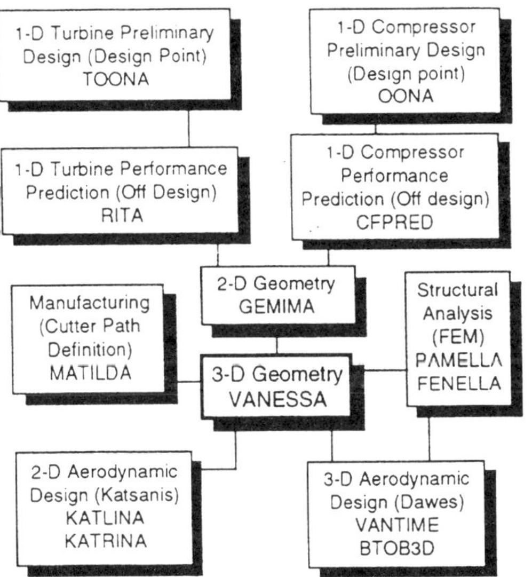

Fig.1 Schematic of design system

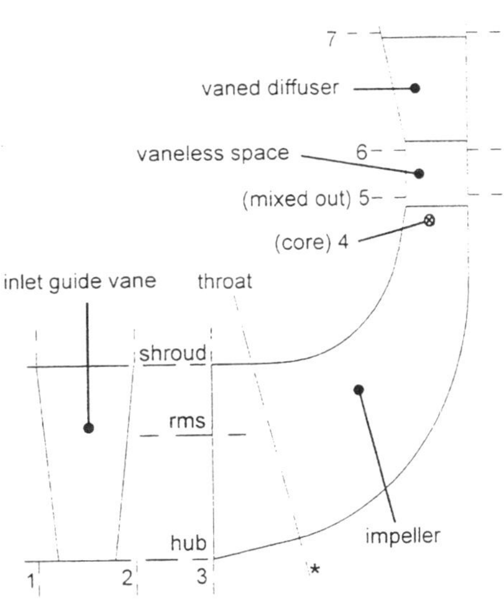

Fig.3 Compressor calculating stations

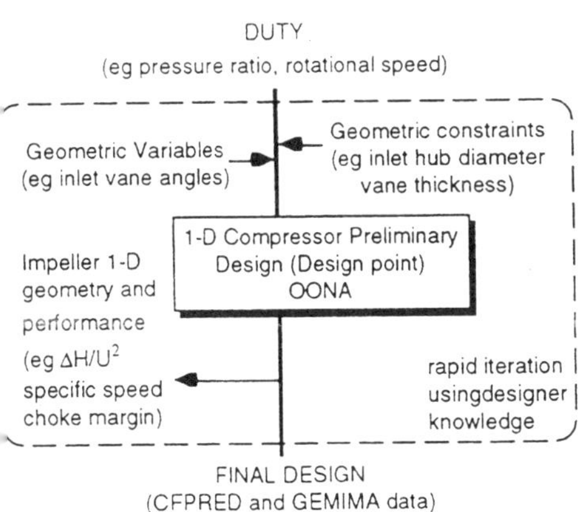

Fig.2 Preliminary design module

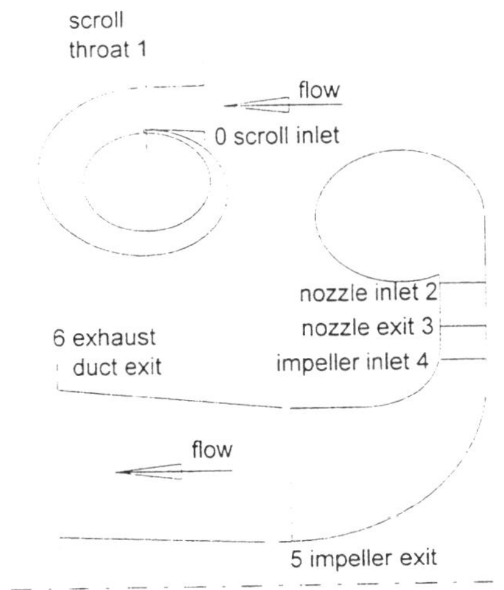

Fig.4 Turbine calculating stations

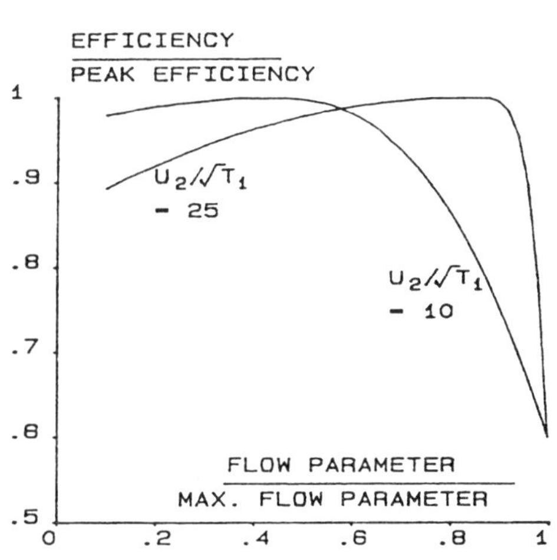

Fig 5 Impeller characteristic

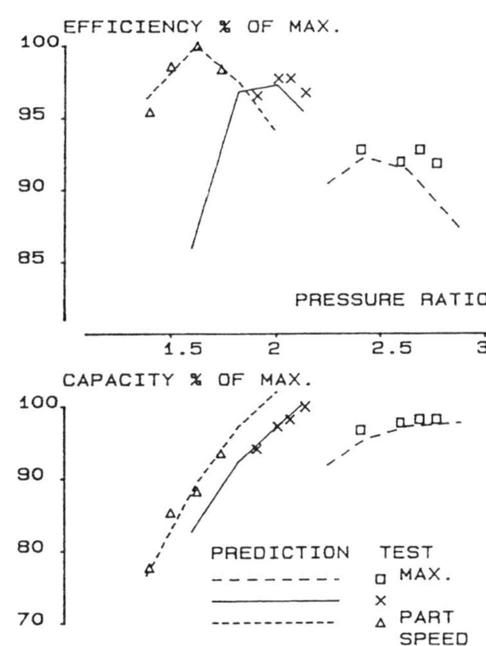

Fig.7 Comparison of predicted and measured turbine performanc

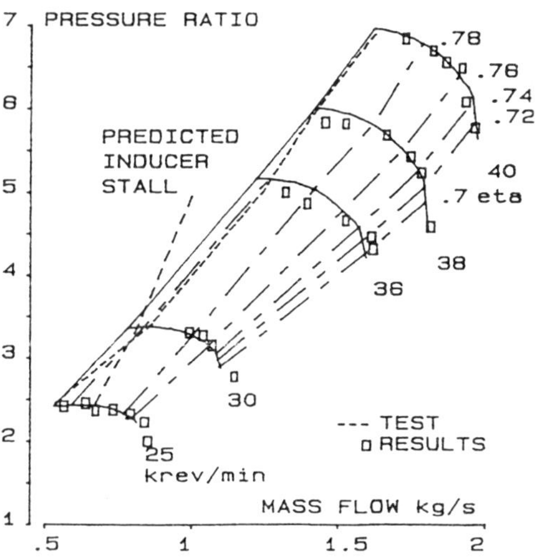

Fig.6 Comparison of predicted and measured performance

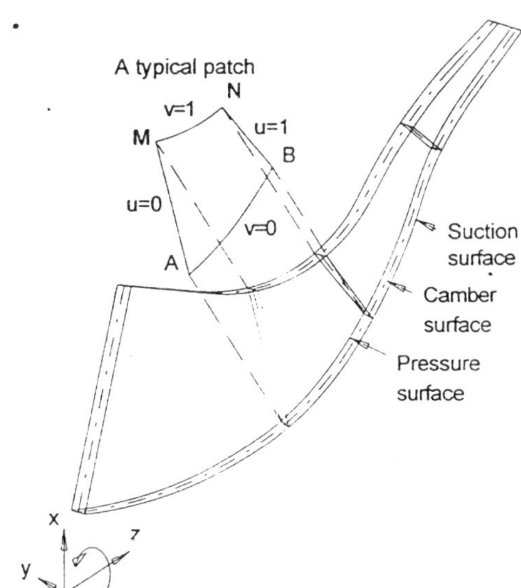

Fig 8 Analytic patch representation of impeller van

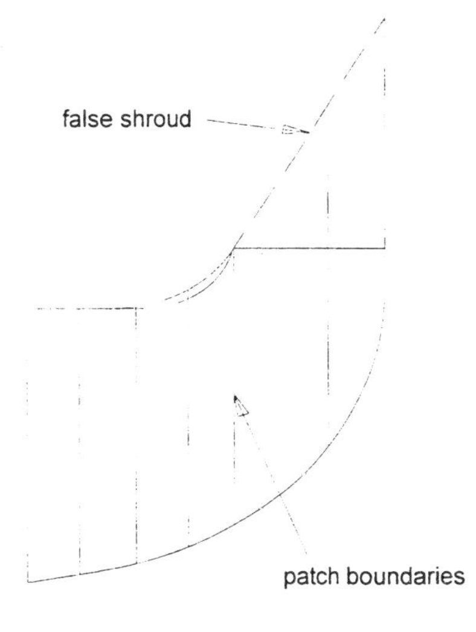

Fig.9 Typical turbine vane meridional view

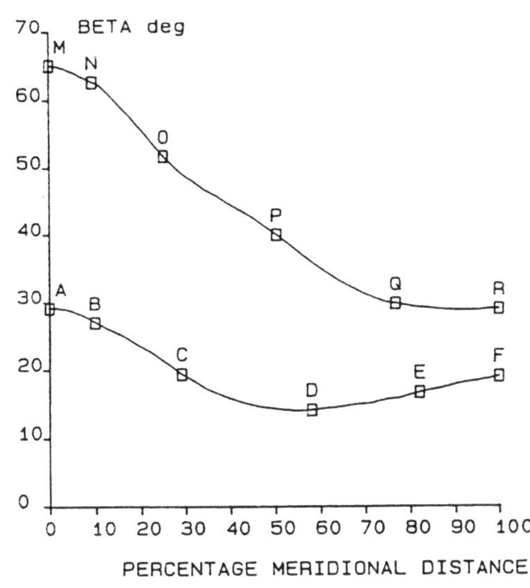

Fig.11 VANESSA output of BETA distribution

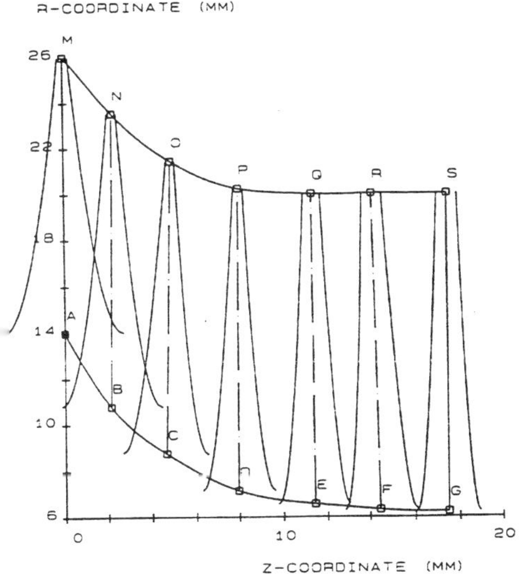

Fig.10 VANESSA output - vane thickness distribution

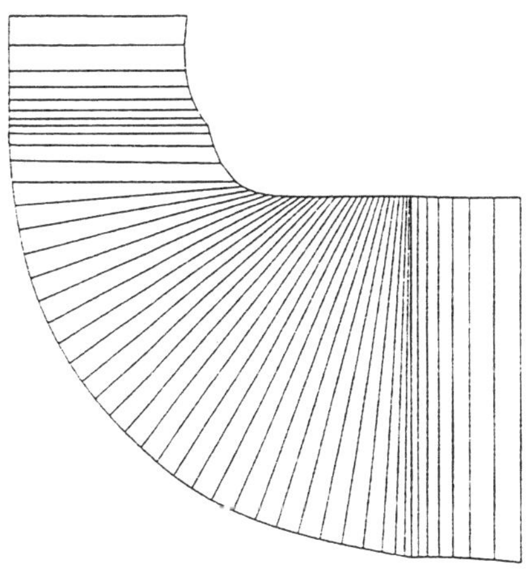

Fig.12 VANTIME streamwise mesh of turbine wheel

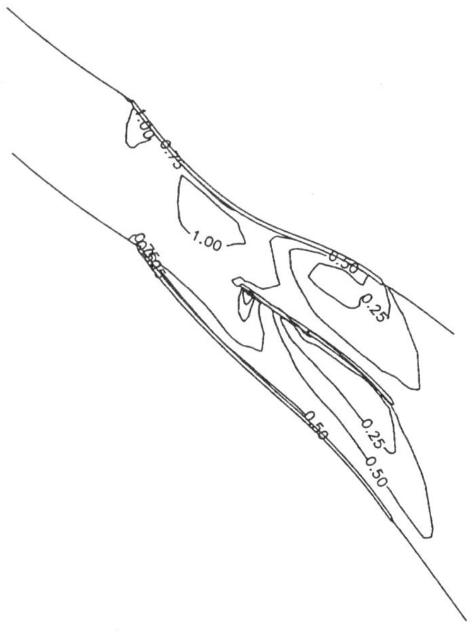

Fig.13 BTOB3D Mach number contours on a surface near the impeller shroud

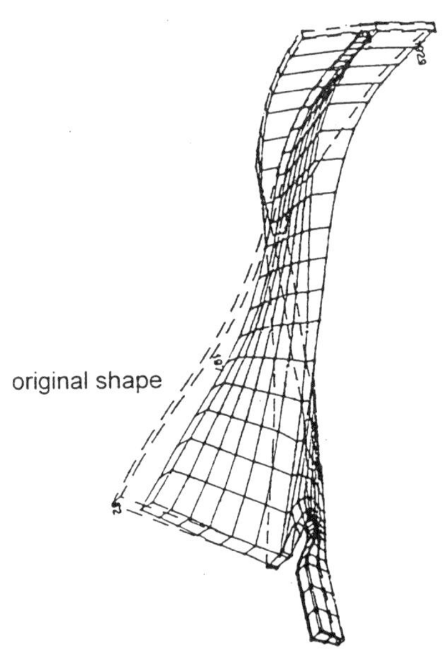

Fig.14 FENELLA dynamic analysis - Vane mode 1

C499/026/96

Active support systems for integrated rapid prototyping

R A FIORINI PhD and **G F DACQUINO**
Politecnico di Milano, Italy

SYNOPSIS

Rapid Prototyping plays a key role in reducing the time-to-market by allowing limited numbers of new products to be designed and built quickly. To achieve higher productivity and optimal design in mechanical engineering and manufacturing, the numerical simulation concept has to be combined with an efficient, user friendly, multi-physical integrated development environment, capable of maximizing or minimizing any measure of the target system performance while satisfying a wide variety of design constraints: an Active Support System.
In the case of finite computational resource systems, it is necessary to choose the best compromise in allocating the computational load for each physical domain according to the available global computational resources, in order to avoid a severe degradation of the overall required accuracy for the final global solution.
The present contribution describes the main development considerations of Subsystem Granularity Concise Descriptors for Active Support Systems. An example of their use for Multi-Physical Virtual Prototyping, with autonomous decision capability on computational load optimization, is shown.

Keywords: Numerical Simulation, Virtual Prototyping, Integrated Design and Manufacturing, Active Support Systems

NOTATION

$\mathbf{z}$	position vector
$\mathbf{v}$	fluid velocity vector
$\mathbf{A}_\iota$	vector potential
$\mathbf{r}$	system point mass
$\mathbf{a}$	viscous damping coefficient
$\mathbf{f}$	spring constant
ρ	fluid density

μ_f	fluid dynamic viscosity
ε_f	fluid elastic modulus
g	acceleration of gravity
μ_e	magnetic permeability
σ	electric conductivity
ε_e	dielectric constant
R, A, F	generic equation parameters
h, m, n	generic equation parameters
H, M, N	generic equation parameters
$\mathbf{E_k}$	kinetic energy
$\mathbf{E_p}$	potential energy

1 INTRODUCTION

Rapid Prototyping plays a key role in reducing the time-to-market, by allowing limited numbers of new products to be designed and built quickly. The goal of research in Rapid Prototyping is to develop and integrate the tools and technologies needed for rapid and efficient design and manufacturing of products, processes, and systems. New tools and technologies coupled with innovative services and an updated infrastructure, that allows distributed prototyping over high speed networks, hold the promise for integrated systems for Rapid Prototyping.
From an operative point of view, Rapid Prototyping can be divided into Virtual Prototyping and Physical Prototyping. Virtual Prototyping includes modelling, numerical simulation, model validation, and design tools and techniques, while Physical Prototyping includes elements of machine design, process design, and CAD/CAM integration.

Virtual Prototyping is the substitution of computer models for physical products, processes, and systems and their use in numerical simulations. Virtual prototypes can model products or processes, up to a complete factory, at several levels of abstraction. Examples include the 3D solid models used in designing mechanical systems, engines, actuators and transducers. To substitute for Physical Prototypes, Virtual Prototypes must accurately reflect the properties of materials and objects, allowing a range of design and evaluation activities. This implies accurate multi-physical integrated representation and rendering of objects, high speed access to remotely located virtual prototypes or models, and high-performance computation to carry out operations on the virtual prototypes.

To achieve higher productivity and optimal design in mechanical engineering and manufacturing, the modelling concept has to be combined with an efficient, user friendly, multi-physical integrated development and numerical simulation environment, capable of maximizing or minimizing any measure of the target system performance while satisfying a wide variety of design constraints: an Active Support System [1].

Most advanced devices have to be analyzed under a few distinct and sometimes strongly coupled physical domains, namely electrical, magnetic, mechanical, hydraulic, fluid dynamic, combustion and dynamic. Usually, the related 3D dynamic simulation and performance analysis require extensive mathematical calculations for each physical domain and then further computational resources are needed for data fusion and synchronization, in multi-physical simulations. In order to save computational resource, it is possible to reduce the detail complexity of analyzing some physical phenomena, but the solution loses precision in the process. If you study a simplified approximation of the real problem, the solution will also be an approximation.

In the case of finite computational resource systems, it is necessary to choose the best compromise in allocating your computational loads for each physical domain, according to the available computational resources, in order to avoid a severe degradation of the overall required accuracy for the final global solution.

The present contribution describes the main development considerations of Subsystem Granularity Concise Descriptors for Active Support Systems. An example of their use for Multi-Physical Virtual Prototyping, with autonomous decision capability on computational load optimization, is shown.

2 VIRTUAL MULTI-PHYSICAL PROTOTYPING

The ultimate goal of the different description techniques and numerical procedures developed over the years is to find the solution to a complex problem by solving a simpler one, that closely represents the actual problem.
In the application of finite computational resource systems for simulating complex multi-physical coupled problems, one sensitive activity is the choice of the description detail level to assign to each physical subsystem domain, in order to obtain the final overall desired accuracy in the computed results.
The numerical simulation effort has to deal with different physical subsystem organizational levels as one moves from the macrocosm to the microcosm.
The final global representation of the physical simulated behaviour is function of the numerical simulated physical behaviour of those organizational levels. The simulated behaviour of those organizational levels, in turn, depends on their specific physical properties and physical couplings or decouplings, as offered by the description precision of the related subsystem domains. So the finer you are able to describe the subsystem coupling-decoupling properties the better your overall final numerical simulation.

Unfortunately, by using finite resource machines, limitations on the resolution of the coupling-decoupling description are unavoidable. So the coupling-decoupling description resolution can be the main limiting factor on the overall accuracy of the simulation effort.
This is a reason why the realistic global physical behaviour simulation of non-stationary multi-physical mechanical systems is still an art rather than a science. In fact, for a given overall computational resource and for a desired global modelling accuracy, a reasonable computational resource allocation to each subsystem would be defined by an objective partition function rather than by the personal "sensitivity" of each expert user.

Quite often to describe nonlinear dissipative fields, there is the need to deal with coupled implicit nonlinear algebraic equations that can be identified automatically, in the case of a structured approach for the selection of solution procedures.
The coupled-decoupled "energy domains", which loom out of these problems with their related physical scale boundaries, usually lead to a natural identification of power or energy field characteristics: conservative, dissipative (non-conservative), storage, source, transformer, gyrator, etc..

Furthermore, a structured quest for the behaviour analogies of physical quantities in different physical subsystems, in terms of flux and characteristic variables, would be extremely convenient.
As an example, let us consider fluid and hydrodynamic subsystems to put into evidence the numerical simulation constraints and limitations of the traditional engineering approach.

When they are coupled to magnetic, mechanical and thermal subsystems the first computational challenge is given by the term $(\mathbf{v} \cdot \nabla)\mathbf{v}$ in the Navier-Stokes equation, where $\mathbf{v}$ is the fluid velocity vector. The simplified parabolic structure of the Navier-Stokes equation is modified by the presence of the convective term into an elliptic one. A variety of fluid regimes can be described at different precision levels and boundary conditions. The associated computational load can be quite heavy even for a relatively coarse global solution accuracy.

The specific subsystem descriptions and the applied numerical techniques define the computational simulation load of the global solution.

As an example, let us consider the numerical simulation of the physical behaviour for a 3D axial-symmetric viscous incompressible fluid in stationary regime.
A usual problem description can be given by the classical formulation in terms of streamline function and vorticity. The solution of the flow field needs the computation of three numerical fields: streamline, vorticity and pressure. In transient regime the computation of those three numerical fields can force a high number of numeric iterations for a von-neumann computer.
For this specific example a problem description using mixed variables (i.e. streamline function and vorticity) can minimize the fluid field computational load [2].
Today, a variety of numerical procedures and techniques are ready available on workstations and mainframes. The following consideration are formulated on the assumption that finite element methods are used [3], but they can be immediately extended to other methods with no lack of effectiveness.
From the numeric simulation point of view, the mixed variables description allows to obtain a reasonable operational compromise between the overall simulation desired accuracy and the associated computational load.

But what about the computational load of the boundary coupling-decoupling with different physical subsystem domains?
To overcome those intrinsic approach constraints and limitations, the use of specific problem concise mathematical descriptors, called Granularity Descriptors, could be quite convenient. In fact, it is possible to formulate physical scale boundaries granularity descriptors from the information gathered in the problem definition step, given by the user (i.e. geometrical specifications, material properties, boundary conditions, dynamic parameters data, etc.).

The use of concise descriptors can give an estimation of the final simulation global accuracy obtainable by the user specific subsytem descriptions, with no need to solve the whole problem computationally.

3 PHYSICAL SCALE BOUNDARIES GRANULARITY DESCRIPTORS

Let us focus our attention on the simulation of an electromagnetic actuator device for automotive liquid injection. Electromagnetic actuators, presently, are used in a wide variety of on-off control injection applications where very fast response is required. In all these applications, the primary goal is to control the movement of a small mass, typically less than

10 grams, through a distance of 0.1 to 0.5 mm. in approximately one millisecond or less, in a temperature operating range between 243 K and 333 K, with a fluid supply pressure range between 0.1 and 10 MPascal, using voltage or current driving pulses. The desired main result is the computation of the fluid flow (mm^3/pulse) as a function of pulsewidth, together with all the other physical related operational quantities.

In this case, it is possible to identify at least four different physical subsystems immediately, namely: electric, magnetic, mechanic and hydraulic. Furthermore, those subsystems are coupled together by physical and geometrical boundaries and constraints.

So the global simulation has to deal with different subsystem organizational levels.

Assuming, as a simplification, that the basic laws of physics are fully understood, this is like asking a biologist to model the properties of a cell starting from the rules of quantum chemistry. The inherent complexity of the undertaking is forbidding!

Such a complexity is inherent in the very nature of the object of investigation. It arises from the "characteristic size" of the interaction level of different subsystem domains, each one with its different characteristics and space-time parameters.

However, if the "characteristic size" of the global multi-physical interaction is much larger than the "characteristic size" of the interacting subsystems, then the final result is determined only by the slowest components, interacting along a stationary action path, according to the quantum physics Correspondence Principle [4].

By definition, the stationary action path is given by the Least Action Principle. In 1744 Maupertuis [5] showed that the classical physical trajectories are stationary action paths and his considerations were the basis for the subsequent formulation of the Least Action Principle: the reference concept of all the modern techniques of the Calculus of Variations[6].

According to this interpretation, it is possible to consider that the fast dynamic subsystems are forced to "syncronize" themselves to the slowest ones giving as a final result the classic trajectory at the macroscopic level [4].

Therefore the larger the ratio between the "characteristic size" of the global multi-physical interaction and the "characterictic size" of the interacting subsystem domains, the more precise will be the final simulation of the global prototype.

In order to be able to formulate the Granularity Descriptors for the present example, it is necessary to consider the basic equations for each subsystem domain.

For the Mechanic subsystem, starting from the generic equation structure:

$$r\ddot{z} + a\dot{z} + f z + d = 0, \tag{1}$$

by eliminating the known term, with no lack in generality, and considering only isotropic and homogeneous materials for simplicity, we arrive to:

$$\nabla^2 z = \left(-\frac{a}{f}\right)\frac{dz}{dt} + \left(-\frac{a}{f}\right)\left(\frac{r}{a}\right)\frac{d^2 z}{dt^2} \tag{2}$$

where:

z	=	position vector	[m];

r	=	system point mass	[kg];
a	=	viscous dumping coeff.	[Ns/m];
f	=	spring constant	[N/m];

under the assumption that it is possible to describe the Mechanical subsystem behaviour by the "field" of the position vector **z**.

For the Hydrodynamic subsystem we can write the fluid equation using the Lagrangian approach:

$$\nabla^2 \mathbf{v} = \frac{\rho}{\mu_f \, \mathbf{g}} \frac{\mathbf{D v}}{\mathbf{dt}} + \frac{\rho}{\varepsilon_f \, \mathbf{g}} \frac{\mathbf{D^2 v}}{\mathbf{dt^2}} \tag{3}$$

where:

$\mathbf{v}$	=	fluid velocity vector	[m/s];
ρ	=	fluid density	[N/m^3];
μ_f	=	dynamic viscosity	[Ns/m^2];
ε_f	=	fluid elastic modulus	[N/m^2];
$\mathbf{g}$	=	acceleration of gravity	[m/s^2].

Finally, for the ElectroMagnetic subsystem that creates the actuator magnetodynamic field we can write:

$$\nabla^2 \mathbf{A_e} = \mu_e \, \sigma \, \frac{\mathbf{d A_e}}{\mathbf{dt}} + \mu_e \, \varepsilon_e \, \frac{\mathbf{d^2 A_e}}{\mathbf{dt^2}} \tag{4}$$

where:

$\mathbf{A_e}$	=	vector potential	[Weber/m];
μ_e	=	magnetic permeability	[H/m];
σ	=	electrical conductivity	[S/m];
ε_e	=	dielectric constant	[F/m].

Now, the above subsystem description equations present the same mathematical structure and it is possible to draw a few considerations about the physical parameters that characterize each subsystem domain.

In the Mechanic subsystem equation the physical parameters are "point" parameters, as they are associated to a "point" or "localized" variable. The main description interest lies on the information connected to the system characteristic position.

On the other hand, in the Hydrodynamic and ElectroMagnetic subsystem equations the physical parameters are "average" parameters, as they are associated to an "averaged" or "delocalized" variable. The main description interest lies on the information connected to the system characteristic velocity.

Building from those considerations it is possible to arrive to the hypothesis that the "Interaction Field" describing the interaction between localized and delocalized subsystems can be expressed by the following mathematical structure:

$$\nabla^2\left(z\dot{z}\right) = \frac{A^2}{F^2}\left(\frac{dz}{dt}\right)^2 + \frac{A^2}{F^2}\frac{R^2}{A^2}\frac{d^2z}{dt^2}\frac{dz}{dt}. \qquad (5)$$

The above equation describes a nonlinear field which the component superposition principle cannot be applied for. Therefore linear combination of solutions cannot be used to generate the general analytic solution. But, posing:

$$z = e^{-t} \qquad (6)$$

we obtain:

$$\nabla^2\left(z^2\right) = \left(-\frac{A^2}{F^2}\right)\left(\frac{dz}{dt}\right)^2 + \left(-\frac{A^2}{F^2}\right)\left(-\frac{R^2}{A^2}\right)\left(\frac{d^2z}{dt^2}\right)^2. \qquad (7)$$

This equation shows a quadratic relation between the "Interaction Field" and the position coordinate z. The "Interaction Field" is symmetrical in respect to the position coordinate z, just like for the well known "Energy Field".

Now, let's start from the followin generic relation structure:

$$\pm z = \frac{m}{h}\frac{dz}{dt} + \frac{n}{h}\frac{d^2z}{dt^2} \qquad (8)$$

taking its square, using the constraint (6), $z = e^{-t}$, and taking the Laplacian, we can arrive to:

$$\nabla^2\left(z^2\right) = \left(\frac{m}{h}\right)^2\left(\frac{dz}{dt}\right)^2 + \left[\frac{n(n-2m)}{h^2}\right]\left(\frac{d^2z}{dt^2}\right)^2. \qquad (9)$$

By comparison with the previous "energy field" equation (7), the two expressions can be an identity when the following relations are satisfied:

$$\left(\frac{m}{h}\right)^2 = -\frac{A^2}{F^2} \quad \text{and} \quad \left[\frac{n(n-2m)}{h^2}\right] = \frac{R^2}{F^2}. \qquad (10)$$

Therefore, either:

$$h = \pm F, \quad m = \pm iA, \quad n = \pm iA \pm \sqrt{R^2 - A^2}$$

or:

$$h = \pm iF, \quad m = \pm A, \quad n = \pm A \pm \sqrt{R^2 + A^2},$$

where $\mathbf{i}$ is the imaginary unit, are solution to equations (10).

In this way, it is possible to formulate the description of the interaction between "localized" or "localizable" systems and "delocalized" or "delocalizable" systems by means of the equivalent "Energy Field" motion equation [7].

That equation can express either the information associated to the system characteristic position and then its mathematical structure is given by:

$$\pm z = \frac{M}{H}\frac{dz}{dt} - \frac{N}{H}\frac{d^2 z}{dt^2}, \tag{11}$$

or the information associated to the system characteristic velocity and then its mathematical structure becomes:

$$\pm \dot{z} = \frac{h}{m} z - \frac{n}{m}\frac{d^2 z}{dt^2}. \tag{12}$$

Now, by multiplying the above two expressions (11) and (12), it is possible to formulate the following relation structure:

$$\left(z\,\dot{z}\right) = \frac{1}{2}\left[\frac{M}{m}\left(\frac{h-n}{H}\right)\left(z\,\dot{z}\right) + \frac{m}{M}\left(\frac{H-N}{h}\right)\left(\dot{z}\,\ddot{z}\right) + \frac{N}{m}\left(\frac{h-n}{H}\right)\left(z\,\ddot{z}\right) + \frac{n}{M}\left(\frac{H-N}{h}\right)\left(\ddot{z}\right)^2\right] \tag{13}$$

and imposing the usual constraint (6), $z = e^{-t}$, we arrive to:

$$\left(z\right)^2 = +\frac{1}{2}\left[\frac{M}{m}\left(\frac{h-n}{H}\right) + \frac{m}{M}\left(\frac{H-N}{h}\right)\right]\left(\dot{z}\right)^2 - \frac{1}{2}\left[+\frac{N}{m}\left(\frac{h-n}{H}\right) + \frac{n}{M}\left(\frac{H-N}{h}\right)\right]\left(\ddot{z}\right)^2. \tag{14}$$

This equation can be rewritten in concise notation as:

$$\left(z\right)^2 = P_i\left(\dot{z}\right)^2 - P_r\left(\ddot{z}\right)^2 \tag{15}$$

where:

$$P_i = \frac{1}{2}\left[\frac{M}{m}\left(\frac{h-n}{H}\right) + \frac{m}{M}\left(\frac{H-N}{h}\right)\right], \quad \textit{and} \quad P_r = \frac{1}{2}\left[+\frac{N}{m}\left(\frac{h-n}{H}\right) + \frac{n}{M}\left(\frac{H-N}{h}\right)\right]. \tag{16}$$

Comparing equation (13) with (5), and equation (14) with (7), it is possible to enunciate the following "Synchronicity Principle":
Given a nonlinear "Interaction Field" described by:

$$\nabla^2\left(z\dot{z}\right) = -E_k\left(\dot{z}\right)^2 - E_p\left(\dot{z}\ddot{z}\right), \tag{17}$$

a level of conservative stationary interaction can emerge if and only if the "Interaction Field" is "synchronic" to a local minimum of the associated "Energy Field":

$$\nabla^2\left(z\right)^2 = E_k\left(\dot{z}\right)^2 - E_p\left(\ddot{z}\right)^2; \tag{18}$$

otherwise the nonlinear "Interaction Field" describes a dissipative field.
Futhermore, by comparing equation (7) and (18):

$$E_k \equiv -\frac{A^2}{F^2}, \quad \text{and} \quad E_p \equiv -\frac{R^2}{F^2}. \tag{19}$$

So, the generic system can be described by a "localized or localizable field", or by a "delocalized or delocalizable field", or by an "interaction field".
Now, it is possible to define the level of conservative stationary interaction of a system at rest as the System Granularity Descriptor given by:

$$SG \equiv \left|\frac{1}{E_p}\right| \equiv \frac{F^2}{R^2}, \tag{20}$$

and the level of the associated stationary interaction band as the Band Granularity Descriptor given by:

$$BG \equiv \left|\frac{1}{E_k}\right| \equiv \frac{F^2}{A^2}. \tag{21}$$

By using the above two system Granularity Descriptors for each interacting subsystem domain, it is possible to give an estimation of the final simulation global accuracy obtainable by the used specific subsytem description level, with no need to solve the whole problem computationally.

4 A PRACTICAL EXAMPLE

By referring to the present example, the Global System can be divided into four subsystems, interacting to one another.
In general, there are at least N(N-1) distinct physical interactions between N interacting physical subsystems, forming a compound global system.
By using the Synchronicity Principle, those physic interactions come down to 1/2[N(N-1)] base interactions: in this case, six base interactions.
But the Magnetic interactions, for the present device virtual prototyping, are more important than the Electric ones. So can be convenient to consider a unified description for the Electric and Magnetic subsystems by one strong coupled Electro-Magnetic subsystem only. In this way the number of subsystem base interactions comes down to three, namely ElectroMagnetic--Mechanic (EM--ME), Hydraulic--Mechanic (HD--ME), and ElectroMagnetic--Hydraulic (EM--HD). Now, by considering the physical media that can be involved in general physical interactions [Empty Space (ES) or Matter (MA)], and the Type of Interaction (Self-Interaction or Cross-Interacion), it is possible to compute the following Table of Subsystem Granularity Descriptors. This **Master Table** is the concise description of all the self-interactions and all the cross-interactions occurring to each interacting physical subsystem and it represents **the Minimal Parameter Set (MPS) needed for the global design optimization of the desired device**.
From this Master Table, by supposing to prototype a global virtual actuator device description with a characteristic granularity in the order of 10^{-2} [m], it is possible to obtain the Effective Spatial Description consistency constraint **ESD** and the Effective Dynamical Description consistency constraint **EDD**, for the driving ElectroMagnetic--Mechanic cross-interaction:

$$\mathbf{ESD} = \left(\mathbf{BG}^2 / \mathbf{SG}\right) * 10^{-2} \leq 1.16 * 10^{-5} \quad [\text{m}] \quad \text{and} \quad \mathbf{EDD} = \left(\mathbf{BG} / \mathbf{SG}\right) \geq 3.80 * 10^{-5} \quad [\text{s}].$$

MASTER TABLE
SUBSYSTEM GRANULARITY DESCRIPTORS

SELF-INTERACTION			CROSS-INTERACTION		
SYS TYPE	MEDIA	DESCRIPTOR	SYS TYPE	MEDIA	DESCRIPTOR
ME-	MA	SG=3.97*10^{03}	EM-	ES	SG=1.11*10^{04}
-ME	MA	SG=3.97*10^{03}	-HD	ES	BG=3.72*10^{-5}
ME-	MA	BG=9.98*10^{-1}	EM-	MA	SG=2.03*10^{07}
-ME	MA	BG=9.98*10^{-1}	-HD	MA	BG=6.22*10^{02}
HD-	MA	SG=1.65*10^{06}	HD-	MA	SG=1.60*10^{05}
-HD	MA	BG=2.18*10^{-4}	-ME	MA	BG=1.97*10^{03}
EM-	ES	SG=8.99*10^{16}	EM	ES	SG=5.47*10^{02}
-EM	ES	BG=4.77*10^{15}	-ME	ES	BG=1.82*10^{-6}
EM-	MA	SG=1.07*10^{09}	EM-	MA	SG=8.05*10^{05}
-EM	MA	BG=1.70*10^{01}	-ME	MA	BG=3.05*10^{01}

In other words, only subsystem spatial descriptions, satysfying the **ESD** constraint on their boundaries, can offer an effective contribution to the global system behaviour dynamical simulation with **EDD** constraint, at the desired global characteristic granularity description. Therefore, automatic mesh generation algorithms must satisfy the **ESD** constraint in order to obtain meaningful global dynamic simulation results. The empirical approaches, used by many authors [8, 9, 10, 11] in literature, are quite often inaccurate, mainly for the inconsistent physical description of, at least, one of the interacting subsystem domains in the global simulation.

REFERENCES

[1] Fiorini, R.A., Sistemi di Supporto Attivo, 1994, Edizioni CUSL.

[2] Gallagher, R.H., Oden, J.T., Taylor, C. and Zienkiewicz, O.C. eds., Finite Elements in Fluids, 1975, John Wiley & Sons.

[3] Zienkiewicz, O.C., Morgan, , K., Finite Elements and Approximation, 1983, John Wiley & Sons.

[4] Feynman, R.P., Statistical Mechanics: A Set of Lectures, 1967, McGraw-Hill, 72-96.

[5] Abraham, R., Marsden, J.E., Foundations of Mechanics, 1978, The Benjamin/Cummings Publ. Co., 246-251.

[6] Finzi, B., Meccanica Razionale: Dinamica, vol.II, 1966, Zanichelli.

[7] Peat, F.D., Synchronicity, 1987, Bantam Books.

[8] Schechter, M.M., Fast Response Multipole Solenoids, SAE, Detroit, 1982, n.820203.

[9] Karidis, J.P., Turns, S.R., Optimization of Dynamic Actuator Performance of High Speed Valves, Proc. Am. Contr. Conf., vol.4, S.Diego, 1984, pp.1548-1555.

[10] Kushida, T., High Speed, Powerful and Simple Solenoid Actuator DISOLE and Its Dynamic Analysis Results, SAE, Detroit, 1985, n.850373.

[11] Edelson Halpert, J., An experimental system may help clean up auto emissions, New York Times, Sunday, January 15, 1995.

C499/046/96

A new approach to the modelling of engine cooling systems

P MENEGAZZI and **J TRAPY**
Techniques of Energy Application Department, Institut Français du Pétrole, France

SYNOPSIS

Optimizing the internal cooling circuits has become vital for the design of modern engines. Although three dimensional (3D) Computational Fluid Dynamics (CFD) proves to be a powerful tool, well suited to this kind of application, a very long time is still spent in the mesh generation step. This is certainly a major obstacle to its systematic use in the engine development process.

This paper proposes an original approach enabling the 3D grid generation of complex internal circuits where traditional methods prove to be unusable within reasonable time. The unstructured meshes of tetrahedra obtained in this way are usable for fluid-dynamics calculation with N3S, a finite element code. As an illustration of the method, a complete example is presented from the mesh generation step up to the CFD calculation.

1 INTRODUCTION

In recent years, most engine developments have been directed towards producing more reliable cars with lower fuel consumption and pollutant emissions. In addition, there is an increasing demand for engines with higher output leading to the use of turbochargers and multivalve technology.

All these trends raise the thermal load of the engine components and particularly of the cylinder head, where the local heat flux is maximum and where hot spots may appear. The efficiency of internal cooling circuits in term of reliability, is closely related to the local characteristics of the flow field: placing the cooling jackets only in the locations where they are needed will eliminate hot spots and allow reductions in coolant flow requirements and water pump horsepower.

Recent works suggest that numerical modelling is now able to provide a reliable and comprehensive description of flows in complex 3D geometries like cooling circuits [1], [2]. Some of them even show that it has become a valuable tool for their optimization [3], [4]

The numerical modelling methodology is usually characterized by four main steps:

1. Describing the geometry to be modelled
2. Meshing the geometry

3. Solving the equations on the mesh

4. Analyzing the solution (postprocessing)

Generally, the geometry to be modelled is first described using a Computer-Aided Design (CAD) tool. The mesh is then generated by importing the CAD data to an adapted software. In most cases this whole procedure turns out to be efficient. However two main restrictions should be mentioned :

- the CAD description doesn't exist in every case : many industries have a large catalogue of traditional parts created without CAD tools. Moreover, current CAD are far from providing the tactile and visual advantages of traditional media such as wood for example so computer-aided designs sometimes begin with a physical object (reverse engineering). Lastly, it may be quite interesting to analyze some parts created by others and whose CAD description is generally not available.
- in the case of very complex 3D geometries like the internal cooling circuits, the CAD description can consist of hundreds of CAD surfaces like NURBS, whose handling is very difficult within a meshing software.

In both cases, generating the mesh of the model becomes quite a critical task in terms of time consumption, particularly for structured grids, and can take several months, that is to say up to 80 % of the whole processing time [5],[6]. Of course, this strongly limits the applicability of 3D modeling within an industrial environment.

2 CREATING THE COMPUTATIONAL MESH

In order to reduce the time required for the generation of the computational mesh in the case of complex 3D internal circuits, a novel approach, based on the digitization of the object to be described, is currently being developed at IFP. An overview of the general methodology is given in this section.

2.1 Describing the geometry

As mentioned before, the first problem to deal with is getting a description of the object to be meshed. When the object exists, several methods like mechanical touch probes or laser range scanners, enable to acquire the external object shape. However, in the case of internal geometries like cooling circuits, these traditional methods are unusable. Then, we opted for X-Ray Computed Tomography (X-Ray CT) scan digitization, a non-invasive technique for imaging the whole structure of 3D objects.

The X-Ray CT consists of making X-Ray projections from different angular positions around the object. Each of these projections yields a one dimensional absorption profile. These profiles are then used to reconstruct slices through the object. Having reconstructed several of these slices sequentially along the third axis (figure 1), they can be stacked into a true 3D image of the object.

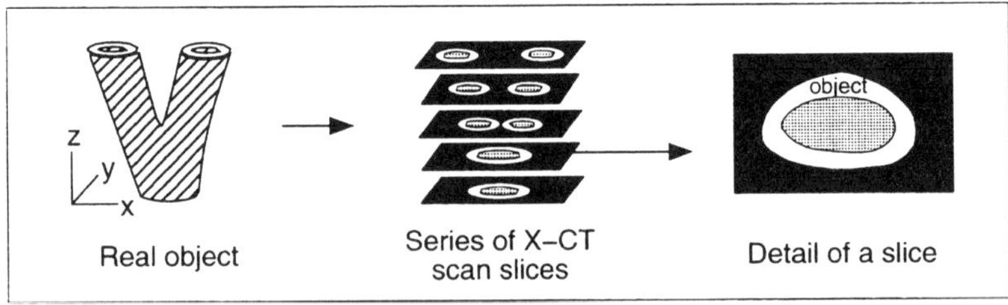

Fig 1 From the real object to the slices

The main parameters for the scan acquisition are the size of the acquisition field, the size in pixels of the slice images (generally 256x256 or 512x512), the thickness of the slice and the distance between two slices. The values of these parameters depend on the type of scan device, the size of the object to be inspected as well as the precision required for the images. A concrete example of X-Ray CT acquisition is proposed in section 3.

2.2 Reconstructing the geometry

The previous operation provides a series of sections in parallel planes. For each section, the following procedure is performed (figure 2):

1. Extraction and orientation of the contours of the object
2. Approximation to each contour by closed polygons (segmentation step),
3. Editing of the polygons (add, delete points, segments, etc.) if necessary.

Once all the segmented contours are obtained, the reconstruction of the object is carried out: the method consists of the construction of a volume whose boundary is a polyhedron with triangular faces lying on the contours. The surface mesh is obtained by processing two slices at the same time and by doing a Delaunay's triangulation [7], [8].

The reconstruction procedure is performed using a particular release of the commercial software GEOLIS, developed and distributed by CRIL INGENIERIE.

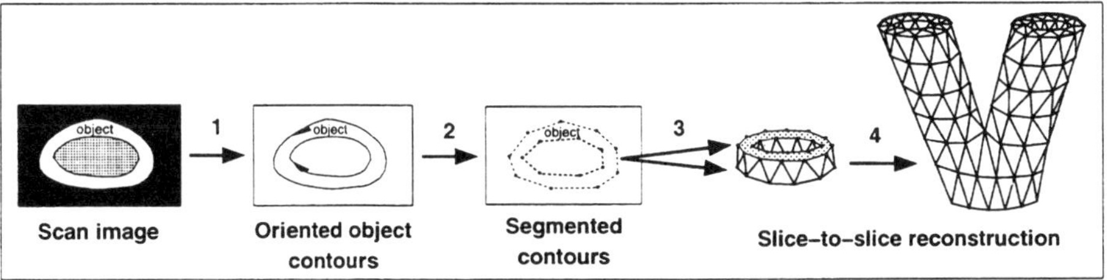

Fig 2 Sketch of the reconstruction process.

This methodology, widespread in medical imagery, mainly intends to provide in a short time, 3D images of complex internal objects without destroying anything. In some industrial applications, it can be very instructive to get such information, for example to create a visual data base of 3D objects which is easy to handle and to exploit.

2.3 Calculation meshes

The surface meshes of triangles obtained after the previous steps, mainly designed for visualization, need particular conditioning before being used in numerical application. This is done by specific softwares developed at IFP: the meshes are checked and regularized in size and in quality on the entire surface. The unstructured meshes of tetrahedra are then generated in the internal volume using an automatic meshing algorithm, GHS3D, based on the VORONOI/DELAUNAY algorithm [9], which is available in the commercial meshing software, SIMAIL (SIMAIL is developed and distributed by SIMULOG). After referencing the surface nodes needed to impose the boundary conditions, the final 3D meshes can be used for CFD calculations.

We can note that this conditioning step is certainly the most penalizing in terms of time consumption. Other ways are currently investigated to integrate as far as possible the mesh calculation constraints into the 3D reconstruction procedure.

3 APPLICATION EXAMPLE

The object of this section is not to make a comprehensive analysis of a flow in a given cooling circuit but to provide a practical illustration of the implementation of the methodology described in the previous section, within a computational procedure.

The following example is extracted from a general study made at IFP on the comparison of several cylinder head cooling circuits of existing V6 engines. In this study, we considered the case of a non-isothermal turbulent incompressible flow coming from the engine block circuit and being distributed to the 22 inlets of the cylinder head circuit. As the thermal problem is assumed not to be coupled to the hydrodynamic one, the temperature is considered as a passive effluent and the physical properties of the working fluid, considered constant in temperature, are those of water at 20°C.

3.1 Mesh generation

Before being scanned, the aluminum alloyed engine head is prepared as follows: the cooling circuit is filled with water to be easily distinguished from the other internal circuits on the images. Moreover, the heavy metalled parts (valves guides, valves, ...), being too absorbent for the X-Ray beam, are extracted so as not to create disturbances on the slice images. Figure 3 shows a short sequence of slice images as well as the detailed image obtained from the engine head. The acquisition field is 315 mm and the reconstruction matrix (that is to say the slice image) is 512x512 pixels which results in a pixel size of 0.615 mm. About 200 slices are necessary to describe the whole object. The duration of the acquisition is about 3 hours with the scan device available at IFP.

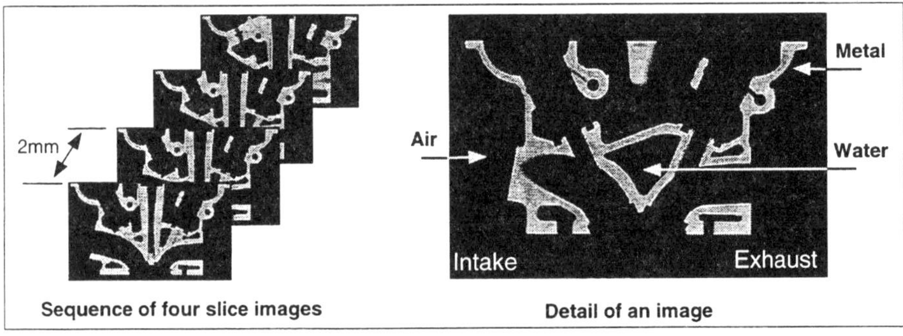

Fig 3 X-ray CT scan sections of the engine head

Using the procedure described in the preceding section, the coolant circuit is reconstructed. In Figure 4 an exhaust side view of the reconstructed circuit is presented as well as a view from below showing the above mentioned 22 inlets coming from the cylinder block and being calibrated by the resizable head gasket holes.

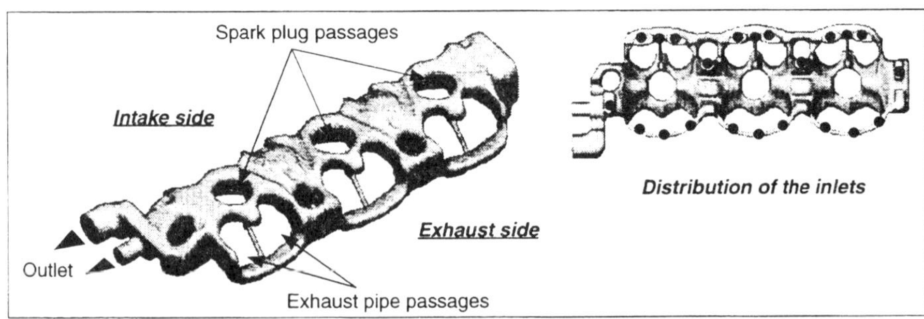

Fig 4 Three dimensional images of the circuit.

After the conditioning step including the surface mesh regularizing and the surface nodes referencing, the final computational mesh of tetrahedra, shown in Figure 5, is generated. For that kind of circuit, the whole procedure can typically take less than two weeks.

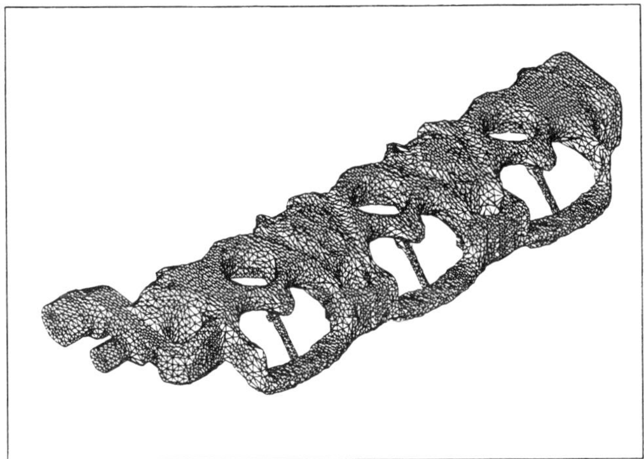

Fig 5 Computational mesh of tetrahedra

The main features of the 3D tetrahedral mesh are given in Table 1 below.

Table 1 Features of the computational mesh

Number of tetrahedra	91979
Number of P1-nodes	22677
Number of velocity nodes	149951
Volume (l)	1.116
External surface (m^2)	0.206

3.2 Numerical method

For the CFD calculation, a commercial computer code, N3S, was used. The important features of the code are presented here.

N3S is a 3D finite element (FE) code, developed by Electricité De France (EDF), for simulating incompressible turbulent flows in industrial applications such as internal flows or turbomachinery problems. The equations governing the fluid motion are the Navier-Stokes equations. For thermal flows the energy equation gives the evolution of the temperature. Turbulent flows, characterized by high Reynolds number, are simulated by the Reynolds-averaged Navier-Stokes equations together with the classical $k - \epsilon$ model in which k denotes the turbulent kinetic energy and ϵ the turbulent dissipation rate.

The previous equations are generally completed by boundary conditions that depend on the type of boundary to be considered:

. for the inlet of the fluid domain, forced constrained conditions (Dirichlet) are used on all the variables.

. for the outlet, vanishing normal stress for the velocity and vanishing flux condition (homogeneous Neumann) for scalar quantities are used.

. for walls, models based on the generalization of the boundary layer on a flat plate is used. For velocity, the normal component satisfies an impermeability condition completed by a friction condition on the tangential stress. Moreover, it is assumed that an equilibrium between turbulent production and dissipation exists near the wall. This enables us to express k and ϵ at the wall. Flux condition and Dirichlet condition can also be imposed for temperature.

Time discretization is realized using a fractional step method: at each time step, we solved successively an advection step by means of a characteristics method, a diffusion step for the scalar terms and a generalized Stokes problem using a preconditioned Uzawa algorithm for velocity and pressure.

The space discretization is based on a standard Galerkin finite element method. The unstructured grids consist exclusively of triangles for two dimensional problems or tetrahedra for three dimensional ones, with mixed formulation for the velocity and the pressure in order to have the Stokes problem well posed. The elements available are P1-P2 or P1-isoP2 elements, mostly used in industrial cases for allowing the velocity matrix to be mass-lumped without diminishing the global spatial precision.

Further details on these algorithms are available in reference [10].

3.3 Boundary conditions

The scavenging of hot coolant by cold flow is modelled using two kinds of boundary conditions:

- Hydrodynamic conditions : a velocity corresponding to a flow rate of 100 l/min for a 4500 rpm engine speed, is imposed at the outlet, with standard Dirichlet conditions on k and ϵ at the inlet. The inlet pressure is imposed.
- Thermal conditions : Dirichlet conditions for the temperature are used.
  - inlet coolant temperature imposed at 90°C
  - wall temperature imposed at 110°C everywhere.

  Moreover, at the initial time, the fluid is supposed to be at 100°C.

As mentioned above, however unrealistic, these thermal conditions are only designed to provide results for comparing several circuits under similar conditions.

3.4 Computational details

The calculation has been performed on the Fujitsu VP2400, a batch vector computer with 512 Mbytes of core memory.

As the thermal problem is assumed not to be coupled to the hydrodynamic one, the resolution can be performed in two successive steps. First the stationary velocity field is computed by solving the Reynolds-averaged Navier-Stokes equations coupled to the $k-\epsilon$ model. Then the thermal steady state problem is solved using the energy equation.

For the hydrodynamic calculation, the duration of the time step is about 65 s and 20 hours are necessary to reach the steady state. In comparison, the duration of the time step for the thermal problem (7s), is rather small but, due to the dominating diffusive effects in the stagnant areas of the circuit, several CPU hours are yet required to reach the thermal steady state.

3.5 Some results

3D simulations usually provide a great amount of data always difficult to be exploited. Fortunately, the use of adapted softwares on 3D graphic stations greatly eases the visualization of these data and the understanding of the physical phenomena they are supposed to describe. Even though it remains hard to reproduce them outside of the interactive software context, we have gathered in this section a few characteristic visual results as well as quantitative ones useful for the flow analysis.

Visual results

Figure 6 shows particle paths going from the inlet up to the outlet of the fluid domain. This kind of visualization is very convenient to highlight the preferential paths of the flow and particularly the influence of the inlet distributions on the structure of the flow.

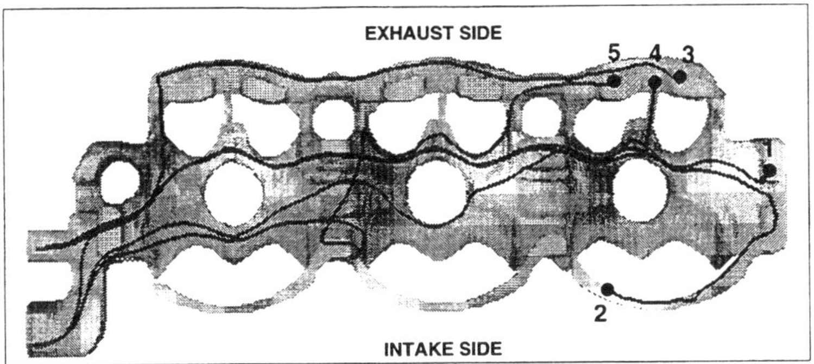

Fig 6 Particle paths from five inlets of the circuit

Figure 7 presents a top view of the velocity and the temperature fields on the external surface of the circuit. We can note a funnel effect characterized by a longitudinal velocity gradient which yields an heterogeneous flow around the spark plug passages along the circuit. The velocity field also highlights the stagnant areas of the circuit mainly located on the intake side.

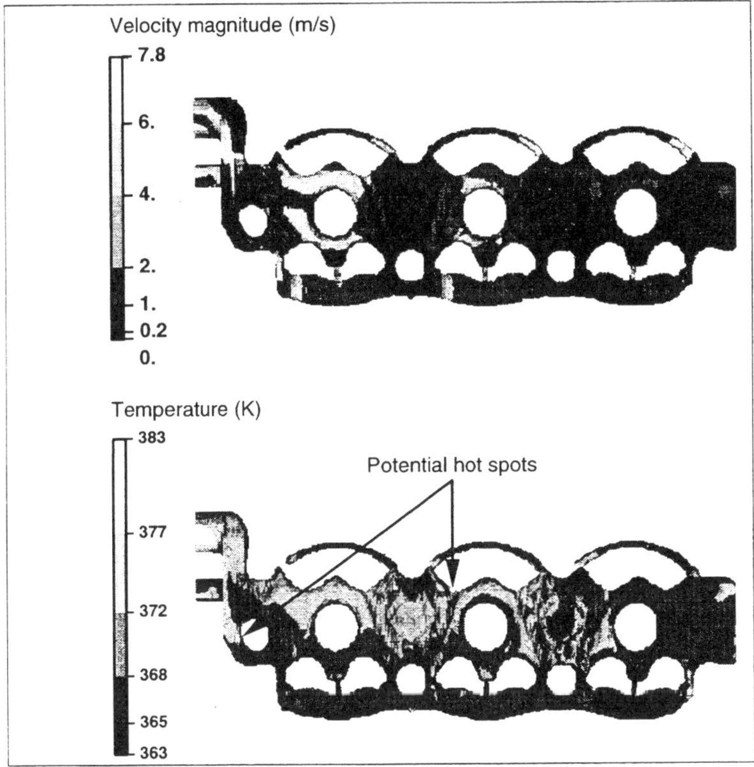

Fig 7 Top view of the velocity and temperature fields on the external surface of the fluid domain

The temperature field is particularly useful to locate the potential hot spots related to a small local velocity as well as those related to a local recirculating flow which confines a small quantity of fluid near the wall and that the velocity field is not able to show.

Global results
Table 2 presents a few global features of the circuit obtained numerically. The results concerning the power exchange give a good idea of the circuit efficiency in term of cooling ability while the time to reach the steady state stands for the ability of the circuit to limit the duration of the engine warm-up.

Table 2 Global numerical features of the circuit

ΔT (outlet - inlet)	4.5
Total extracted power (kW)	31
Extracted power per surface unit (kW/m^2)	151
Bulk temperature (K)	366.3
Mean flow velocity (m/s)	0.43
Time to reach the steady state (s)	1.8

As mentioned above, all these results are mostly useful when comparing several circuits to one another.

Future modelling work will concentrate on extending the thermal analysis to the whole cylinder head including the solid part. This will be possible with the new release of N3S enabling the coupled use of two distinct meshes for the fluid and the solid, to compute the global thermal field.

4 CONCLUSION

In this paper, we have described an original procedure aiming to significantly reduce the mesh generation task in the case of complex 3D geometries when classical methods prove to be unefficient in reasonable time.

The method, based on the X-Ray CT digitization of the object to be modelled, proves to be particularly efficient in applications like the characterization of internal engine circuits. This ability is clearly illustrated through a typical example in which the coolant flow in a cylinder head is investigated using CFD tool.

Although promising results have been already obtained, there are a lot of developments in extending the current approach. First of all, the duration of the meshing procedure can be greatly reduced particularly by a higher automation of the mesh conditioning task and by taking into account the numerical constraints on the meshes in the reconstruction procedure. The problem of integrating the CAD tool in the process is also under consideration so as to extend the applicability of such an approach in industrial design.

References

[1] ARCOUMANIS, C. NOURI, J.M. and WHITELAW, J.H. Coolant Flow in the Cylinder Head/Block of the Ford 2.5L DI Diesel Engine, 1991, SAE paper 910300.

[2] SANDFORD, M.H. and POSTLETHWAITE, I. Engine Coolant Flow Simulation - A Correlation Study. 1993, SAE paper 930068.

[3] PIEDAGNEL, X. and VOCHER, L. Optimization of the cooling configuration of a 3 valve/cylinder IDI engine using 3D computer modelling, 3rd EAEC international conference, Strasbourg, 1991, 368-373.

[4] JAE-IN LEE and NAM-HYO CHO Numerical Analysis of Gasoline Engine Coolant Flow, 1995, SAE paper 950274.

[5] BORETTI, A.A LISBONA, M.G. MILAZZO, P. and NEBULONI, P. The Use of Computational Fluid Dynamic Based Tools in Engine Design, 2nd International Conference on Fluid Mechanics, Combustion Emissions and Reliability in Reciprocating Engines, Capri (Italy), 1992, 69-80.

[6] O'CONNOR, L. Giving A Boost To Engine Design, Mechanical Engineering, May 1992, 44-50.

[7] BOISSONNAT, J.D. Shape reconstruction from planar cross-sections, Computer Vision, Graphics and Image Processing, 1988, 44, 1-29.

[8] GEIGER, B. Construction et utilisation des modèles d'organes en vue de l'assistance au diagnostic et aux interventions chirurgicales, 1993, (Thèse de l'Ecole des Mines de Paris).

[9] FRANCEZ, L. GEORGE, P.L. Description of a 3D mesh generator, European conference on new advances in computational structural mechanics, Giens (France), 1991.

[10] CHABARD, J.P. METIVET, B. POT, G. THOMAS, B. An efficient finite element method for the computation of 3D turbulent incompressible flows, Finite Element in Fluids, Hemisphere publishing corporation, 1992, 8.

C499/021/96

Model based development of engine control algorithms

H J DEKKER MSAE and **W L STURM**
TNO Road-Vehicles Research Institute, Delft, The Netherlands

SYNOPSIS

Model based development of engine control systems has several advantages. The development time and costs are strongly reduced because much of the development and optimization work is carried out by simulating both engine and control system. After optimizing the control algorithm it can be executed by a real-time control system which is hooked up to the engine.

The real-time control system is also able to simulate the engine model in real time. This opens the way for Hardware In the Loop Testing of existing engine controllers, sensors and actuators.

For the optimization of the closed loop control system for a Variable Geometry Turbocharger (VGT), fitted on a Heavy Duty (HD) Diesel Engine, simulated time responses as well as Nyquist diagrams were used. An improvement in transient engine behaviour was demonstrated, both with simulations and with measured responses on a transient test bench.

1. INTRODUCTION

Present and future emission legislation puts quite some pressure on the development of combustion engines. An important role in reducing the emissions is played by modern technologies like electronically controlled fuel injection, variable geometry or wastegated turbochargers, EGR systems etc. When applying these systems (stand alone, or in combination with each other) it is difficult to arrive at proper control algorithms. Besides, it is a rather time consuming job to find the optimal adjustments for obtaining predefined development targets, both in steady state and transient conditions.

TNO carried out a research project to decrease HD diesel engine emissions. This work resulted in an engine which produces 2.4 g/kWh NOx and 0.105 g/kWh particulates in the European 13-Mode test. The BSFC penalty was limited to 2.5%. Characteristic EURO-4 emission requirements are expected to be: 2.5 g/kWh NOx and 0.10 g/kWh particulates (approximate year of implementation is 2004). These results were obtained at a 4 valve, 12 litre engine, with a rated power of 315 kW at 2000 RPM. The engine was equipped with a TNO developed EGR system, a VGT turbocharger and an electronically controlled high pressure diesel injection system (1).

Because of the complexity of the control system, TNO uses a mathematical engine model for the development of the control system. Because much of the development and optimization is carried out by simulating engine and control system, the development time and costs are considerably reduced. Simulation also provides detailed information about the interaction between engine and controller, which supports the development of a proper functioning control system. In addition to this it is possible to study new engine concepts, without building the real engine (eg. different turbochargers, EGR systems).

The advantage of model based development of engine control algorithms will be demonstrated in the following, using some results from the development of the low emission HD diesel engine.

2. DESCRIPTION OF THE METHOD USED

2.1 Model Based Development of Engine Control Algorithms

When developing the engine control system using model based control, the development process consists of three phases:

Phase 1: Engine Model Development
A mathematical, dynamic model of the engine is developed.

Phase 2: Control Algorithm Development
A mathematical model of the control system is developed, modified and tested using simulation of both the engine and the control algorithms.

Phase 3: Implementation
Real-time code is generated out of the controller model, and is transferred to a real-time controller, which is connected to the sensors and actuators of the engine. The engine is controlled running either on a test bench or in a vehicle (hardware in the loop testing).

The MATRIXx Product Family was used during the design process (2). This product enables modelling in a graphical oriented environment, as well as generating and running real-time code.

2.2 Application of Real-Time Engine Model

In addition to generating real-time code of the engine controller, it is also possible to generate real-time code of the engine model itself. This enables Hardware In the Loop testing of existing engine controllers, actuators and sensors (fig. 1).

Moreover it is possible to use the real-time engine model as a non-linear observer in a more complex control system.

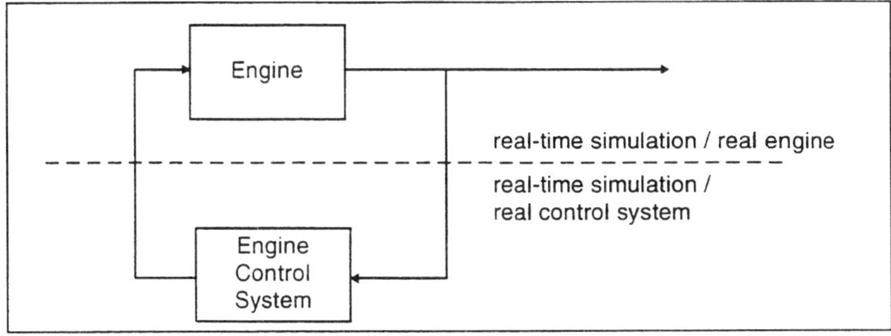

Fig 1 Hardware In the Loop Testing

3. MODEL DEVELOPMENT

The engine model that is described in the following is that of a Heavy Duty turbocharged diesel engine. This engine was equipped with electronic diesel injection and a variable geometry turbocharger. Although at a later stage an EGR system was added (3) it is not incorporated in the work presented here. The model describes the steady state as well as the transient performance of engine torque, boost pressure, exhaust pressure, turbocharger speed, smoke emission etc.

3.1 Model Setup

A mean value model was used to describe the transient behaviour of the different engine parameters (4), (5). This means that the mean value during one engine cycle is used, rather than calculating the engine parameters at every crank angle. This level of detail is sufficient for modelling effects that are of interest when evaluating the control properties during transient conditions of an engine.

The basic equations of the model are fundamental thermodynamic principles, mass and energy balances etc. In addition a limited number of empirical relations is used. The empirical data can be obtained from measurements of the specific engine. Usually a 13 mode test provides sufficient information. Transient data can be used to fine tune the model, but this is not necessary. It is also possible to generate data with a more detailed 1D gasdynamical model, in order to find the empirical relations.

3.2 Model Structure

The dynamical model was made in Xmath, using Systembuild's graphical representation. Each engine component is represented as a block. The interaction between the different engine components is represented by arrows to and from the blocks. The model as a whole is represented as a block diagram with several levels of detail (fig. 2).

The modular structure makes it possible to build up a library of engine components, and to use these components in different engine models.

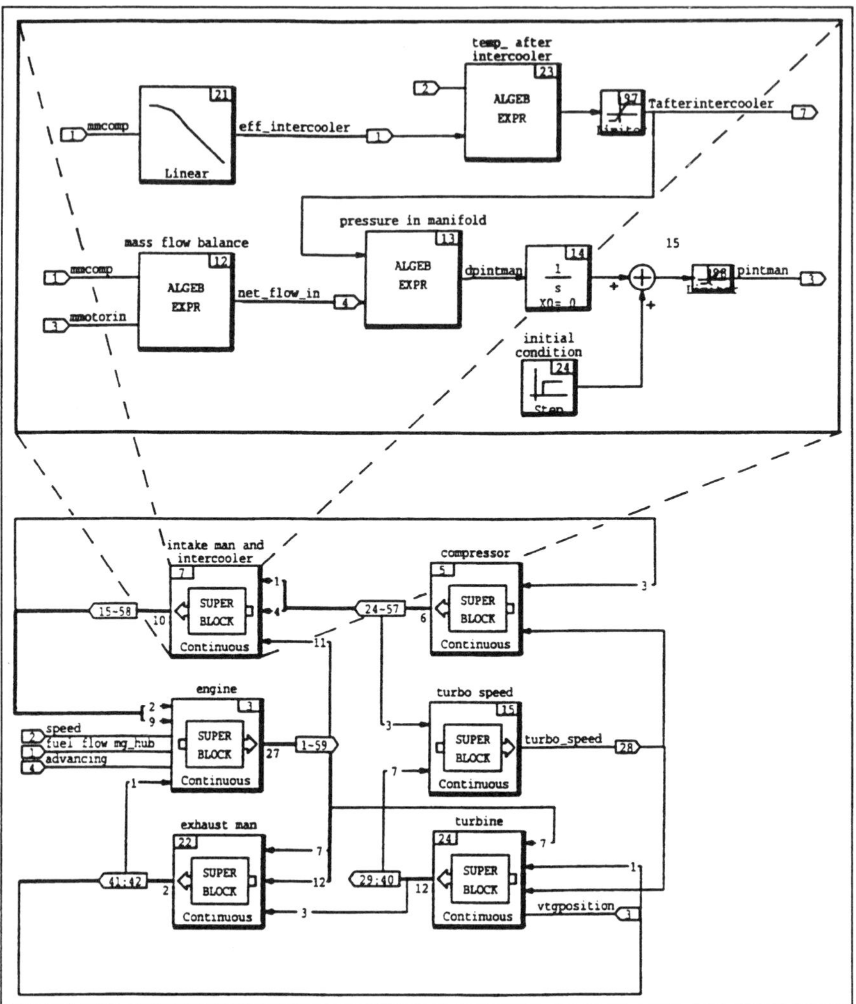

Fig 2 Hierarchical model structure

3.3 Transient Validation

The validation of the model was carried out on TNO's Heavy Duty Transient Engine Dynamometer (Schenk DYNAS 690). In order to evaluate transient behaviour of a truck engine, seven rather simple transient tests were identified as to be representative for real engine behaviour in a truck. These tests would provide sufficient reproducibility for comparing different control strategies, notwithstanding the fact that the DYNAS 690 has the capability of simulating the complete drive line of a vehicle by pre-programming any driving cycle.

The transient tests comprise some low gear responses at which the engine speed follows a ramp from eg. 1200 to 2000 RPM in a certain period of time. A high gear response was approximated by a load acceptance test from 0 to 100 % load, while the engine speed is kept constant.

During the validation experiments the position of the guide vanes was electronically controlled. The control algorithm was a look-up table defining the guide vane position as function of load and speed (ie. open loop control).

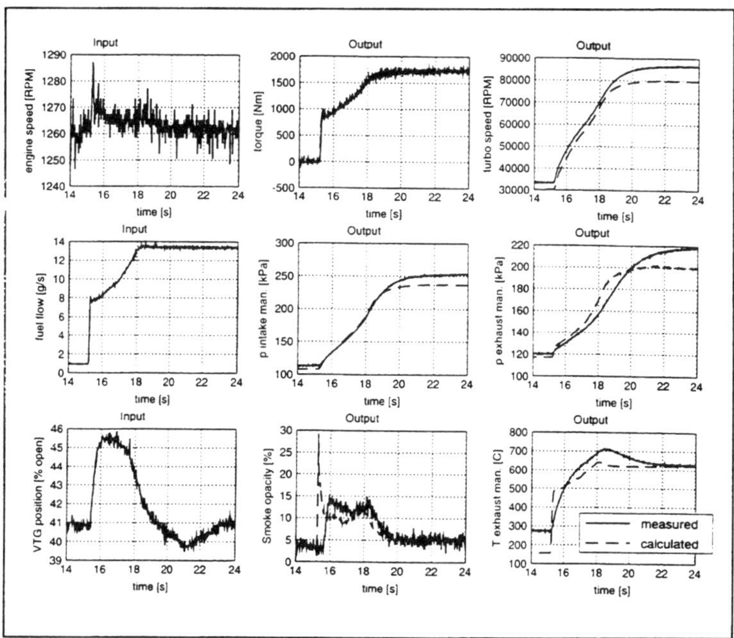

Fig 3 Validation of transient engine model

Figure 3 shows the validation results of a load acceptance test at 1260 RPM. These pictures show that the model provides a rather good description of the transient engine behaviour. Differences between model predictions and measured responses in all engine conditions are less than 10%.

The next chapter demonstrates the use of this model in the development of improved control algorithms for this engine, in order to obtain a better transient response.

4. CONTROL SYSTEM DEVELOPMENT

This section describes the model based development of a closed loop VGT control algorithm for a Heavy Duty turbocharged diesel engine. Closed loop control usually is preferred to open loop control because it is possible to guarantee that the parameter under control meets it's specifications. In case of controlling a VGT it is common practice to use a look-up table which defines the desired boost pressure dependent on engine load and speed. In each steady state engine operation condition the guide vanes are directed to the position that gives the specified boost pressure, even at higher altitudes. Other examples of closed loop VGT control are found in (6) and (7).

In addition to this the transient engine response is significantly improved when the VGT is closed loop controlled. The electronic control system then realises the desired pressure as quickly as possible. It is important to arrive at a proper control algorithm for the VGT, in order to prevent overspeeding the turbocharger, a too high boost pressure during transients etc.

4.1 Design of the control system

In the design of the control system simulations play an important role. By simulating the engine in combination with it's control system the interaction between the control system and the engine is investigated. Control problems can be identified, and solutions can be generated. This kind of investigation was carried out in order to find a proper control algorithm for closed loop VGT control. The result is schematically given in figure 4.

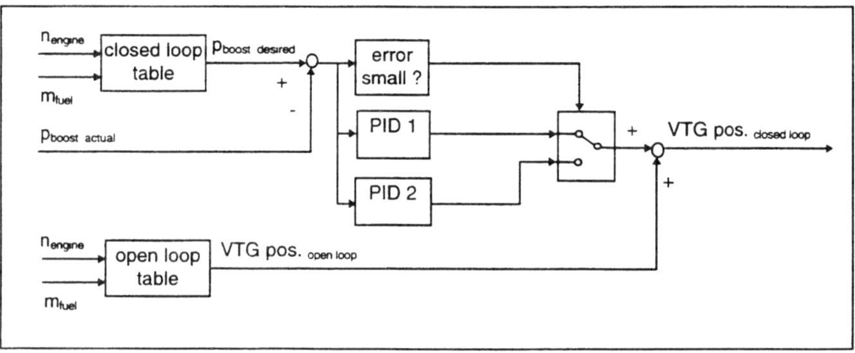

Fig 4 Schematic of closed loop VGT control system

When the difference between desired and actual boost pressure is small, the first PID controller is used. When the error is large, the controller switches to a second PID controller. Because small and large errors are discriminated, overboosting and turbocharger overspeed are prevented. A fast response is obtained by the PID controller for large errors while at the same time overboosting is avoided by making the controller for small errors relatively slow.

The fuelling is also electronically controlled. The desired fuel flow is dependent on pedal position and engine speed, while it is limited during transients in order to prevent high smoke emissions. The maximum fuel flow depends on boost pressure and engine speed. This implies that a faster response of boost pressure due to closed loop VGT control results in a faster increase of fuel flow and engine torque.

The effect of the VGT actuator on the transient engine response was assessed from simulating the engine in two situations:

- equipped with the current actuator
- equipped with an ideal actuator, i.e. direct coupling between the control signal and the actual guide vane position.

Comparing both situations showed a minor improvement of transient engine response when an ideal actuator was used. Based on these simulations it was decided that there is no need to replace the current actuator. The improvement of the response was not sufficient to justify the effort and costs of selecting and applying a new actuator.

4.2 Simulated time and frequency responses

Fig. 5 shows some responses of a load acceptance test at 1260 RPM, for different controller settings. These responses were used to find the optimum controller gains. It shows that at large controller gains there is overshoot in the pressure response. This is not acceptable, since the maximum allowable cylinder pressure could be exceeded when the boost pressure is higher than the specified pressure. Also the damping is very poor, resulting in an oscillating guide vane position, which causes wear of the turbocharger.

Smaller gains give a slower response, resulting in poor transient engine performance. The optimum controller gains give the best transient performance without overboosting the engine and overspeeding the turbocharger.

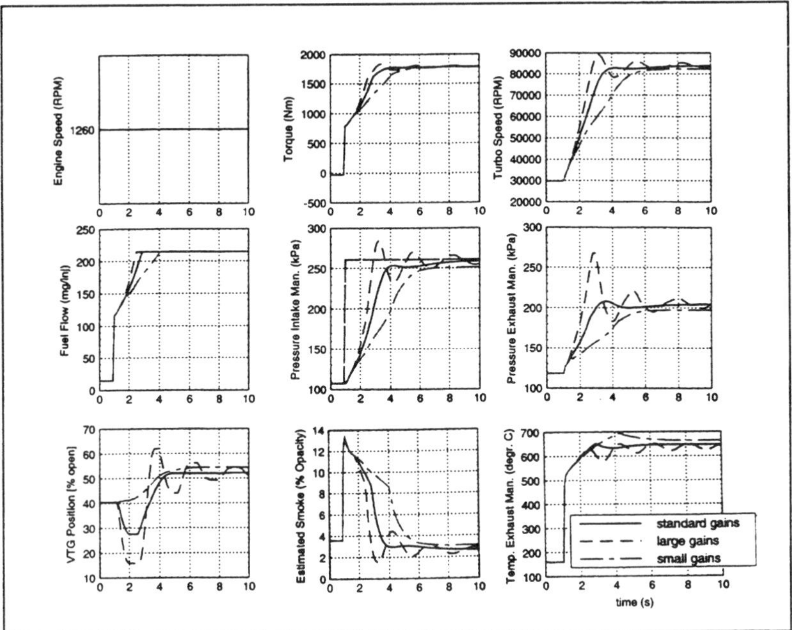

Fig 5 Simulated response for different controller settings

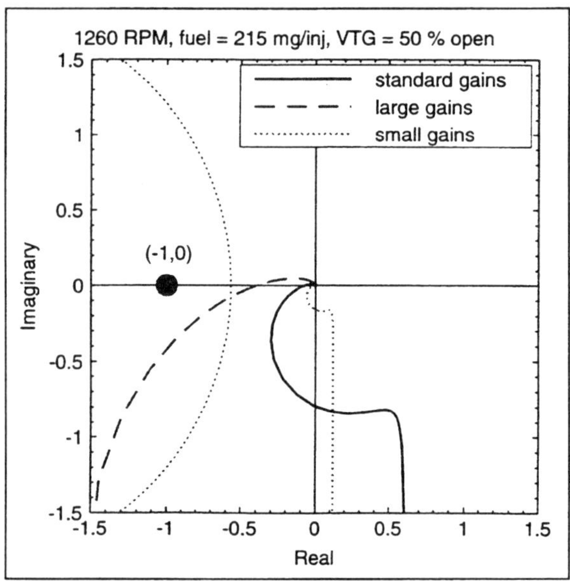

Fig 6 Nyquist diagram of the closed loop system

In addition to time responses, Nyquist diagrams were used to evaluate the stability of the closed loop system. When determining a Nyquist diagram, the non-linearized engine model must be linearized in a certain operating point of the engine. Repeating this for more points provides information about the margin between a stable and an unstable control algorithm. This shows which operating points are critical with respect to stability and damping.

Figure 6 shows the Nyquist diagrams corresponding to the different controller settings showed in fig. 5. There is a relatively high margin between the M=1.3 circle and the Nyquist curve of the standard controller. Touching the M=1.3 circle would mean an overshoot of approximately 23% which is not acceptable as mentioned earlier. The Nyquist curve of the controller with large gains, enters the M=1.3 circle, which means overshoot and poor damping. This is in accordance with the time responses in fig. 5.

4.3 Measured results

When the simulated time responses and the stability margins based on the engine model are acceptable, the developed control algorithm can be implemented in a real-time controller. Then the optimization program on the test bench can start. Figure 7 shows two measured responses, one with an open loop VGT control strategy and one with a closed loop VGT control strategy. The results indicate that the closed loop control strategy gives better engine response.

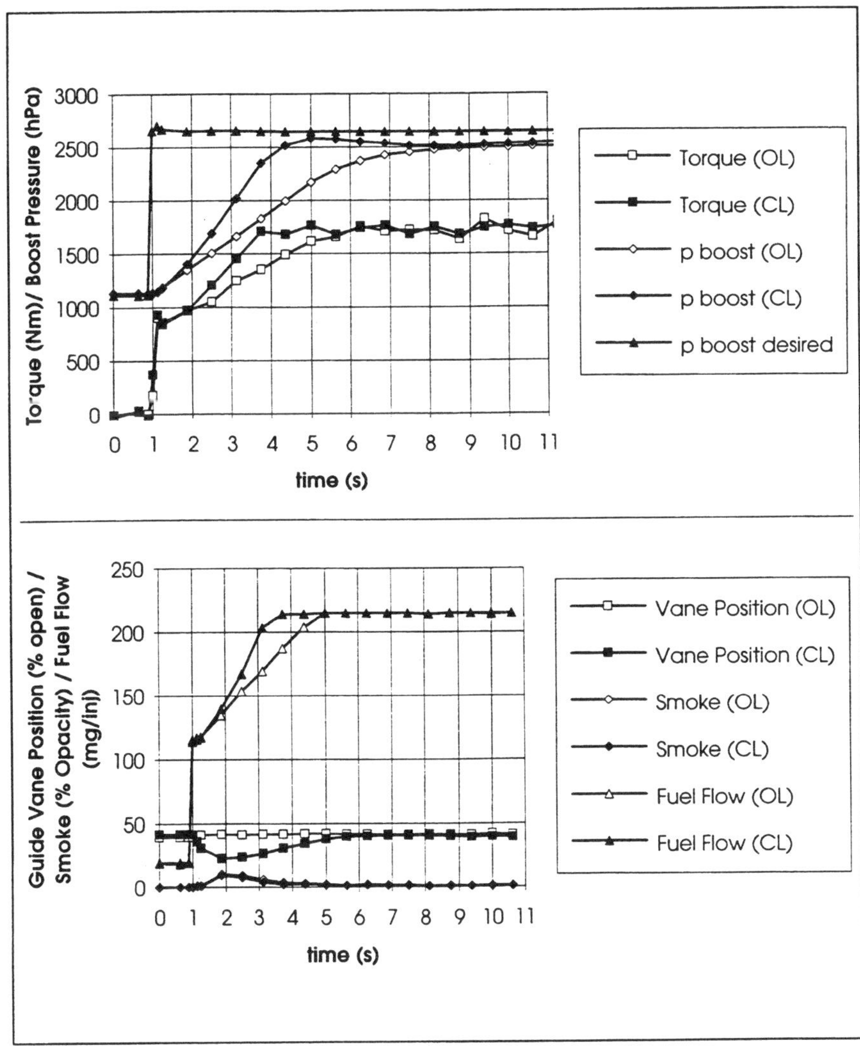

Fig 7 Measured load acceptance test at 1260 RPM.

They also show that the measured response is very similar to the simulated response. This means that almost all the development and optimization work of the closed loop VGT control algorithm can be carried out with simulations. As a result the time needed to optimize the control algorithm on the real engine, placed on the test bench, is very short, despite of the complexity of the control algorithm. This means a considerable reduction on the expenses for developing the control system.

5. CONCLUSIONS

- The dynamic engine model developed using physical equations and supplemented with a limited amount of empirical data provides a reliable and accurate prediction of transient engine behaviour.
- The model is a useful tool for developing control algorithms. It saves quite some optimization work on the test bench, and it gives more insight in the interactions between engine and control system. Because of the growing number of electronic controlled components that are used on engines, the importance of this tool will increase.
- A well designed closed loop control algorithm for a VGT can significantly improve the transient behaviour of a turbocharged diesel engine, without problems like overboosting and turbocharger overspeed.
- The model can be used to evaluate engine components in advance (eg. VGT actuator), giving useful information about the optimum dimensions and the control strategies needed. Different components can be compared, and the one that best fits the requirements can be used in a test program.

REFERENCES

(1) BAERT R.S.G. et al., New EGR Technology Retains HD Diesel Economy with 21st Century Emissions, to be published at SAE in 1996.

(2) INTEGRATED SYSTEMS INC., MATRIXx Product Family Technical Specifications, 1995

(3) DEKKER H.J. et al., Simulations and Control of a HD Diesel Engine Equipped with New EGR Technology, to be published at SAE in 1996.

(4) HENDRICKS, E., A Compact, Comprehensive Model of Large Turbocharged, Two-Stroke Diesel Engines, SAE paper 861190.

(5) JENSEN J.P. et al., Transient Simulation of a Small Turbocharged Diesel Engine, IMechE, November 1990.

(6) PILLEY et al., Optimization of Heavy-Duty Diesel Engine Transient Emissions by Advanced Control of a Variable Geometry Turbocharger, SAE paper 890395.

(7) WATSON N. et al., A Variable-Geometry Turbocharger for High Output Diesel Engines, SAE paper 880118.

C499/029/96

Computer control design and simulation of a gas turbine aeroengine

E ST. JOHN-OLCAYTO BEng, SM (mit), **J WILKIE** BSc, PhD, AMIEE, and **M GRIMBLE**
Industrial Control Centre, University of Strathclyde, UK

Synopsis

This paper outlines the motivation and use of advanced control techniques for an aeroengine application. This is primarily stimulated by three features: firstly advances in measurement and actuation technology introduce a multivariable aspect to the problem, secondly, the increasing demands of efficiency, safety and response of aeroengine systems and thirdly the availability of computer software tools that make possible the analysis and design of such systems.

Notation

A_j	Nozzle Area
$IGVPOS$	Inlet Guide Vane Position
W_f	Fuel Flow
$NHPC$	Percentage HPC Speed
$PRCT2$	Bypass Pressure Ratio
HPC	High Pressure Compressor
$MIMO$	Multi-Input, Multi-Output
$SISO$	Single-Input, Single-Output
$\|\cdot\|_\infty$	L_∞ Norm
Δ	Uncertainty Matrix
j	$\sqrt{-1}$
s	Complex Variable of the Laplace Transform
ω	Frequency
$\underline{u}$	Control Vector

1 Introduction

In recent years there have been many advances made: both in the mechanical design of aeroengines and in the possibilities for advanced control design and simulation of these systems. This paper considers the complementary motivation and role of such aspects. Firstly, variable geometry components on modern aeroengines, such as variable area nozzle and inlet guide vanes motivate research into modern control design methods which can cope with the demands of such highly complex and multivariable systems. The prospect of improving aeroengine efficiency, whilst avoiding areas of instability such as compressor surge, encourage the examination of new control techniques. The advances made in robust control in the last ten years and its introduction in many applications form an attractive basis for control design. There are advantages in using such techniques which include robustness to design model deficiencies. Lastly, the growth of graphical simulation tools promotes flexible, interactive design. These packages permit controller testing on nonlinear engine simulations, reducing the time required on expensive test–bed trials.

This paper examines the need for robust, multivariable control and its application to an aeroengine. $NHPC$ is required to track the pilot demand for thrust control. Additionally, to maintain adequate compressor surge margin, $PRCT2$ is required to follow a demand calculated as a function of the demanded $NHPC$. This is achieved through the continuous control actuation of W_f, A_j and $IGVPOS$. The control design makes uses of weighting functions to shape the frequency response of the closed loop system. By choosing these functions appropriately, the effects of disturbances (such as power off–take and inlet distortion) modelling uncertainty can be reduced.

The nonlinear dynamics of an axial, twin spool, bypass aeroengine were obtained from an older simulation and incorporated into Simulink. The controllers, designed using Matlab, were readily integrated into the simulation, together with routines to limit auxiliary variables such as temperature and spool acceleration.

2 Aeroengine Design

2.1 Measurement and Actuation

In recent years there has been an increase in the sensor and actuator technology available for all types of plants, including aeroengines. For example, the use of pyrometric techniques in temperature sensing ([1]) has often led to improved monitoring and control. Applied to an aeroengine, improved temperature sensing can provide significant information on the life of turbine blades as well as more accurate measurements for the control system to act upon.

Similarly, the increased number of fully manipulatable actuators within a closed loop feedback scheme provides the possibility of improved performance from systems. The improvement in measurement and actuation has provided the impetus to reform the control problem from the traditional role of SISO design to MIMO design.

2.2 The Aeroengine Control Problem

Aeroengine control can be briefly defined as maintaining engine thrust whilst not causing the engine to pass into surge. Adequate thrust control is normally achieved by controlling the spool speed of the engine as a good indication of the thrust. However, as can be seen from the compressor map Figure 1, the lines of constant spool speed exist for varying pressure ratios - therefore, another control objective is to maintain an appropriate pressure ratio across the compressor. Specifying the spool speed and compressor pressure ratio places the operating point of the aeroengine on the compressor map.

It is noted that it is more efficient to operate closer to the surge line, but at the same time, this represents a hard limit on the safe operation of the aeroengine. In practice a soft limit is obtained by developing a steady state working line at some distance from the surge line. This margin of error is termed the surge margin and relates the current working pressure ratio to the actual surge pressure ratio.

Under SISO control it is not possible to control simultaneously $NHPC$ and $PRCT2$. In practice, $NHPC$ (and hence the thrust) is regulated whilst auxiliary variables such as temperature or acceleration are monitored. In the case of excessive rises in, say, temperature, the SISO control would be switched to control this variable until it was brought back within an acceptable range.

In this study, 3 variables were manipulated by the control scheme (W_f, A_j, $IGVPOS$). Traditionally the only fully continuous input was W_f. The inlet guide vanes were open–loop scheduled against the rotational spool speeds. Likewise the nozzle area was scheduled to operate at pre–defined positions for specified operating conditions. The advances in current technology and expertise in control design has meant that the aeroengine can be operated with three fully manipulatable controls.

The problem faced by designers with many measurable variables is in choosing appropriate input–output pairings. This has been tackled in many ways. For example, the process industries have developed many control structure selection techniques, such as the relative gain array, [2], and have been successfully applied. Some of these have been applied to this aeroengine problem, [4]. In this study, a nonlinear simulation (see section 3.2) provided information on the effects that each controls had on the output variables.

2.3 Uncertainty and Aeroengine Dynamics

Current aeroengine demands dictate that they operate at higher pressure ratios and hence, lower specific fuel consumption. However, this means that their compressors operate closer to the surge line than those of their predecessors and it is likely that future units will continue the trend. Deliberate tracking of the compressor working line, then, is becoming more of a necessity in order to prevent aeroengine damage.

This requirement has to take into account the lack of knowledge the designer has of the actual dynamics. This knowledge is affected by:

1. Unit to unit variation:
   Even though aeroengines are constructed to very high degrees of precision, every unit is unique,

2. Ignorance of plant dynamics:
   Due to the complex nature of aeroengines, it is difficult to obtain complete dynamical descriptions,

3. Intake geometry:
   Simulations for control system design often assume that the flow into the aeroengine is 1-D, parallel to its axis and undistorted.

4. Changing operating conditions:
   The pressure distribution across the compressor face can vary with time, for example, when the aircraft is at an angle of incidence or when the inlet air is turbulent.

2.4 Robustness and H_∞ Control

To cope with the issues raised in the previous section, aeroengine controllers are required to make the closed-loop system *robust*. This means that the closed-loop system is guaranteed to be stable given the bounds of the ambiguity in the dynamics while maintaining the specified performance. Achieving both of these objectives result in a control that possesses both *stability robustness* and *performance robustness*.

Many frequency-domain methods have been proposed to design controllers resulting in closed-loop systems with the above properties. One such method is *H-Infinity* (H_∞) control design. H_∞ design is achieved through the minimisation of a cost function. Furthermore, it makes clear statements about system stability in the face of unknown, but bounded plant uncertainty. These statements have their foundations in the *Small Gain Theorem* (SGT). [5]. For this reason, H_∞ control appears to be well suited to aerospace systems whose qualifying procedures are stringent.

Although developed for general nonlinear systems, the SGT can be recast into a linear system framework. To illustrate how the SGT can be used for linear systems analysis and in particular, aeroengine control, consider the situation shown in Fig 2 a). Suppose a linear representation $G_N(s)$ of the aeroengine dynamics have been obtained about a steady-state operating point. These dynamics may have been the result of a system identification procedure on the real plant, a linearisation of a nonlinear dynamical description or may be a combination of information drawn from both methods.

Since $G_N(s)$ is only an approximation to the true dynamics, it is referred to as the *nominal* model (indicated by the subscript N). Even though the actual dynamics are nonlinear, the difference between the actual and nominal plants is assumed to be a linear system $\Delta(s)$. Although at first questionable, this assumption is reasonable if $\Delta(s)$ captures the *bounds* of the modelling discrepancy. These bounds are measured using $|| \cdot ||_\infty$, which is the maximum value of the multivariable frequency response magnitude. Furthermore, let $K(s)$ be a compensator designed for $G_N(s)$ such that the nominal closed loop system is stable. The nominal closed-loop system is expressed in the block form of Fig 2 b) so that the SGT can be invoked. $P(s)$ is the transfer function matrix seen by

In the context of linear systems, the SGT asks:

> Given bounds on $\Delta(s)$ and a nominal, linear representation of the plant, what are the restrictions on the $P(s)$ (and therefore the compensator $K(s)$) such that the actual closed-loop system remains stable?

In the frequency domain, the SGT states:

$$\| P(j\omega \|_\infty < \frac{1}{\| \Delta(j\omega) \|_\infty} \qquad (1)$$

H_∞ design attempts to produce a $K(s)$ such that the above inequality is satisfied.

While a guarantee of stability, the above uses only the upper bound of the uncertainty ($\| \Delta(j\omega) \|_\infty$). Therefore, a design procedure using this test may yield an overly conservative controller. However, some knowledge of the bounds on $\Delta(j\omega)$ is usually available and may be exploited to produce more realistic designs. This knowledge is reflected by specifying a weighting function that indicates frequency regions in which there is least confidence. Another weighting function is used to specify the performance by reformulating the disturbance rejection requirement as a stability robustness problem.

3 Nonlinear Simulation tools

3.1 Nonlinear simulation

Modelling and simulation are often termed as inseparable procedures. Therefore, there is no longer a need to spend as much time on the construction of processes or expensive test-bed trials. Simulations can benefit the user by increasing the understanding of the process, predicting system responses, evaluating control systems or optimising overall system behaviour.

Simulation languages developed through the 'CSSL' type structure (Continuous System Simulation Language). They were designed to help engineers/designers by simplifying the format required to apply numerical techniques in order to extract transient system responses from a set of nonlinear differential equations(or mixed differential/difference/algebraic equations). As they developed, they incorporated more features such as linearisation at steady state conditions, the ability to 'trim' models or the calculation of frequency response analysis.

The current generation of simulation languages are menu-driven simulation tools with substantial libraries of standard components represented graphically by icons. This has made the role of simulation more accessible to those who would have avoided simulation techniques due to the need to learn a specialised language.

3.2 Aeroengine Simulation

A 15–state, nonlinear description of the aeroengine, obtained by reverse engineering an existing FORTRAN programme, was incorporated into Simulink. Some use was made of standard Matlab routines, but most modules were coded in 'C' and incorporated into Simulink for increased speed. The main advantages of the Simulink simulation are:

- The graphical interface to Matlab which aids usage and permits modifications to be made easily,
- Usage of Matlab functions within the simulation.
- Use of Matlab for analysing simulation results,

The simulation was used extensively within this project for the evaluation of controllers and the determination of various dynamic models. It provided far greater insight into the engine behaviour than could have been achieved otherwise.

4 Results

Several H_∞ controllers were designed at operating points on the working–line. A sampling period of 30 ms was selected and the controllers were discretised using Tustin's method with prewarping at 10 rad s^{-1}. To construct a full range controller, the poles and zeros of the local, discrete–time controllers were parameterised as functions of the spool speed. The $PRCT2$ demand is generated as a function of $NHPC$ demand.

To test the resulting full–range controller, a step increase of 3% in $NHPC$ was demanded. The output responses of $NHPC$ and $PRCT2$ are shown in Fig 3. The tracking of the $NHPC$ demand is good but takes a long time to settle (>4 s). There is a slight undershoot in the $PRCT2$ response (it should increase as $NHPC$). This was due to an inadequate profile generator.

5 Implementation Issues

5.1 Actuator Saturation

Actuator nonlinearity is probably the most important factor limiting control system performance. W_f and A_j are effected by actuators whose dynamics are approximated to be of first order with similar time constants. However, each actuator has different rate and magnitude limits. When one or more of these limits is attained the actuator saturates and the system is essentially in open–loop. Not only is the desired control no longer achieved but integrator and slow dynamics within the controller wind–up making linear analysis no longer applicable. Various anti–windup schemes have been suggested including Hanus conditioned controllers and variable gain conditioning (a thorough review is given in [3]).

The problem of windup and actuator saturation is accentuated in MIMO systems since control direction must be taken into account. The concept of control direction is illustrated in Fig 4. The diagram shows the perturbation control signals for a two–output controller. The control $\underline{u}$ is a vector $[u_1 \ u_2]^T$. The rectangle indicates the upper and lower bounds of the signal that may be input to the plant. The 'direction' is specified as the tangent of the angle that $\underline{u}$ makes with the u_1 axis:

$$d = \frac{u_2}{u_1}$$

When saturation occurs the control vectors are distorted and the desired direction is not obtained the controllers are lost. In this case, the direction d_k, can be used to reduce the control vector's size by the required amount. For example, if u_1 saturates (thereby attaining its maximum value of $u_{1_{max}}$, the output vector can be altered to:

$$\underline{u} = [u_{1_{max}} \ u_2 d]^T$$

thereby preserving the direction. Further logic is required to cover special cases of saturation.

5.2 Auxiliary Variable Limiting

In addition to the primary control requirements, there are subsidiary requirements, mentioned in section 2, to control auxiliary variables such as various temperatures and spool speed acceleration. This is often achieved by means of 'switching' controllers once the limit of the auxiliary variable has been exceeded. An alternative approach is to use the primary controller for all tasks but to filter $NHPC$ demand when one or more of the auxiliary variables reaches its respective limit.

This method requires an internal model of the relationship between the control inputs signals, such as fuel flow, and the auxiliary variable, such as turbine blade temperature. Furthermore, an adaptive filter is used to filter the pilot's $NHPC$ demand. The model is used as a one–step ahead predictor of the variable in question. During a manoeuvre, the filter is adapted until an $NHPC$ demand that does not cause the limits to be exceeded is found. The advantage of this scheme is that the stability of the closed loop system is unaffected since it is the command reference signal that is being altered. This scheme was used to limit the HPC acceleration to $\pm 5\%$ of $NHPC$ as shown in Fig 5.

5.3 Summary

The motivation for and development of robust multivariable controllers for modern aeroengines was presented. The controller presented here regulated the high pressure compressor speed and the bypass pressure ratio by varying the fuel flow and nozzle area. The robust aspects of the design were highlighted and results were shown which illustrate the performance of the controller. The advantages of having a nonlinear simulation of the simulation which could be used to develop various models of sections of the system were outlined. Lastly some of the current implementational strategies were outlined. Figure 5 shows the implementation of the final design adopted: the demand for a value of $NHPC$ with the associated feedback from the auxiliary one–step ahead predictor, the corresponding pressure ratio demand and the iterative schemes for actuator saturation/anti–windup.

6 Acknowledgements

The authors would like to acknowledge Lucas Aerospace Ltd. and the Science and Engineering Research Council for their support. The help of the Defence Research Agency is also gratefully acknowledged.

References

[1] K. Eckersdorf L. Michalski and J. McGhee. *Temperature Measurement.* Wiley Series in Measurement Science. Wiley, 1991.

[2] J. W. Maciejowski. *Multivariable Feedback Design.* John Wiley and Sons, Inc., 1984.

[3] S. Ronnback. *Linear Control of Systems with Actuator Constraints.* PhD thesis, Lulea University of Technology, Sweden, 1993.

[4] R. Samar and I. Postlewaite. Multivariable controller design for a turbofan engine. Technical Report 93–58, Leicester University Department of Engineering, October 1993.

[5] G. Zames. On the input–output stability of time–varying nonlinear feedback systems part 1: Conditions derived using concepts of loop gain, conicity, and posivity. *IEEE Transactions on Automatic Control*, 11(2):228–238, 1966.

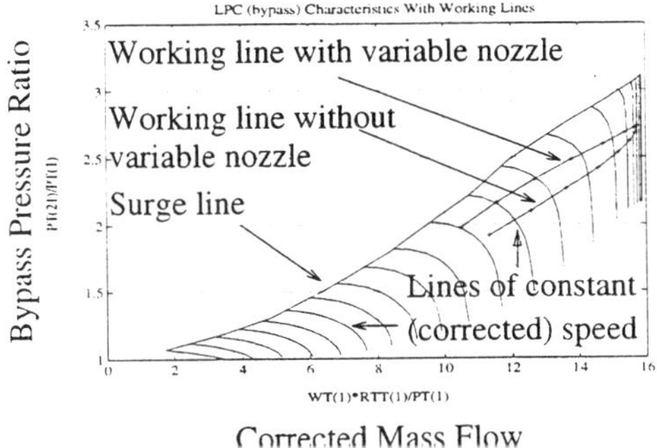

Figure 1: LPC Bypass Characteristics Showing (Partial) Working Lines Using Variable Area and Fixed Area Nozzle

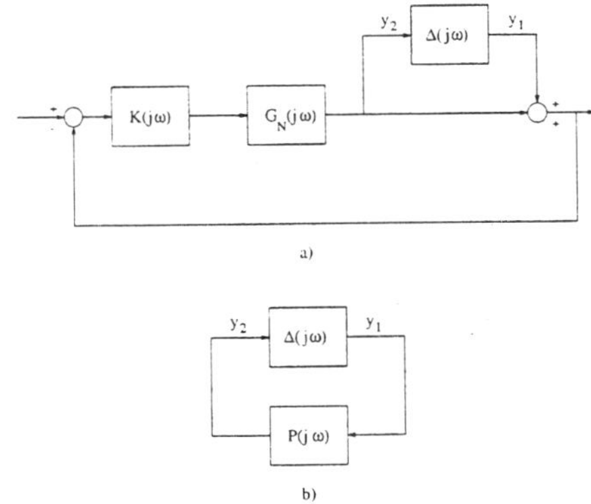

Figure 2: Feedback System with Uncertainty Matrix: a) Standard Form, b) Block Form for H_∞ Design

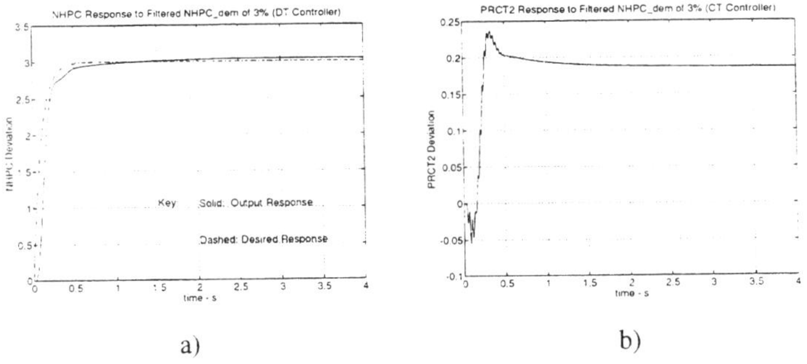

a) b)

Figure 3: Primary Output Responses of the Aeroengine To Step in $NHPC$ Demand: a) $NHPC$ Response, b) $PRCT2$ Response

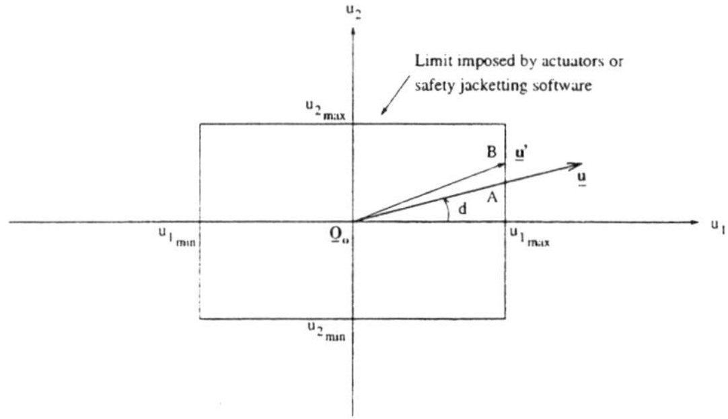

Figure 4: The Effect of Multivariable Saturation

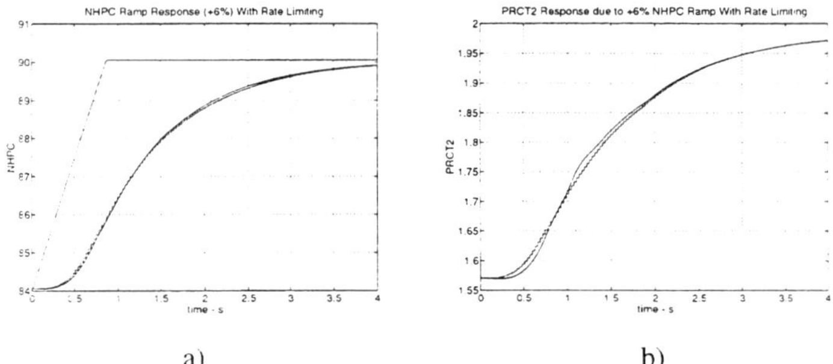

a) b)

Figure 5: Primary Output Responses of the Aeroengine To Ramp Demand: a) $NHPC$ Response, b) $PRCT2$ Response

:499/044/96

Application of an advanced diagnostic technique on nedium speed marine diesel engines

A KOUREMENOS MASME, **D T HOUNTALAS**, and **A D KOUREMENOS**
Iechanical Engineering Department, NTUA, Greece
N KOTSIOPOULOS BSc, MSc, PhD
ellenic Air Force Academy, Attiki, Greece

ABSTRACT

. great deal of advances have been obtained by introducing computers in the field of internal combustion engines. Special fforts have been made in using computers for monitoring and diagnosis of diesel engines. The main scope of the present ork is to give the results obtained by the application of an engine diagnosis method developed by the authors on a edium speed marine diesel engine. The method is based on a detailed simulation model of the engine and on easurements taken using a fast data acquisition system. The results obtained reveal the condition and deliver suggestions r repairs and optimum tuning. The use of this method on propulsion marine engines (large and medium scale ones) is of eat importance since their sizes do not allow the use of trial-error methods. Further trials can be conducted using the ngine simulation model, to predict the performance of the engine when various operating conditions are changed.

OMENCLATURE

	= Area (m^2)
d	= Discharge coefficient (-)
	= Function
s	= Load parameter (-)
	= Mass (kg)
	= Mass flow rate (kg/s)
	= Pressure (N/m^2)
	= Time (s)
	= Temperature (K)
	= Velocity (m/s)
	= Turbulent velocity (m/s)
	= Mean velocity (m/s)
	= Weighting matrix
	= Matrice
,Z	= Matrices

reek letters

	= Matrix
	= Matrix
h_{is}	= Isentropic specific enthalpy rise or drop (m^2/s^2)
P	= Pressure difference (N/m^2)
	= Air cooler effectiveness (-)
s	= Isentropic efficiency (-)

ıbscripts

	= Air
	= Burned
	= Compressor
l	= Calculated
:p	= Experimental

f = Fuel
g = Gas
T = Turbine

Abbreviations
CA = Crank angle
T/C = Turbocharger
TDC = Top dead centre

Dimensionless Groups
Nu = Nusselt Number
Re = Reynolds Number

1 INTRODUCTION

In the past various methods have been proposed for engine monitoring and fault diagnosis [1,2,3,4,5]. These are mainly statistical requiring a great deal of data for the specific engine under consideration and usually provide the condition of the engine or the fault with uncertainty. In the present work a method developed by the authors is used based on a thermodynamic simulation model for engine operation [6] and on real time measurements taken from the engine. Its advantage is that it can be applied on any type of engine regardless of its configuration or geometry while requiring only a small amount of data. The method has been applied in the past on large scale two-stroke low speed diesel engines with good results [7].

In the present work the results obtained from the application of the method on a medium speed four stroke turbocharged marine diesel engine are presented. The engine was a vee-type turbocharged 18 cylinder marine engine with a maximum power output of 5500 PS at a speed of 530 RPM. The task in these medium speed engines compared to the low speed two-stroke ones is more difficult due to their higher engine operating speed since their operation is much more sensitive to adjustments. The last results from the less time available for the various processes taking place in the combustion chamber during combustion i.e. fuel atomization, evaporation and mixing of fuel vaour with the surrounding air. For this reason a small error in the injection angle may have a serious effect on engine operation. Another problem is the experimental procedure, since in such engines the pressures and temperatures in the combustion chamber are very high affecting seriously the sensors used (pressure transducers). Another difficulty introduced is the high engine operating speed which makes the TDC determination problem more difficult.

The results from the application of the method on a medium speed marine diesel engine are encouraging and reveal the advantage of the present method over other existing statistical ones [1,2,3,4,5]

2 DESCRIPTION OF THE DIAGNOSIS METHOD

The main task of the proposed method is to determine the condition of the engine cylinders and the various subsystems.The method is based on a detailed engine simulation model [6,7] used to describe engine operation and real time measurements taken from the cylinders and the engine's subsystems.

The condition of each engine subsystem is described through the use of constants that represent geometrical data of the engine. These constants are used in the engine simulation model and are unknown parameters that have to be determined. For their estimation a constants determination procedure developed by the authors is used [7,8] .

Constants β_j are determined so that the model predicts accurately the measured output values Y_{exp} given the input conditions. This process is repeated twice, first using the engine shop or sea trial data which provides the reference constants β_o and then using data measured in the field at the specific engine conditions which provides the current constants values β By comparing these sets of constants between them the condition of the engine is revealed. An error or fault exists when the following relation occurs,

$$\left|\frac{\beta - \beta_o}{\beta_o}\right| \times 100 \geq 3 \quad (1)$$

The three percent value error is used because the error in pressure measurement is ± 1% and the cycle from cycle variation in the cylinder pressure diagram of medium speed diesel engines is very small (the use of 20 cycles to calculate the mean engine cycle is adequate).

3 THE SIMULATION MODEL-BRIEF DESCRIPTION

A detailed simulation model developed by the authors is used to describe the operation of the engine and its subsystems [6,7,8,9,10,11,12]. This model is used for diagnosis purposes by the method described above but it is also used after having determined the sets of constants to find the optimum operation point of an old engine. This is especially important for medium speed diesel engines that have been working for a long period since their tuning for optimum operation is different from the one provided by the manufacturer which corresponds to a new engine condition.

Detailed information regarding the simulation of the engine and its subsystems and the related constants is given in the appendix and in previous publications of the authors [6,7,8].

The simulation model used considers the following subsystems.

a) Engine Cylinder.
b) Fuel Injection System.
c) Inlet and Exhaust Manifolds.
d) Turbocharger and Air Cooler.

3.1 Engine Cylinder Simulation

The simulation of the engine cylinder operation requires the modeling of the following mechanisms :

a) Heat transfer

In the present work a turbulent kinetic energy viscous dissipation rate k~ε model [6,9,13,14,15] is used to determine the characteristic velocity for heat transfer calculations. Details are given in the Appendix.

b) Blowby

Blowby has a serious effect on the compression and combustion expansion pressure diagram [9,10]. In a previous work [16] the authors have presented a detailed method for blowby calculation considering the rings motions in their grooves. In the present work a simpler method is used assuming an equivalent blowby area giving a blowby rate equal to the actual rate of mass loss.

c) Combustion

The combustion mechanism in the present work is described using a two-zone model details of which are given in previous publications of the authors [6,7] and in the Appendix.

d) Gas exchange

The filling-emptying method [7,9,10,17,18,19] is used to estimate the pressure-temperature versus time history in the two manifolds (inlet-exhaust). The method gives very good results for constant pressure turbocharging systems [17,18,19] and good results for pulse turbocharging systems like the one under consideration.

3.2 Simulation of the Fuel Injection System

In the present analysis a model developed by the authors is used, that has been found to give good results [20]. A brief description is provided in the Appendix.

3.3 Turbocharger-Air Cooler

To have a correct simulation of the turbine and the compressor their characteristic charts are required [17,18,19]. But we know from practice these are difficult to obtain because, T/C manufacturers usually do not provide this kind of data for commercial use. For engine diagnosis in the present work, the method of operation similarity is adopted [21] which makes use of the operational data provided in the engine sea or shop trials.
The data required are,

a) Pressure before and after the compressor.
b) Pressure before and after the turbine.
c) Air temperature before and after the compressor.
d) Exhaust gas temperature before the turbine.
e) Rotational speed of the turbocharger.

The air cooler is modeled in a much simpler way, based on experimental data from the engine shop tests or sea trials. The pressure drop and the effectiveness are given as functions of the mass flow rate through it [7,17,18,19].

4 DESCRIPTION OF THE CONSTANTS DETERMINATION PROCEDURE

Let us assume that a number of constants "j" are to be determined while the available number of measured values are "i". A least squares method is used aiming to the minimization of the following function,

$$f_{err}(\beta_{1,2,\ldots,j}) = \sum_i \left(Y_{exp,i} - Y_{cal,i}\right)^2 \tag{2}$$

where "β" is the matrix of the unknown constants and Y_{exp} , Y_{cal} are the matrices of the experimental and calculated values respectively. The problem is thus reduced to the minimization of the sum of squares function f_{err} which is non linear in parameters [22,23]. The algorithm for the solution has as follows,

$$\beta^{k+1} = \beta^k + Z^k X^{T(k)}\left(Y_{exp} - Y_{cal}^{(k)}\right) \quad \text{where}$$
$$Z^{-1(k)} \equiv X^{T(k)} W X^{(k)} \tag{3}$$

In this relation β represents the constants or parameters vector to be determined. k is the iteration number. X is the sensitivity matrix defined by.

$$X = \begin{bmatrix} X_{11} & . & X_{1j} \\ . & . & . \\ X_{i1} & . & X_{ij} \end{bmatrix} = \begin{bmatrix} \frac{\partial Y_{1,cal}}{\partial \beta_1} & . & \frac{\partial Y_{1,cal}}{\partial \beta_j} \\ \frac{\partial Y_{i,cal}}{\partial \beta_1} & . & \frac{\partial Y_{i,cal}}{\partial \beta_j} \end{bmatrix} \tag{4}$$

and W an weighting matrix defined in a way that takes into account the importance of each parameter by either including it or excluding from the current calculation step (values 1 and 0 respectively).

5 LIST OF CONSTANTS TO BE DETERMINED

5.1 Cylinder

The sets of constants affecting the cylinder pressure diagram are divided into two main categories, the ones that affect the compression stroke and the ones that affect the combustion and expansion stroke also.

5.1.1 Constants Affecting the Compression Stroke

The set of constants or parameters to be determined are :

a) Initial Pressure.
b) Equivalent Cylinder Ring Clearance, δr_c .
c) Heat Transfer Coefficients c_1 and c_2 .
d) Cylinder Wall Temperature.
e) Effective Compression Ratio of the Cylinder, CR.
f) Cylinder TDC position (considered to be a constant).

In Figure 1 a comparison is given between the measured and calculated compression part of the cylinder pressure for cylinder No.1 at 500 rpm speed after having determined the previous set of constants-parameters. From this process as shown in Figure 1 the engine TDC is obtained avoiding the use of an initial sensor (TDC pickup device) which is difficult to use in field applications.

5.1.2 Combustion - Exhaust Period

The constants or parameters to be determined are :

a) Ignition Delay Coefficient (affected by fuel cetane number), a_{del}
b) Fuel Mixing Rate Coefficient, a_{mix}
c) Combustion Rate Coefficient, K_b .
d) Amount of Injected Fuel per Stroke and Cylinder, m_{finj}.

Constant a_{del} is determined in order to have a theoretical ignition delay equal to the experimental one while constants a_{mix}, K_b and m_{inj} are obtained in order to have the least possible deviation between the calculated and measured cylinder pressure diagrams, from the point of ignition up to exhaust valve or port opening.

A good initial value for m_{finj} is obtained from the following equation.

$$m_{finj} = \frac{Q_{g,cum}}{H_{calor}} \tag{5}$$

where H_{calor} is the lower calorific value of the fuel taken equal to 42000 kJ/kg, which is a good estimate for the type of fuel used. The cumulative gross heat release $Q_{g,cum}$ is obtained from the net heat release rate [9] and the calculated heat losses to the cylinder walls obtained through the simulation model from the estimated wall temperature and heat transfer coefficients.

5.2 Fuel Injection System Constants

The constants or parameters affecting the fuel injection system are.

a) High pressure pump clearance, δr_p .
b) Injector effective flow area, A_{inj}
c) Delivery valve leakage area, A_{dval}

The values of the first two constants are determined in order to have the least possible deviation between the measured and calculated pressure traces in the connecting pipeline at the point where the pressure transducer is mounted. The third is determined in order to have a residual pressure value equal to the measured one.

5.3 Engine Subsystems Constants

The constants used to describe the operation of the turbocharger and the air cooler are,
a) Constants "a and b" of the air cooler.
b) The effective discharge area of the turbine nozzle, A_{eff}.

Constants "a and b" are determined so that the calculated pressure losses and air cooler effectiveness match the measured ones while constant A_{eff} is determined to have a mean pressure before the turbine equal to the experimental one.
In Figure 2a comparison is given between the calculated and measured cylinder pressure diagrams for cylinder No.1 after all constants have been determined. It reveals the accuracy of the simulation model used and the constants determination procedure.

6 ANALYSIS OF THE EXPERIMENTAL PROCEDURE

The measurements taken from the various subsystems are :

6.1 Engine Cylinder

From the engine cylinder pressure measurements are taken using a piezotron type air-cooled transducer.The sampling rate is one degree of crank angle [24] and about twenty consequtive cycles are taken from which a mean engine cycle is obtained after processing .

6.2 Fuel Injection System

Pressure measurements are taken from the high pressure fuel pipeline close to the injector using a high pressure piezotron sensor. The sensor was installed on the fuel pipe by making the proper modifications.

6.3 Pressure-Temperature Measurements from the Engine Subsystems

Finally temperature and pressure measurements are taken from the various subsystems of the engine, the turbine, the compressor and the air-coolers using conventional instrumentation.

7 APPLICATION OF THE METHOD

The method has been applied on a medium speed four stroke diesel engine used to power a vessel in the area of Greece. The main purpose was to determine the condition of the engine (wear) and to examine if its present tuning was correct. The specific vessel was not able to reach its maximum cruising speed. The reason for this was that the engine could not reach its maximum operating speed of 530 RPM under full load suffering thus from a low power output. This was also justified by the fact that the propeller and the other systems of the vessel were in excellent condition since the trials were made after a major repair. The main data of the engine under examination are provided in Table 1 of the Appendix.
During the sea trials two sets of measurements were taken :
a) One at 325 rpm and zero propeller pitch in the harbour.
b) And a second at 500 rpm and 100 % propeller pitch at sea.

8 RESULTS-DISCUSSION

In Figure 3 are given the measured pressure diagrams for all engine cylinders at 500 rpm speed. As shown there exists a great non-uniformity between the individual cylinders where pressure deviations up to 15 bar are observed. It is difficult to determine the cause for these deviations since they can be the result of combined faults or wrong settings. To overcome the problem the proposed diagnosis method was used from the application of which we were finally able to determine the cause for these deviations and the low engine power output.
In Figure 4 are given the results obtained from the constants determination procedure for the compression stroke. The maximum compression pressures for all cylinders were obtained from the measured firing cycle through the constants determination procedure of the compression stroke, thus it was not necessary to shut-off the fuel to the individual cylinder. From the results given in Fig.4 it is obvious that all cylinders have an increased compression pressure compared to a new engine condition. The reason for this as shown in Table 2 was that all cylinders had a compression ratio higher than the one given by the manufacturers data. The cause for this was revealed during the inspection of the engine and was the replacement of the original cylinder head gasket with a thinner one. The variation of CR among the cylinders caused by the fact that each one had a separate gasket contributed also to the variation in their combustion pressure values.
The results obtained from the constants determination procedure of the compression-expansion stroke are given in Fig.5. As revealed there exists a high deviation over 25% of the fueling rate among the cylinders (fuel flow rate from each pump). The cause for this was the different fuel rack position of the cylinders fuel pumps. Thus the need for adjusting the fuel pumps is obvious.
Another cause for the pressure deviations among the cylinders was the difference in injection timing. In Figure 6 an illustration is given of the method used in the present work to determine, the static injection timing and the dynamic injection timing. The results of this analysis are given in Fig.5 revealing a great difference among the cylinders. Only a few are close to the proper static injection timing value.

The condition of the fuel injector nozzle was determined by comparing the mixing coefficient a_{mix} described in the Appendix with its reference value obtained from the engine shop trial data (new engine condition). Values over 100% do not reveal better operation but higher atomization of injected fuel that may be caused, as an example, by a higher injector opening pressure which does not nessecerally favour engine operation. The results from this analysis are given in Fig.5 for all cylinders. The need for adjustments is obvious. Concerning the operation of the high pressure fuel pumps no specific problems were observed since their piston-cylinder clearances were found to be close to the values for a new fuel injection pump.

All these resulted in a wide variation of the power output of the engine cylinders as shown in Table 2, which was obtained by processing the measured cylinder pressure-crank angle signals using the calculated TDC position. This is an important advantage of the method proposed in this work since it is able to distinguish between multiple faults that have similar overall effects

Having determined the fuel consumption of each cylinder and its power output it is a simple task to determine its specific fuel consumption. The results are given in Table 2. From this it is revealed that almost all cylinders have a specific fuel consumption considerably higher than the one referring to a new engine condition. The reasons for this have been given above, and are the increased compression ratio which is higher than the optimum one, the error in the injection timing, the error in the fueling rate and the injector's condition.

Finally considering the operation of the turbocharger and the air cooler no special problems were detected since at the operating condition examined, 500 rpm, all parameters were found to be close to the normal ones. For this reason and due to lack of space no detailed results are given in the present work.

An overall verification of the diagnosis results predicted by the present method was obtained since the vessel was able to obtain its maximum cruising speed after the various repairs-adjustments and by the fact that the injectors and gaskets condition was predicted accurately. The last was varified during the maintanance work. Of course this would be made even more clear if measurements were taken after repair but this was difficult in the present case because the vessel was sailing abroad.

9 CONCLUSIONS

A diagnosis method initially developed for large scale two-stroke low speed diesel engines [7] has been applied on medium speed four stroke ones. The difficulties when determining the operating condition of such engines is greater compared to the low speed ones due to their higher engine operating speed. They are more sensitive to adjustments of the various subsystems since as, an example, a small error in the injection timing may cause severe problems in their operation while the effect on low speed ones is considerably reduced.

The method has been applied on a medium speed marine diesel engines and the results are encouraging. From this analysis the condition of each engine subsystem is obtained revealing an error in the compression ratio of the engine and a wear of the injector nozzles. Also errors are found in its settings; specifically the injection timing and the fuel mass flow rate to the various cylinders. All these were responsible for the wide variations in maximum combustion pressure and for the low power output of the engine cylinders that resulted in a decrease of the cruising speed of the vessel.

The method is promising but requires further tests on a wide range of medium speed diesel engines. The authors are looking forward to applying the method on further engines in order to determine the necessary repairs or adjustments and repeat the measurements after repair in order to have a validation of the present method. This has already been done for low speed engines and the results have been reported in a previous publication [7] revealing a good accuracy of the method.

APPENDIX A : DETAILED DESCRIPTION OF THE SIMULATION MODEL

1 Blowby

In the present work blowby is modeled using an equivalent cylinder ring clearance "δr_c ", i.e. an equivalent blowby area. The mass flow rate is then calculated using compressible isentropic flow relations [6,9].

The equivalent blowby area A, in this case, is equal to

$$A = \pi D \delta r_c \qquad \text{(A-1)}$$

2 Heat transfer

A turbulent kinetic energy viscous dissipation rate k~ε model [6,13] is used to determine the characteristic velocity fo heat transfer calculations.

The characteristic velocity for heat transfer calculations is determined from,

$$u_{car} = \left(\bar{u}^2 + u'^2\right)^{1/2} \qquad \text{(A-2)}$$

and the heat transfer coefficient from the well known relation [9,10,14,15],

$$N_u = c_1 \, \mathrm{Re}^{c_2} \qquad \text{(A-3)}$$

Constants c_1 and c_2 are calculated from the constants determination procedure

3 Combustion process

The combustion model used in the present work is a two-zone one originally developed for IDI diesel engines [6,7]. It is based on both mechanisms that control combustion i.e. evaporation of injected fuel and mixing of the fuel vapour with air entrained into the combustion zone to form a combustible mixture.

The injected fuel is divided into packages where the fuel droplets are assumed to have a diameter equal to the Sauter Mean Diameter [9].

For evaporation, the model of Borman and Johnson is followed described in detail in [6]. The fuel vapours mix with the surrounding air to form a combustible mixture. The mixing rate is controlled by turbulent diffusion, the evaporated fuel and the air entrained are divided in two portions; a macromixed one and a micromixed one [6]. The corresponding rates are given by,

$$\dot{m}_{fmic} = D_t(u)\left(m_{fmac} - m_{fmic}\right) \quad \text{(A-4a)}$$

$$\dot{m}_{amic} = D_t(u)\left(m_{amac} - m_{amic}\right) \quad \text{(A-4b)}$$

$$D_t(u) = a_{mix} u \quad \text{(A-4c)}$$

where a_{mix} is a constant to be determined and u the relative velocity of the burning zone element with respect to the surrounding air.

Ignition delay is determined from the relation [9,25],

$$S_{pr} = \int_0^t \frac{1}{a_{del} P_g^{-2.5} \Phi_{eq}^{-1.04} \exp(5000/T_g)} dt = 1 \quad \text{(A-5)}$$

where "Φ_{eq}" is the equivalence ratio of the fuel air mixture. The constant a_{del} is calculated from the constants determination procedure.

The combustion rate of fuel is modeled using an Arrhenius type equation of the form [6,9,10],

$$m_{fb} = \left\{ K_b \frac{m_{fmic} - m_{fb}}{T^{0.5}} e^{-E_c/T} P_{O_2}, \text{ if (AFR)} > \text{(AFR)}_{st} \right.$$

$$m_{fb} = \left\{ K_b \frac{m_{fmic} - m_{fb}}{(AFR)_{st} T^{0.5}} e^{-E_c/T} P_{O_2}, \text{ if (AFR)} \leq \text{(AFR)}_{st} \right. \quad \text{(A-6)}$$

where K_b is a constant to be determined, E_c the activation energy, AFR the air fuel ratio and P_{O2} the partial pressure of oxygen.

4 The fuel injection system

At present a brief description is given revealing the philosophy of the model and the constants involved, while a detailed description can be found in a previous publications [7,20. The simulation considers the following control volumes,

- High pressure pump chamber
- Delivery valve chamber
- Delivery pipe from pump to injector
- Injector main volume
- Injector chamber under the needle

The pressure history in each control volume, except the fuel pipe, is obtained from the compressibility of the fuel [7,20] and the rate of its volume change. The volume flow rate of fuel through orifices, various openings, valves or ports with an area of A_j is given by,

$$\dot{Q}_j = A_j C_{dj} \left(\frac{2\Delta P_j}{\rho_j} \right)^{0.5} \quad \text{(A-7)}$$

The axial displacement of the delivery valve and the injector needle are calculated from Newton's second law of motion, by considering the forces acting on each element.

The unsteady flow equations, inside the pipe connecting the pump and the injector, are solved using the method of characteristics [7,20].

The cylinder-piston clearance of the high pressure fuel pump "δr_p ", the injector holes discharge area "A_{inj}" and the delivery valve leakage area "A_{dval}" are the constants that have to be determined.

5 Turbocharger

The simulation of the turbocharger is achieved using the method of operation similarity [21]. We calculate, using a least squares method, the constants of polynomial curves fitting the following functions,

$$\eta_{is_C} = f_1(\phi) \quad \text{(A-8)}$$

$$\eta_{is_T} = f_2(\phi) \quad \text{(A-9)}$$

$$k_{is} = f_3(\phi) = \Delta h_{is} / U^2 \quad \text{(A-10)}$$

where $\phi = \dot{m}/(\rho AU)$ is the flow coefficient.

6 Air Cooler

The air cooler effectiveness and pressure drop are functions of the mass flow rate of air m [7,17,18,19] and are derived from the following expressions,

$$\varepsilon = 1 - b\dot{m}^2 \quad \text{(A-11)}$$

$$\Delta P = a\dot{m}^2 \quad \text{(A-12)}$$

"ε" is the effectiveness given by,

$$\varepsilon = \frac{T_{a,in} - T_{a,out}}{T_{a,in} - T_{c,in}} \quad \text{(A-13)}$$

and indices "a, c, in, out" denote respectively : air, cooling medium, inlet and outlet from air cooler. Constants "a and b" are determined from the constants determination procedure.

APPENDIX B: ENGINE DATA AND DIAGNOSIS RESULTS

TABLE 1 Engine Data

Engine Type	:DI 4 Stroke
Bore	:400 mm
Stroke	:460 mm
Connecting Rod Length	:950 mm
Compression Ratio	:13
Number of Cylinders	:18
Number of T/C	:2
Number of Air Coolers	:2

TABLE 2 Engine Performance Data

Cyl. No. a/a	Power PS	B.S.F.C. g/PS/h	Cyl. No. a/a	Power PS	B.S.F.C. g/PS/h
1	214	186	10	296	156
2	241	158	11	217	208
3	271	155	12	288	153
4	278	158	13	316	153
5	240	176	14	169	197
6	235	167	15	259	176
7	292	151	16	326	149
8	247	162	17	264	165
9	243	162	18	208	153
Normal Values	**250**	**150**		**250**	**150**

REFERENCES

1. Fredriksen, P.S. Advanced condition monitoring system based on microprocessor components. Ship Operation Automation, 1976.
2. Terano, T. Trouble diagnosis system of marine engine-its system analysis Ship Operation Automation, 1976
3. Holtrop, J. and Menner, G.J. A statistical power prediction method. Int. Shipbuilding Progress. Volume 25, No 290, Oct. 1978.
4. Bruel and Kjaer. Machine health monitoring using FFT frequency analyzer type 2031 or 2033 with Desk-Top calculator. 1980.
5. Collacot, R.V. Mechanical fault diagnosis and condition monitoring. Vol.I-II, Chapman and Hall, 1977.
6. Kouremenos, D.A., Rakopoulos, C.D. and Hountalas, D.T. Thermodynamic analysis of indirect injection diesel engines by two-zone modeling of combustion. Trans. of the ASME, Journal of Engineering for Gas Turbines and Power Vol. 112, pp. 138-149, 1990.
7. Kouremenos, D.A., Hountalas, D.T. and Kotsiopoulos, P.N.. Computer simulation of turbocharged marine diesel engines and its application for engine and turbocharger diagnosis. 5th International Conference on Turbocharging and Turbochargers. Institution of Mechanical Engineers, Paper C484/008/94, pp.13-20, London, 1994.

8. Kouremenos, D.A., Rakopoulos, C.D. and Hountalas, D.T. A model calibration and constants determination procedure based on the elaboration of the measured pressure crank angle diagram in a divided chamber diesel engine. ASME-WA Meeting, Dallas TX., Nov. 25-30, AES-Vol. 21, pp. 21-31, 1990.
9. Heywood, J.B. Internal Combustion Engine Fundamentals. McGraw-Hill Book Co., New York, 1988.
10. Benson, R.S and Whitehouse, N.D. Internal Combustion Engines. Pergamon Press, Oxford, 1973.
11. Benson, R.S. and Baruah, P.C. Some further tests on a computer program to simulate internal combustion engines. SAE Transactions 730667, 1973.
12. Benson, R.S. A computer program for calculating the performance of an internal combustion engine exhaust system. Proc. Instn. Mech. Engrs., Vol.182, Part 3L, pp.91-108, 1967-68.
13. Launder, B.E. and Spalding, D.B. Mathematical Models of Turbulence. Academic Press, London & New York, 1972.
14. Kamel, M. and Watson, N. "Heat transfer in the indirect injection diesel engine. SAE Paper 790826, 1979.
15. Annand, W.J.D. "Heat transfer in the cylinders of reciprocating internal combustion engines. Proc. Inst. Mech. Engrs, Vol.177, pp.973-990, 1963.
16. Kouremenos, D.A., Rakopoulos, C.D, Kotsos, K.G. and Hountalas, D.T. Modeling the blowby rate in a reciprocating internal combustion engine. Proc. 16th IASTED IMS Int. Conf., Paris, pp. 465-468, June 1987.
17. Watson, N. and Janota, M.S. Turbocharging the Internal Combustion Engine. MacMillan Press, London, 1982.
18. Marzouk M. Simulation of turbocharged diesel engines under transient conditions. Ph.D. Thesis, Imperial College, University of London, 1976.
19. Watson N. and Marzouk M. A non-linear digital simulation of turbocharged diesel engine under transient conditions. SAE paper 7710123, 1977.
20. Kouremenos, D.A., Rakopoulos, C.D., Hountalas, D.T. and Kotsiopoulos, P.N. A simulation technique for the fuel injection system of diesel engines. ASME-WA Meeting, Atlanta GA., Dec. 1-6, AES-Vol 24, pp.91-102, 1991.
21. Vavra, M.H. Aero-Thermodynamics and Flow in Turbomachines. Robert E. Krieger Publ. Co., New York, 1974.
22. Beck, J.V. and Arnold, K.J. Parameter Estimation in Engineering and Science. John Wiley & Sons, New York, 1977.
23. Sage A.R. and Melsa, J.L., System Identification. Academic Press, New York 1971.
24. Marzouk M. and Watson N. Some problems in diesel engine research with special reference to computer control and data acquisition. Proc. Instn. Mech. Engrs., Vol.190, No.23/76, 1976.
25. Kadota, T., Hiroyasu, H. and Oya, H. Spontaneous ignition delay of a fuel droplet in high pressure and high temperature gaseous environments. Bulletin JSME, Vol. 19 (130), 1976.

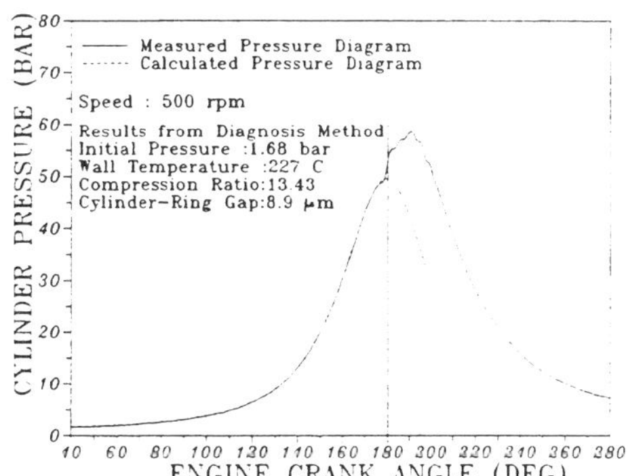

Fig.1 Comparison between calculated and measured compression stroke diagrams after having determined the constants at 500 rpm.

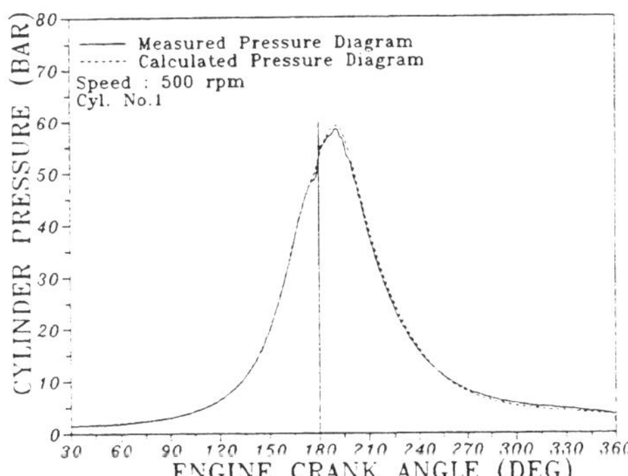

Fig.2 Comparison between calculated and measured pressure diagrams after having determined the models constants at 500 rpm.

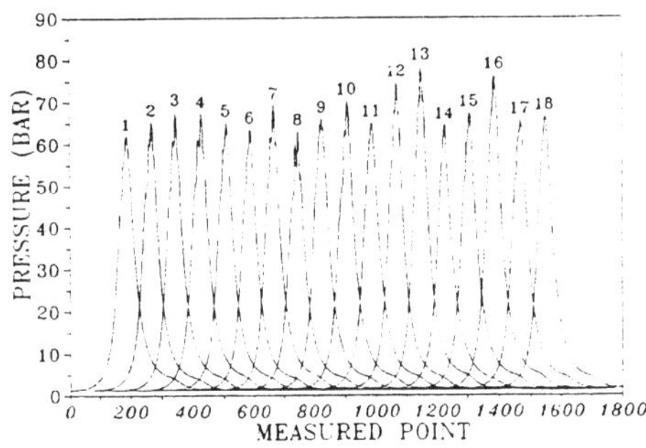

Fig.3 Meaasured cylinder pressure diagrams at 500 rpm for all engine cylinders

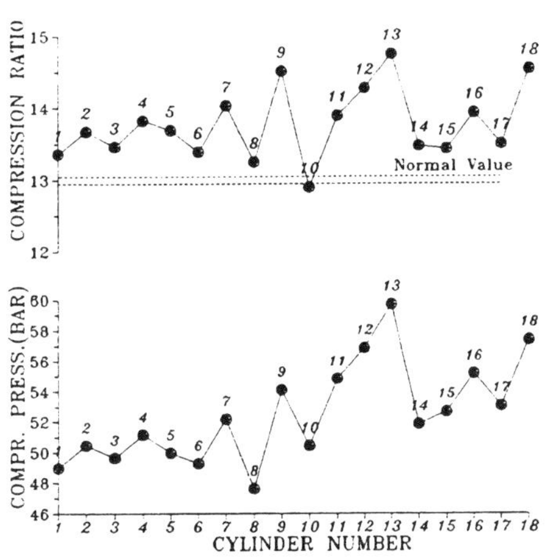

Fig..4 Results obtaiined from the constants determination procedure of the compression stroke

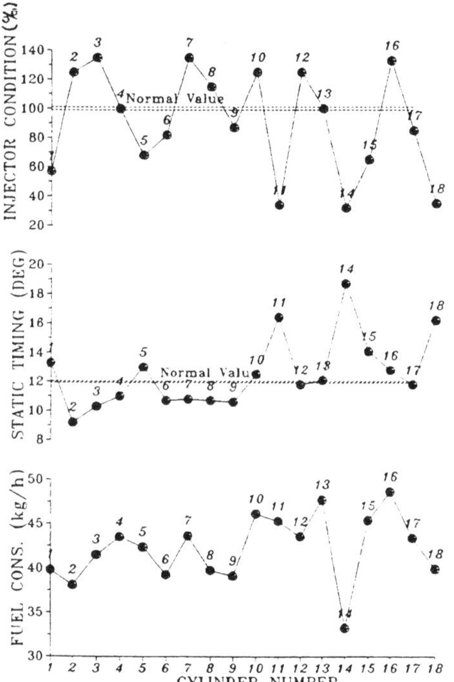

Fig.5 Results obtained from the constants determination procedure of the combustion expansion stroke

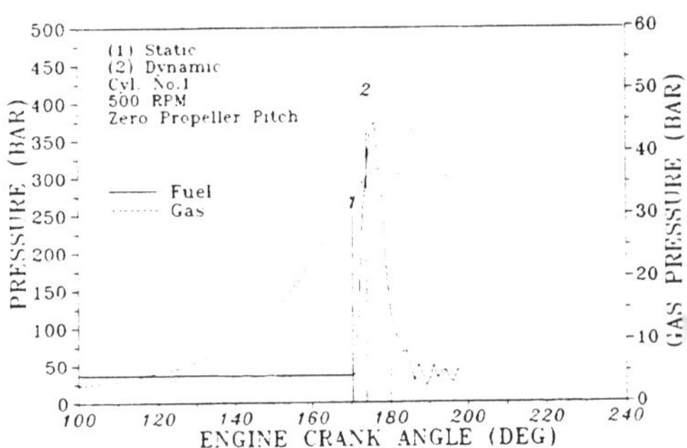

Fig.6 Determination of the static and dynamic injection timing using the measured fuel injection system pressure diagram

C499/012/96

Calculation of one-dimensional unsteady flow in internal combustion engines – how long should it take?

R J PEARSON BEng, MSc, PhD and **D E WINTERBONE** FEng, MSAE, FIMechE
Department of Mechanical Engineering, UMIST, UK

Abstract

The computer time taken to calculate the unsteady gas flows in intake and exhaust systems seems to be very dependent on the code being used. Some programs take hours on a powerful workstation to evaluate the flows in the engines, whereas others could perform the same task in minutes on a microcomputer. This paper presents comparative run times for codes based on the method of characteristics and the two-step Lax-Wendroff Scheme with various flux limiters. The ability of the numerical schemes to conserve mass flow between the ends of the exhaust system are assessed and the effects of the different flux limiters on pressure predictions are compared.

1. INTRODUCTION

The calculation of one-dimensional unsteady flow in the intake and exhaust manifolds of internal combustion engines is a mature area[1]. Graphical methods[2,3] were superseded more than thirty years ago by the mesh method of characteristics[4] (MMC) which enabled the calculation of non-homentropic flow on a digital computer. MMC was the dominant technique for calculating unsteady flows in engine manifolds up to the mid 1980's[5-10] but since then second-order conservative finite difference schemes (with flux limiters) have become predominant for the pipe algorithm, with MMC being retained at the boundary meshes[11-20]. These algorithms have been widely used by the computational fluid dynamics community. Chapman et al[21] and Morel[22] have adapted the FRAM algorithm[23] for engine calculations whilst some workers have tried schemes which are at the forefront of research in numerical analysis[24,25]. Others[26] have adapted older techniques[27] to produce computational algorithms.

Whilst there remains scope for improvements in modelling various complex boundary features and devices such as junctions[28] catalytic converters[29] and silencing elements[20] the basic features of one-dimensional models are well established. Indeed the equations and solutions algorithms for many of the usual boundaries required in such codes have been described by Benson[1] and it has been found that when the same boundary conditions are used different pipe algorithms give similar results[13,16,18,19,20,30,31]. It is reasonable to expect, therefore, that there should be some consensus on the computational effort required by any variant of this type of code. This conjecture has been vindicated in some instances [13,16,18,19,20,30,31] but the authors have recently encountered cases where extremely long run times have been quoted for relatively simple data sets, in some instances even when homentropic flow is assumed. This paper presents computer run times for a variety of different engines using variants of the UMIST comprehensive engine simulation code. Mass flow conservation for different pipe calculation algorithms are given and the performance of the flux corrected transport (FCT) and total variation diminishing (TVD) flux limiters is compared.

2. GOVERNING EQUATIONS

2.1 Conservation Law Form

The continuity, momentum, and energy equations for the one-dimensional flow of a single phase compressible fluid in a pipe can be written in symbolic vector form as

$$\frac{\partial W}{\partial t}+\frac{\partial F(W)}{\partial x}+C(W)=0, \tag{1}$$

where

$$W=\begin{bmatrix}\rho\\ \rho u\\ \rho e_o\end{bmatrix}, F(W)=\begin{bmatrix}\rho u\\ \rho u^2+p\\ \rho u h_o\end{bmatrix},$$
$$C(W)=\begin{bmatrix}\rho u\\ \rho u^2\\ \rho u h_o\end{bmatrix}\frac{d}{dx}\ell nS+\begin{bmatrix}0\\ \rho G\\ -\rho q\end{bmatrix} \tag{2}$$

and

$$G=\frac{1}{2}u|u|f\frac{4}{D}. \tag{3}$$

This presentation of the equations is called the *conservation law form* since the equations can be obtained directly from the integral conservation equations applied to a fixed control volume. The elements of the conserved vector W represent the mass, momentum and total stagnation internal energy per unit volume, respectively. It is usual to close this equation set by assuming that the fluid is an ideal gas with constant specific heat capacities, obeying the equations

$$\frac{p}{\rho}=RT \tag{4}$$

and

$$e=C_vT. \tag{5}$$

3. METHODS FOR SOLVING THE GOVERNING EQUATIONS

3.1 The Lax-Wendroff Schemes

This family of numerical methods contains many variants which, for the linear advection equation, reduce to the one-step Lax-Wendroff scheme which is defined for the Euler equations as

$$W_i^{n+1}=W_i^n-\frac{\Delta t}{2\Delta x}\left(F_{i+1}^n-F_{i-1}^n\right)+\frac{(\Delta t)^2}{2(\Delta x)^2}\left[A_{i+1/2}^n(F_{i+1}^n-F_i^n)-A_{i-1/2}^n(F_i^n-F_{i-1}^n)\right] \tag{6}$$

where A is a 3 x 3 Jacobian matrix.

Two-step variants of this scheme which circumvent the requirement to evaluate the Jacobian matrix A are the two-step Lax-Wendroff, or Richtmyer method[34], and the MacCormack method[35]. Including the source terms, the two-step Lax-Wendroff scheme is defined as:

$$W_{i+1/2}^{n+1/2}=\frac{1}{2}\left(W_{i+1}^n+W_i^n\right)-\frac{\Delta t}{2\Delta x}\left(F_{i+1}^n-F_i^n\right)-\frac{\Delta t}{4}\left(C_{i+1}^n+C_i^n\right) \tag{7}$$

and

$$W_i^{n+1}=W_i^n-\frac{\Delta t}{\Delta x}\left(F_{i+1/2}^{n+1/2}-F_{i-1/2}^{n+1/2}\right)-\frac{\Delta t}{2}\left(C_{i+1/2}^{n+1/2}+C_{i-1/2}^{n+1/2}\right). \tag{8}$$

The numerical schemes described above are second-order accurate in time and space and can be interpreted as conservative finite volume schemes. It is a consequence of the Godunov Theorem[32], however, that discretization schemes such as these which are greater than first-order accurate, and have constant coefficients, produce spurious oscillations at discontinuities. The way around this problem is to use *non-linear* difference schemes where the coefficients of the scheme depend on the solution itself.

3.2 Non-linear Difference Schemes

The spurious oscillations, encountered when linear second-order difference schemes are applied to flows containing discontinuities, are caused by the anti-diffusive flux which is required to produce the higher-order accuracy. Limiting this flux by producing sufficient *local* artificial dissipation, using non-linear terms which are based on the total variation diminishing (TVD) concept, enables the spurious oscillations to be largely eliminated. The TSpark code has the option to use a variety of flux corrected transport (FCT) algorithms[36-39] or the Davis TVD[40] algorithm as non-linear flux limiters. These limiters preserve the symmetric nature of the original difference schemes and are therefore easily adapted to calculating the effects of variable gas properties[41,42]. In the work reported in this paper the gas properties are regarded as constant.

The different FCT algorithms[39] have been previously described in the context of engine calculations by Poloni et al[17] and will not be repeated here. The Davis TVD scheme is essentially a local non-linear local artificial dissipation which can be appended to either the two-step Lax-Wendroff or MacCormack Scheme. This scheme is currently favoured by the UMIST group and its implementation in an engine code is described by Pearson[31]. The FCT schemes will be shown to have deficiencies in handling some flows in engine manifolds[31].

4. RESULTS

In this section pressure and temperature predictions for the numerical schemes described above are presented, and computer run times are given. The computer run time for an engine simulation is a function of the engine configuration, the calculation method used for the simulation, and the specification of the computer on which the program is run. As far as calculation effort is concerned, the engine configuration is defined by the number of cylinders which the engine has and the specifications of the exhaust and intake manifolds. The salient features of the manifolds are the number and length of the pipes, the number of junctions (including loss junctions), the number of plenum chambers and the presence of other complex boundaries such as turbocharger turbines and compressors and intercoolers. The flame geometry calculations and the 'in-pipe' calculations present the greatest load to the CPU and thus the number of meshes into which the pipes are discretized is important. Various pipe algorithms are tried using identical boundary conditions. Unless otherwise indicated the closed period calculations for all cases were performed using a two-zone model with equilibrium reactions giving six product species. The flame geometry is modelled by a propagating spherical flame front which interacts with a pent-roof or disc shaped combustion chamber.

A further important factor in determining the computer run time is the speed at which the engine is run. If the calculation time step, Δt, were constant and the simulation was carried out over the same total crank angle rotation then the computer run time would be expected to be inversely proportional to the engine speed. However, all explicit methods for solving equations (1) are governed by the well known stability criterion of Courant, Friedrichs, and Lewy (CFL) which is given by

$$\Delta t = \left(\frac{\Delta x}{|u| + a} \right)_{\min} \quad (9)$$

For a fixed mesh size, Δx, across the speed range (as in the present work) the maximum allowable time step is dictated by the maximum values of the fluid velocity, u, and the speed of sound, a. The value of u is primarily a function of engine speed (at wide open throttle - WOT), whereas a is determined mainly by the engine load and so whilst Δt may decrease (due to increasing u), as the engine speed increases, it does not do so linearly.

The calculation technique and computer specification have obvious influences on the run times achieved. Most of the run times quoted below refer to the codes being run on a Hewlett Packard Apollo 715/50 work station fitted with a 64 bit, 50 MHz RISC processor rated at 62 MIPS and which has 32 MB RAM. The HP,

FORTRAN/9000 (f77) compiler was used at level 2 optimisation. All run times quoted are for the simulation of five engine cycles.

4.1 Comparison of First and Second Order Methods

Fig.1 shows the pressure and temperature predictions at two locations in the exhaust system of the four-cylinder spark ignition engine described by Worth[45] (see schematic). The engine speed is 3500 rev/min. Results using the two-step Lax-Wendroff scheme with FCT, which is second-order accurate, are compared with the first-order mesh method of characteristics.

In the pressure traces at the exhaust valve (fig.1a) it can be seen that the points of extrema (peaks and troughs) of the waves predicted by the method of characteristics are truncated with respect to those predicted by the second-order scheme, and some of the features of the latter results are not resolved at all. This produces a difference in the predicted volumetric efficiencies of only one per cent and the truncation effect is less pronounced further downstream in the manifold (335 mm downstream of junction 1 - fig.1b). Frequency domain analysis of such results however, using fast Fourier transforms, shows that there is a significant reduction in the high frequency information contained in the first-order results[19] throughout an exhaust system and this will obviously affect the noise spectrum predicted by each method.

Examination of the temperature predictions at the exhaust valve (Fig.1c) reveals trends similar to those described for the pressure predictions at this location. Downstream of exhaust junction 1, however (fig.1d), it can be seen that the method of characteristics gives superior resolution of the two contact surfaces produced by the blowdown events of cylinders 1 and 4. The finely resolved contact surfaces are a result of the Lagrangian technique which is used to track the fluid path lines in the method of characteristics; it is apparent, however, that the method introduces spurious sharp dips into the solution just before the contact surfaces; these can corrupt the pressure calculations in some cases.

For the test case shown in fig.1 the results obtained using the two-step Lax-Wendroff scheme without a flux limiter and with the TVD flux limiter were almost identical to those obtained using the FCT flux limiter and they are, therefore, not shown.

Case No.	Engine Speed (rev/min)	Pipe Algorithm	Mass Conservation(%)
1	3500	MMC	0.390
1	3500	LW2	0.034
1	3500	LW2+FCT(DS PAD)	0.385
1	3500	LW2+TVD (lim2)	0.609
2	3000	LW2	0.867
2	3000	LW2+FCT(DS PAD)	-3.296
2	3000	LW2+FCT(DD EAD)	-0.620
2	3000	LW2+TVD (lim2)	0.864
2	7000	LW2	1.133
2	7000	LW2+FCT(DS PAD)	-6.370
2	7000	LW2+FCT(DD EAD)	-4.566
2	7000	LW2+TVD (lim2)	0.185
3	7000	LW2	0.242
3	7000	LW2+FCT(DS PD)	-25.74
3	7000	LW2+FCT(DD EAD)	-16.81
3	7000	LW2+TVD (lim2)	0.934

Table 1: Exhaust mass conservation for various pipe algorithms

The mass conservation achieved by the various numerical schemes for this engine is summarised in table 1 (case 1). This quantity represents the percentage difference between the sum of the cycle-averaged mass flow through the exhaust valves and the cycle averaged mass flow rate at the end of the exhaust pipe. It can be seen that when the two-step Lax-Wendroff scheme is used with no flux limiter, the mass conservation is significantly better than that of the method of characteristics (MMC). When either the FCT (in this case diffusion via smoothing with phoenical anti-diffusion - DS PAD - as defined by Niessner and Bulaty[39]) or the TVD flux limiters are applied the ability of the resulting numerical scheme to conserve mass is reduced to the same order as the MMC scheme, but the discrepancy remains within one per cent.

The run times required by these different pipe algorithms is shown in table 2. All times given in this table relate to a code with a two-zone combustion model unless otherwise stated. It is apparent that, without any flux limiter, the LW2 scheme is slightly faster than the MMC scheme. Previous work had shown speed advantages of factors of 2 or more using MMC[17,18]. These earlier results were obtained using an LW2 algorithm which was written to be as computationally efficient as possible. In the present work the algorithms used have been modified in a number of ways so as to facilitate greater flexibility for introducing alternative sub-models into the code and they are consequently slower than the older algorithms. The use of the FCT limiter with the LW2 algorithm gives rise to only a small penalty in the overall run time of the program - note that in all instances the flux limiter (FCT or TVD) is applied only to the calculations for the exhaust system.

4.2 Comparison of Different Flux Limiters

In this section TVD and FCT flux limiters are compared, when used in conjunction with the two-step Lax-Wendroff scheme. There are several ways of carrying the FCT technique[39], using different combinations of diffusion via smoothing (DS) or damping (DD) and phoenical (PAD), explicit (EAD), or naive anti-diffusive fluxes. Results from the DS PAD and DD EAD variants have been given here as they represent the worst and best combinations respectively.

Case 2 in tables 1 and 2 represents a 2.2 litre capacity which was fitted with variable valve timing and a variable geometry intake manifold with large variations in cross-sectional area. The exhaust system was an four-into-two-into-one design. From table 1 it is clear that, at 3000 rev/min, all the numerical algorithms, except the DS PAD variant of the FCT scheme, conserved the mass flow between the ends of the manifold to within one per cent. At an engine speed of 7000 rev/min, however, both the FCT variants give a mass flow discrepancy of around five per cent; this could produce substantial errors in turbocharger performance predictions, noise predictions (volume flow rate at the tail pipe exit is used to calculate radiated noise), and catalyst performance calculations. In addition to the mass flow discrepancy there are significant distortions of the predicted pressure and temperature traces (see later).

The total program run times required when the different pipe algorithms are used are shown in table 2. These run times are significantly longer than those for case 1. The fact that the intake manifold for the engine used in case 2 is much more complicated than that used in case 1 accounts for some of the increased run time but a major proportion results from the calculation of the pipe diameter at every mesh point in the code used for case 2. This was done to enable complex pipe shapes to be simulated without the need to introduce a series of two-pipe constant pressure junctions. Even with this modification, however, the computer run times are of the order of a few minutes on the workstation. This rule still remains true for medium speed diesel engines. A diesel engine code (Mark 15) using the LW2+FCT (DS PAD) pipe algorithm (without the modification to calculate the pipe diameter at each mesh point) required only 3m 31s to simulate a large cylinder engine with 27 constant pressure junctions, 8 pressure-loss junctions and a turbocharger turbine boundary (case 4 in table 2). When the TVD option of the code was run on a 486/66 PC in order to simulate the 7000 rev/min speed of case 2, the code required 340 seconds - this is still very acceptable and such times mean that these codes can now be used effectively at the design stage.

It should be noted that the proportion of the computing time spent in the actual pipe boundary calculations (including the valve, nozzle, inlet, junction, throttle body, plenum etc., but not the closed period calculations) is less than 10% for cases 1, 2, and 3 in table 2. This proportion can increase if loss junctions are to be simulated and in the diesel engine described above the loss junction model alone accounts for 18% of the run time. (The heat release calculations in the diesel program require much less computational effort than the two-zone model used in the TSpark code). In cases 1 to 3 the proportion of the total run time which is due to the pipe calculations ranges from 35% to 75% depending on the engine speed and input data file considered. Generally, because they are applied to the exhaust system only, the flux limiter algorithms increase the computer by only a small amount.

Fig.2 compares measured and predicted data for the engine shown in the schematic (case 3 in tables 1 and 2). In fig.2a the predictions (in the exhaust port of cylinder 1) were made using the LW2 scheme with and without the TVD flux limiter. There is very little difference between the predictions made with and without the TVD scheme, beyond some slight truncation of weave peaks by the TVD scheme. This similarity is to be expected in the case shown because there are no spurious oscillations in the LW2 predictions (pressure and temperature) and hence the TVD algorithm is not really required. In fig.2b, however, the FCT (DS PAD) scheme can be seen to produce significant distortions in the results. This variant of the FCT scheme produces this distortion to a greater extent than the other variants, but the results obtained using the DD EAD version, normally the best combination, are not devoid of distortion, as shown in fig.2c. Of course there are engine cases where the flux limiter schemes will justifiably modify the results due to the LW2 scheme alone because of the occurrence of spurious oscillations in the locality of strong pressure or temperature gradients[31].

That the correlation between the measurements and any of the predictions is only moderate in fig.2 is mainly due to the use of a constant pressure model for the complex four-into-one exhaust junction. Fig.2d shows much better predictions in the single intake pipe of cylinder 1. This pipe has a complex port shape.

Consideration of the mass conservation between the ends of the exhaust system of case 3 (table 1) shows that the LW2 and LW2+TVD calculations give better than 1% conservation but the DS PAD and DD EAD variants of the FCT scheme gain almost 17% and 26%, respectively, of the mass which enters at the valves. This very poor performance of the FCT scheme may be due to the presence of the diffuser at the end of the exhaust system (diameter change of 60 mm to 100 mm over 420 mm).

Case No.	Engine speed (rev/min)	Pipe Algorithm	No.of exh. pipes	No.of int. pipes	Total No. of meshes	Run Time (s)
1	3500	MOC	13	10	153	36
1	3500	LW2	13	10	153	34
1	3500	LW2+FCT(DS PAD)	13	10	153	39
2	3000	LW2	7	26	323	215
2	3000	LW2+FCT (DS PAD)	7	26	323	228
2	3000	LW2+FCT (DD EAD)	7	26	323	276
2	3000	LW2+TVD (lim2)	7	26	323	231
2	7000	LW2	7	26	323	129
2	7000	LW2+FCT(DS PAD)	7	26	323	140
2	7000	LW2+FCT (DD EAD)	7	26	323	159
2	7000	LW2+TVD(lim2)	7	26	323	138
3	7000	LW2	5	4	240	142
3	7000	LW2+FCT(DS PAD)	5	4	240	152
3	7000	LW2+FCT (DD EAD)	5	4	240	172
3	7000	LW2+TVD (lim2)	5	4	240	151
4	665	LW2+FCT (DS PAD) (Diesel model)	26	25	333	211

Table 2: Run times for various pipe algorithms

5. CONCLUSIONS

The paper shows that the run time required for the simulation of several different engine configurations is of the order of a few minutes on a workstation or a 486 DX2/66 PC. The numerical schemes used in the present work have only a small influence on the overall run times which are of the order of a few minutes for small high speed automotive engines or medium speed diesel engines. It is suggested that programs based on the numerical methods described in this work, which require significantly longer run times than those quoted have either been coded inefficiently or have unnecessarily severe tolerances for the convergence of some boundary routines.

Mass conservation through exhaust systems of 1% or better can be achieved for the simulation of high speed spark ignition engines using the Davis TVD scheme but the flux corrected transport technique can give much worse results in some situations and can distort the pressure predictions.

6 REFERENCES

1. BENSON, R. S. The thermodynamics and gas dynamics of internal combustion engines. Eds. Horlock, J. H., and Winterbone, D. E., Clarendon Press, 1982.
2. JENNY, E. Unidimensional transient flow with consideration of friction, heat transfer, and change of section.Brown Boveri Rev., 1950, 37, 11, pp. 447-461.
3. BENSON, R. S., and WOODS, W. A. Wave action in the exhaust system of a supercharged engine model. Int. J. Mech. Sci., 1960, 1, pp. 253-281.
4. BENSON, R. S., GARG, R. D., and WOOLLATT, D. A numerical solution of unsteady flow problems. Int. J. Mech.Sci., 1964. 6, pp. 117-144.
5. BENSON, R. S. A comprehensive digital computer program to simulate a compression ignition engine including intake and exhaust systems. SAE paper No. 710773, 1971.
6. GAJENDRA BABU, M. K., and MURTHY, B. S. Simulation and evaluation of exhaust and intake system of a four-stroke spark ignition engine. SAE paper No. 760763, 1976.
7. AZUMA, T., TOKANAGA, Y., and YURA, T.Y. Characteristics of exhaust gas pulsation of constant pressure turbocharged diesel engines. ASME J. Engng. Power, 1980, 102, pp. 827-835..
8. DESANTES, J. M., BOADA, F., and CORBERAN, J. M. Exhaust pipe design method for the optimisation of the scavenging process. SAE paper No. 850083, 1983.
9. WINTERBONE, D. E., ALEXANDER, G. I., and NICHOLS, J. R. Developments in methods of considering wave action in pipes connected to I.C. engines. ASME Conf. on Flows in Internal Combustion Engines - III, Miami, 1985, pp. 71-78.
10. BINGHAM, J. F. Intake system design using a validated internal combustion engine computer model. Inst. Mech. Engrs. Paper no. C25/87, 1987, pp. 111-122.
11. SEIFERT, H. Experiences with a mathematical model for the simulation of operating process in internal combustion engines. MTZ, 1978, 39, 7/8, pp. 321-325.
12. TAKIZAWA, M., UNO, T., OUE, T., and YURA, T. A study of gas exchange process simulation of an automotive multi-cylinder internal combustion engine. SAE paper No. 820410, 1982.
13. AZUMA, T., YURA, T., TOKUNAGA, Y. Some aspects of constant pressure turbocharged marine diesel engines of medium and low speed. ASME J. Eng. Power, 1983, 105, pp.697-711.
14. BULATY, T., and NIESSNER, H. Calculation of 1-d unsteady flows in pipe systems of i.c. engines. ASME Annual Winter Meeting, New Orleans, Louisiana, 9-14 Dec., 1984.
15. MEISNER, S., and SORENSON, S. C. Computer simulation of intake and exhaust manifold flow and heat transfer. SAE paper no. 860242, 1986.
16. POLONI, M., WINTERBONE, D. E., and NICHOLS, J. R. Comparison of unsteady flow calculations in a pipe by the method of characteristics and the two-step differential Lax-Wendroff method. Int. J. Mech. Sci., 1987, 29, No.5, pp. 367-378.
17. POLONI, M., WINTERBONE, D. E., and NICHOLS, J. R. Calculation of pressure and temperature discontinuity in a pipe by the method of characteristics and the two-step differential Lax-Wendroff method. Int. Symp. Flows in Internal Combustion Engines, FED, 62, ASME Annual Winter Meeting, Boston, 13-18 Dec., 1987.
18. WINTERBONE, D. E., PEARSON, R. J., and ZHAO, Y. Numerical simulation of intake and exhaust flows in a high-speed multi-cylinder petrol engine using the Lax-Wendroff method. Instn.Mech. Engrs. International Conf., Computers in Engine Design, Cambridge, Sept. 10-12, 1991.
19. ONORATI, A., WINTERBONE, D. E., and PEARSON, R. J. A comparison of the two-step Lax-Wendroff technique and the method of characteristics for engine gas dynamic calculations using Fast Fourier Transform spectral analysis. 1993,.SAE Paper No. 930428.
20. FERRARI, G., and ONORATI, A. Determination of silencer performance and radiated noise spectrum by 1-d gas dynamic modelling. Proc. FISITA Congress, Beijing, China, 17-22 October, 1994.
21. CHAPMAN, M., NOVAK, J. M., and STEIN, R. A. Numerical modeling of intake and exhaust flows in multi-cylinder internal combustion engines. ASME Conf. on Flows in Internal Combustion Engines, Phoenix, pp. 9-19, 1982.
22. MOREL, T., FLEMMING, M. F., and LaPOINTE, L. A. Characterization of manifold dynamics in the Chrysler 2.2 S.I. engine by measurements and simulation. , 1990, SAE Paper No.900679.

23. CHAPMAN, M. FRAM - Non-linear damping algorithms for the continuity equation. J. Comp. Phys., 1981, 44, pp. 84-103.
24. GIANNATTASIO, P. and DADONE, A. Applications of a high resolution shock-capturing scheme to the unsteady flow computation in engine ducts. Inst. Mech. Engrs. International Conf.,Computers in Engine Technology, Cambridge, Sept. 10-12, 1991, pp. 119-126.
25. BRIZ, G., and GIANNATTASIO, P. Applicazione dello schema numerico `conservation element-solution element' al calcolo del flusso non stazionario nei condotti dei motori a C.I.. Proc. 48th Congresso Nazionale ATI, Taormina, Italy, 1993.
26. BLAIR, G. P. An alternative method for the prediction of unsteady gas flow through the internal combustion engine. 1991. SAE Paper No. 911850.
27. BANNISTER, F. K. Pressure waves in gases in in pipes. Ackroyd Stuart Memorial .Lectures, University of Nottingham 1958.
28. BINGHAM, J. F., and BLAIR, G. P. An improved branched pipe model for multi-cylinder automotive engine calculations. Proc. Inst. Mech. Engrs., 1985, 199, No. D1, pp. 65-77.
29. BARUAH, P. C., BENSON, R. S., and GUPTA, H. N. Performance and emission predictions for a multi-cylinder spark ignition engine with catalytic converter. SAE paper No. 780672.
30. CHEN, C., VESHAGH, A., and WALLACE, F. J. A comparison between alternative methods of gas flow and performance prediction of internal combustion engines. International Off-Highway and Powerplant Congress and Exposition, Milwaukee, Sept. 14-17, 1992, SAE Paper No. 921734.
31. PEARSON, R. J. Numerical Methods for Simulating Gas Dynamics in Engine Manifolds. Ph.D. Thesis, Dept. Mechanical Engineering, University of Manchester Institute of Science and Technology, 1994.
32. ROE, P. L. Characteristic-based schemes for the Euler equations. Ann. Rev. Fluid Mech., 18, pp. 337-365, 1986.
33. LAX, P. D., and WENDROFF, B. Difference schemes for hyperbolic equations with higher order of accuracy. Comm.Pure App. Math., 1964, 17, pp. 381-398.
34. RICHTMYER, R. D. A survey of difference methods for non-steady fluid dynamics. NCAR Technical Note 63-2, National Center for Atmospheric Research, Boulder, Colo., 1962.
35. MacCORMACK, R. W. The effect of viscosity in hypervelocity and impact cratering. AIAA, 1969, Paper No. 69-354.
36. BORIS, J. P., and BOOK, D. L. Flux-corrected transport, I, SHASTA, a fluid transport algorithm that works. J. Comp.Phys., 1973, 11, pp. 38-69.
37. BOOK, D. L., BORIS, J. P., and HAIN, K. Flux-corrected transport, II, generalisations of the method. J. Comp.Phys., 1975, 18, pp. 248-283.
38. IKEDA, T., and NAKAGAWA, T. On the SHASTA FCTalgorithm for the equation $\frac{\partial \rho}{\partial t} + \frac{\partial (v(\rho)\rho)}{\partial x} = 0$. Math. Comp., Oct 1979, 33, pp. 1157-1169.
39. NIESSNER, H., and BULATY, T. A family of flux-correction methods to avoid overshoot occurring with solutions of unsteady flow problems. Proc.GAMM Conf. Num.Meth.Fluid Mech., Paris, 1981, pp.241-250.
40. DAVIS, S. F. A simplified TVD finite difference scheme via artificial viscosity. SIAM J. Sci. Stat. Comput., 1987, 8, 1, pp. 1-18.
41. WINTERBONE, D. E., and PEARSON, R. J. A solution of the wave equations using real gases. Int. J. Mech. Sci., 1992, 34, No. 12, pp. 917-932.
42. PEARSON, R. J., and WINTERBONE, D. E. Calculating the effects of variations in gas composition on wave propagation in gases. Int. J. Mech. Sci., 1993, 35, No. 6, pp. 517-537.
43. COURANT, R., ISAACSON, E., and REES, M. On the solution of non-linear hyperbolic differential equations by finite differences. Comm. Pure App. Math., 1952, 5, pp. 243-255.
44. McAULEY, K. J., BORMAN, G. L., WU, T. ,CHEN, S. K., MYERS, P. S., and UYEHARA, O. A Development and evaluation of the simulation of the compression ignition engine. 1965, SAE Paper No. 650451.
45. WORTH, D.R., Investigation of inlet manifold tuning of an automotive spark-ignition engine. MSc Thesis, Department of Mechanical Engineering, UMIST, 1989.

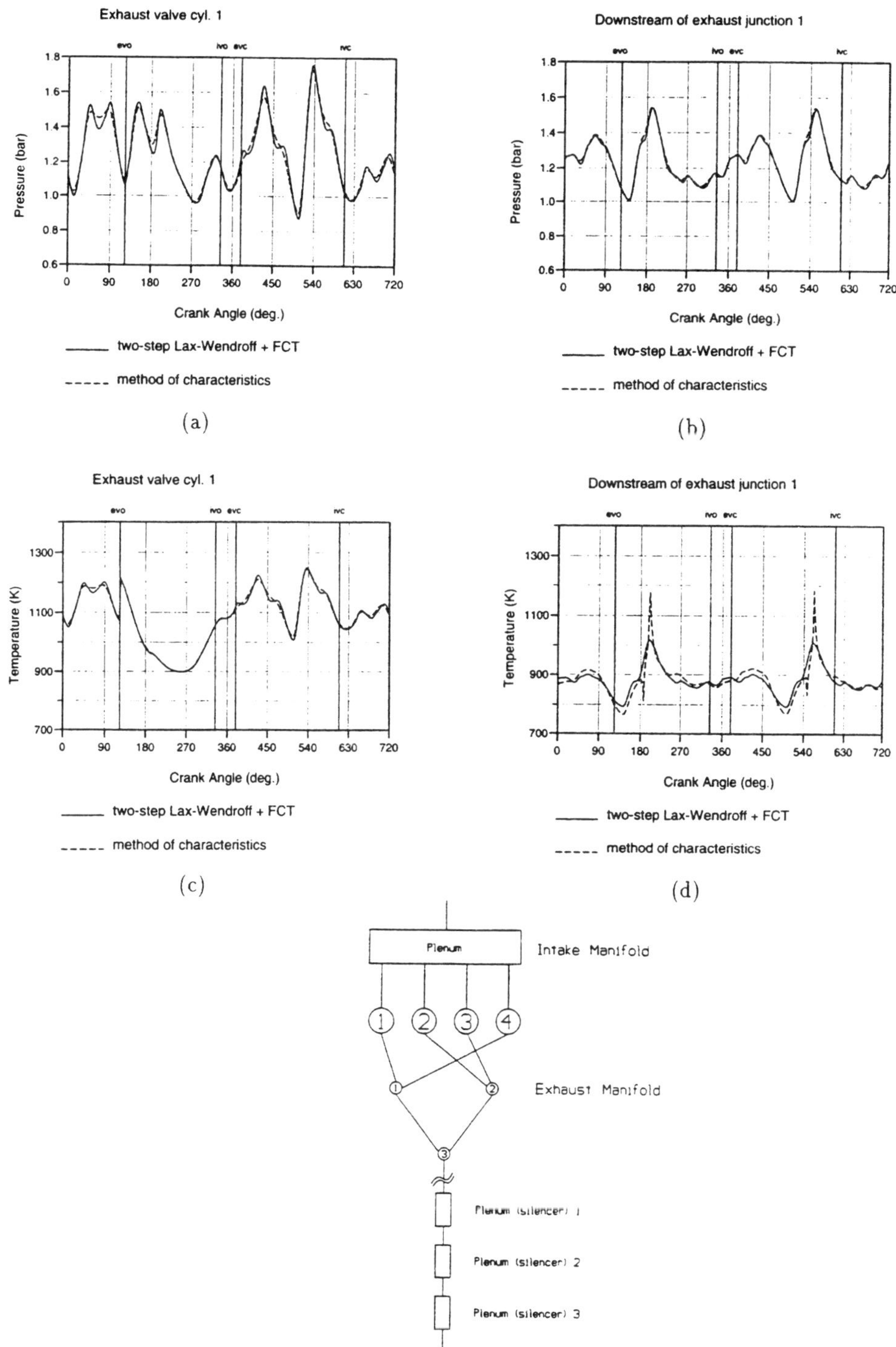

Fig. 1: Comparison of first and second-order methods for prediction of gas dynamics in exhaust system of 1.6 l automotive engine.

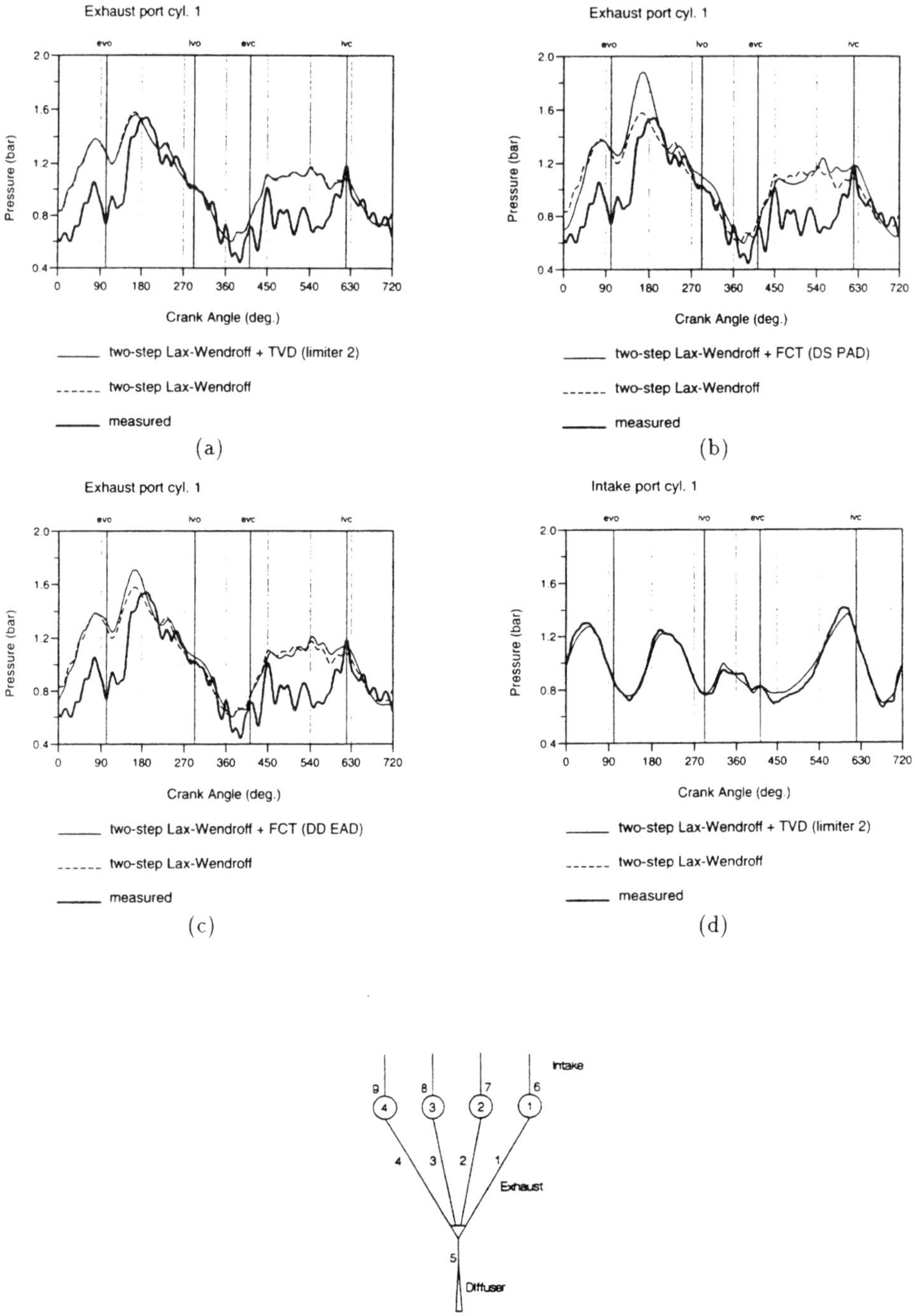

Exhaust port cyl. 1

evo | ivo | evc | ivc

Pressure (bar)

Crank Angle (deg.)

——— two-step Lax-Wendroff + TVD (limiter 2)

- - - - - two-step Lax-Wendroff

▬▬▬ measured

(a)

Exhaust port cyl. 1

evo | ivo | evc | ivc

Pressure (bar)

Crank Angle (deg.)

——— two-step Lax-Wendroff + FCT (DS PAD)

- - - - - two-step Lax-Wendroff

▬▬▬ measured

(b)

Exhaust port cyl. 1

evo | ivo | evc | ivc

Pressure (bar)

Crank Angle (deg.)

——— two-step Lax-Wendroff + FCT (DD EAD)

- - - - - two-step Lax-Wendroff

▬▬▬ measured

(c)

Intake port cyl. 1

evo | ivo | evc | ivc

Pressure (bar)

Crank Angle (deg.)

——— two-step Lax-Wendroff + TVD (limiter 2)

- - - - - two-step Lax-Wendroff

▬▬▬ measured

(d)

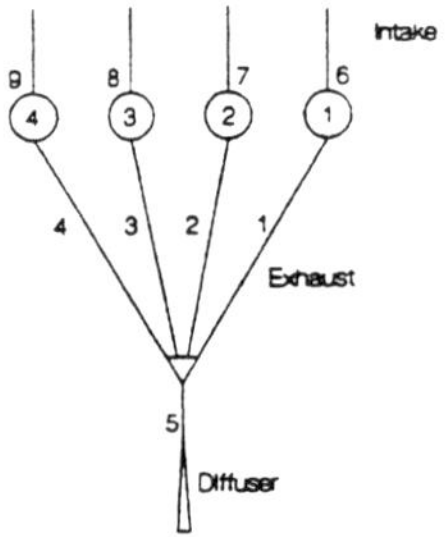

Fig. 2. Comparison of flux limiters for prediction of gas dynamics in exhaust system of racing engine.

C499/017/96

An engineering building block approach to engine simulation with special reference to new application areas

Z BAZARI BSc, MSc, DIC, PhD, MSAE, **L A SMITH** BSc, DIC, PhD, CPhys, MInstP, and
K BANISOLEIMAN BSc, PhD, CEng, MIMechE, MSAE
Lloyd's Register, Croydon, UK
B A FRENCH BSc, MSc, CEng, MIMechE
Ford Motor Company, Basildon, UK

SYNOPSIS

A new generation of diesel engine simulation, based on filling and emptying thermodynamics, multi-zone combustion-emissions modelling, one-dimensional non-isentropic pipe gas dynamics is introduced. A software design and development procedure based on an Engineering Building Block (EBB) approach is adopted to ensure maintainability, re-usability and versatility of the software.

The application of the model in representing complex engine designs, predicting IDI engine performance, pipe wave dynamics for a 4 valve head, turbocharged HSDI engine, employing a variable geometry induction system, and exhaust emissions under transient engine operating conditions are presented.

1. INTRODUCTION

The development of internal combustion engines is increasingly a sophisticated process and is influenced by a variety of technical issues such as emissions legislation, the need for higher engine power density, enhanced drivability and transient response, new materials and electronic engine controls. The use of computer-based systems including engine simulation is expanding throughout the industry to assist in this process.

Simulation of the complete engine cycle goes back to the 1960s following the advent of digital computers in the 1950s. Engine cycle simulation techniques have evolved from early simple air-standard cycle analysis, followed by filling and emptying models [e.g., 1-6], method of characteristics or finite volume procedure for pipe gas dynamic [e.g., 7-9], multi-zone combustion and emissions model [e.g., 10-11] and noise prediction based on pipe gas dynamics [e.g., 9]. The integration of such capabilities into other computer codes for the operation within a comprehensive integrated computer aided engineering environment has been another area of current development [12].

Within the engine industry, simulation is normally used as an aid to design and development activities. The prime objective of simulation is to reduce the cost and time associated with the development of new designs and prototypes. This acceptance took place gradually, initially considering simulation as a way of giving useful prediction trends and later as a complementary tool to engine testing. The availability of low cost high speed digital computers aided this process significantly.

Due to the complexity of engine processes, the demand for enhanced and improved representation of the physical processes within simulation has continued unabated. In recent years, general trends such as increasing power of computers, improved numerical algorithms, enhanced understanding of engine processes due to advances in basic research and thereby more sophisticated models of fluid mechanics, heat transfer, friction, combustion and chemical kinetics have led to more accurate predictive tools.

Concurrent to demand for better representation of processes, the number of physical components on engines has also increased. As an example, FIGURE 1 shows schematically a sequentially turbocharged production engine which includes 12 cylinders, 8 turbochargers, 6 charge coolers, 19 pipes representing intake and exhaust piping systems, poppet and control valves, etc. and associated control systems. Additional engine add-on systems such as exhaust gas recirculation (EGR), variable valve timing (VVT), electronic controlled injection (ECI), exhaust aftertreatment, shafting systems and so on need to be modelled to yield the full system simulation capability.

Also, the continuing demand by the user for enhanced man-machine interface in the form of graphics, provision of capability for alternative design options and engine concepts, and so on have led to software activities beyond the simple process simulation aspects which were the main concern of the simulation engineer in the past.

2. MULTI-DIRECTIONAL DEVELOPMENT ENVIRONMENT

Engine simulation development encompasses several areas of basic expertise including heat and mass transfer, pipe flow dynamics, combustion and emissions, fuel injection system, noise generation and propagation, graphical user interface, data management and software assessment. The enhancement to available simulations is normally carried out by a multi-disciplinary task force in a multi-directional development environment.

The shear complexity of the new generation of engine simulation demands the imposition of a development regime to safeguard the quality and the integrity of the final product and ease the development process of the software by a multi-tasked team of developers.

Also the large software size with added complexity raises the question of how to design and develop a reliable software. From a software development viewpoint, the attributes for a good product can be defined as the functionality (satisfying the stated functional specifications), performance (satisfying the stated performance specifications), usability (meeting the requirements for user interface), reliability (coping with the full range of inputs) and maintainability (can be modified easily). It becomes progressively acceptable that in order to arrive at such desirable attributes, much emphasis needs to be put on the development process and environment in the belief that if this process is controlled, then the quality of the product is ensured. In addition, the requirements for correct prediction must always be met by systematic verification and validation exercises.

This paper deals with the above issues in a systematic way using the MERLIN diesel engine simulation [6, 13]. The simulation is first described with due consideration to the software development strategy and user interface. This is followed by case studies which include IDI engine performance, prediction of pipe gas dynamics and combustion-emissions prediction under transient operating conditions. Further developments are discussed next, followed by conclusions.

3. DESCRIPTION OF MODEL

3.1 Software Development Strategy

The MERLIN software development programme is driven principally by the requirements of research and consultancy activities in the marine and automotive industries, however, much attention has been paid to overcoming the user-unfriendliness usually associated with codes developed in research environments.

The original requirement specification emphasised the need for modularity in engine system model construction. This is achieved by adopting an object orientated Engineering Building Block (EBB) approach to representing and modelling engine components. As such, the simulation philosophy is based on direct representation of well defined physical components by their respective mathematical models. Each component is regarded as a basic EBB which can be used to construct engine systems.

The modularity concept has also been implemented at the software level with the architecture providing for ease of extendability and adherence to software Quality Assurance procedures. Each EBB is represented by a generic model which can be easily linked to other EBBs to build a simulation. This has been adopted to save development time and ensure reusability of the model. The code is written almost exclusively in standard FORTRAN 77 to ensure portability.

Currently, the potential flexibility offered by linking the executable object with user-written modules describing component submodels, boundary conditions, and so on, is being explored. This approach is intended to be taken a stage further with the dynamic linking of simulation to other codes, for example as a boundary data generator for the higher dimensional computational fluid dynamics (CFD) and Finite Element Analysis (FEA) codes.

3.2 Overview of Modelling Technique

The MERLIN engine simulation is a thermodynamic based filling and emptying model which is using the state of the art zero and one-dimensional modelling techniques for solving unsteady fluid flow and conservation equations. The model is essentially an engine system model with the primary objective of predicting overall engine performance and emissions over a wide range of design and operating conditions. The mathematics behind the model are explained by Watson [5] on filling and emptying control volume thermodynamics, Benson [8] on pipe 1-D gas dynamics and Bazari [11] on multi-zone combustion-emissions modelling.

From the modelling point of view each engine component is represented as an EBB. For software development purposes, the EBBs are classified as thermofluid, mechanical or control. The thermofluid EBBs are subdivided into flow controllers, flow receivers (control volume) and pipes. TABLE 1 shows this classification together with the EBBs currently modelled within the simulation.

Table 1 Range of currently available EBBs

Thermofluid	Mechanical	Control
Air filter	Compressor	Governor
Compressor	Crank shaft	Wastegate
Cylinder	Cylinder	
Combustion chamber (constant volume)	Stiff shaft	
EGR Valve	Turbine	
Fluid State (source or sink)	Gear train	
Heat exchanger	Load	
Junction		
Orifice		
Pipe		
Plenum		
Poppet valve		
Port		
Throttle valve		
Turbine		
Wastegate		

The thermofluid EBBs are modelled as flow receivers, flow controllers and pipes to provide efficient numerical solutions to state and conservation equations for a wide range of EBBs.

Flow Receivers - Flow receivers are pressure specifying components and are modelled by the thermodynamic control volume approach. The energy and mass conservation equations are used to calculate the instantaneous rate of change of temperature, mass and pressure for these components at each time step. These are then integrated to calculate the instantaneous level of pressure, temperature and so on. This approach permits the calculation of engine unsteady flow through a large number of quasi-steady time steps.

Amongst the flow receivers, the engine cylinder is considered as a control volume with a moving boundary and has been comprehensively modelled for heat transfer, combustion (heat release), work transfer, gas exchange and emissions calculation. Also a pre-chamber for IDI engines is modelled as a constant volume control volume with both models for heat release and heat transfer.

Flow Controllers - Flow controllers are flow specifying components which are defined within the simulation as those engine components which normally control flow through a change in pressure. These are generally modelled as adiabatic expansion or compression devices with a pressure ratio across them affecting flow and vice-versa. The flowrate-pressure ratio relationship for these components is defined as equation, digitised map and so on. In addition to valves and flow orifices, turbines and compressors (turbocharger in the case of engines) are considered as flow controllers by representing their characteristics in the form of digitised maps.

Pipes - The pipe EBB is represented by the 1-D unsteady flow conservation equations which are solved using the method of characteristic. The treatment is general for non-homentropic, subsonic and supersonic, non-reactive flows. The boundary conditions for the pipe EBB includes fully open, partially open, fully closed, sudden expansion, or junction. Depending on the engine system design and connectivity, the corresponding boundary conditions are assigned to each end of pipe, automatically.

3.3 User Interface

The MERLIN environment allows system building through an interactive session. The interactive session involves provision of the list of EBBs and their connectivity nodes by the user. The code then checks for legal connectivity, availability of EBBs data files and initial conditions file and so on. The system simulation capability of the software has been tested by simulating complex engine system as that shown in FIGURE 1 [13].

A graphical user interface (GUI) is currently under development. Each EBB is represented by an icon with data forms behind each icon containing data points for that EBB. Graphical construction of an engine system, setting up of a database for data management, code execution and post-processing of results of analysis are considered within the GUI.

4. MODEL APPLICATION

4.1 IDI Combustion and Performance

Modelling technique - Zero-dimensional, first law, computer models of the IDI diesel engine combustion have been reported as early as the late sixties [14]. Kamel [15, 16] modelled the combustion in a Ricardo Comet MkVb pre-chamber in the late seventies using experimentally measured data for evaluation of the main and the prechamber heat flux and steady flow tests for calibration of the throat discharge coefficient. The zero-dimensional approach was used to develop a modular divided chamber IDI heat release and combustion model excluding emissions modelling.

The constant volume 'combustion chamber' EBB was used to represent the pre-chamber of the IDI engine linked to the main cylinder via an orifice as shown in FIGURE 2a. Specification of the IDI combustion system geometry in this way is modular and creates flexibility for modelling single or dual throat designs. It also allows re-use of the existing proven models and minimises development time.

Fluid mass transfer between the pre and main combustion chambers driven by the pressure gradient across the connecting passageway, heat transfer between the working fluid and its surrounding metal surfaces in the main and the pre-chamber and work transfer in the main combustion chamber are considered throughout the cycle.

Ignition and combustion in the pre-chamber and the main chamber are considered in both contexts of heat release analysis and combustion. Simultaneous combustion processes in the pre- and the main chambers are allowed to progress with no restriction on mass transfer between them, other than choking of the connecting passageway.

Watson's [17] two part combustion correlation is used to model the fuel burning rate in the main and the pre chamber. Thus, the instantaneous apparent fuel burning rate data for each combustion chamber can be non-dimensionalised and curve-fitted using the two part combustion model. All these capabilities are self-contained within simulation.

Performance Prediction - A four cylinder, 1.8 litre, high speed IDI diesel engine was modelled in a simulated turbocharging configuration as shown in FIGURE 2b by setting the inlet and outlet fluid states to boost and exhaust conditions. This modelling approach was used to isolate the turbocharging system's energy feedback loop from the combustion system, thus allowing the comparisons to be focused on the in-cylinder conditions in order to demonstrate the validity of IDI modelling approach.

Data from 4 measured part-load points on engine map, FIGURE 3, were used to verify the predictions. TABLE 2 shows the predicted overall engine performance and compares these with the measured data.

Table 2 Comparison of the general engine performance data

	1800 (rpm)		2000 (rpm)		2200 (rpm)		2500 (rpm)	
Parameter (unit)	**Siml**	**Expt**	**Siml**	**Expt**	**Siml**	**Expt**	**Siml**	**Expt**
Fuelling (mg/c/c)	16.5	16.6	10.1	10.1	10.8	10.8	17.1	17.0
BMEP (bar)	4.6	5.0	2.1	2.5	2.0	2.5	4.4	5.0
BSFC (g/kw-h)	295	274	404	336	438	352	319	279
Air flow (g/s)	34	35	35	38	41	43	48	56
Boost P. (kPa)	129	125	120	119	126	124	147	150
Boost T. (K)	313	311	311	313	311	309	350	316
Pmax Main (bar)	65	65	61	66	65	63	83	86
Pmax Pre (bar)	65	66	61	66	65	63	83	86
Pexh (kPa)	137	136	141	138	150	148	176	177
V throat (m/s)	65	-	71	-	75	-	83	-

All simulations were run with the measured fuelling levels. As can be seen from TABLE 2 the simulation underpredicted the BMEP in all cases with accuracy being to within 0.6 bar (i.e., 4.5% of the peak BMEP of 13.0 bar).

FIGURE 4 shows the predicted and the measured main chamber pressures specified against crank angle. As can be seen from these diagrams the peak cylinder pressure was predicted with good accuracy, but in all cases the predicted peak cylinder pressure lagged the measured data and occurred later in the cycle.

4.2 Variable Geometry Induction System - Port Deactivation

MERLIN has been used in a study of induction manifold design options for a new generation four-cylinder, four-valve per cylinder, HSDI diesel engine. The engine employs induction manifold 'tuning' to optimise volumetric efficiency over selected parts of the speed range. One major design constraint is the limited space available in the engine bay to accommodate the concomitant 'resonant' pipes and volumes, particularly for turbocharged and intercooled installations and, therefore, emphasis was placed in the study on compactness of design.

Motored steady speed simulations were run with a simplified single cylinder variant of the engine, the objective being to explore the merits of port deactivation. The exhaust system was modelled as a fixed gas state bounding the exhaust valves so as to suppress the gas dynamic influence of the exhaust design on the engine mass flow characteristic.

The initial stage of the study comprised a parametric study on the effect of basic ram pipe geometry (pipe length and cross sectional area) on the volumetric efficiency. FIGURE 5a shows the fluid flow connectivity diagram for the baseline dual ram pipe manifold, designated System I. In the second part of the study the effect of port and port-valve deactivation on volumetric efficiency was examined. MERLIN fluid flow connectivity diagrams for these manifolds, designated System II and System III respectively, are also shown in FIGURE 5a. Note that with System I the respective ram pipe and inlet valve combinations are independent, whereas in System II the ram pipe is deactivated upstream of the valve but allows the remaining ram pipe to breathe through both valves.

FIGURE 5b shows the results of the study on the sensitivity of the volumetric efficiency to ram pipe length for System I. Each curve displays a broadly similar character with an underlying 'low frequency' component and a 'higher frequency' fluctuation superimposed. With increasing pipe length the low frequency part displaces progressively to a lower engine speed. Two underlying, and related, phenomena determine the shapes of these curves. These are the timing and amplitude of the reflected primary compression wave and the presence of a decaying residual wave form in the ram pipe and inlet valve enclosures as discussed further by Prosser [18].

With simple ram manifolds, System I, the diameter of the pipe can also have a pronounced influence on the shape of the volumetric efficiency curve, FIGURE 5c. With a reducing cross sectional area the depression created across the inlet valve must increase and, consequently, so too the amplitude of the expansion wave propagated back into the ram pipe. Since the waveform is non-linear, wave speeds will vary locally and the timing of the reflected pulse will change. The resultant effect on mass flowrate, which is also dependent on local gas density variation, is to increase the peak value and at the same time displace its position to a lower engine speed.

A four-valve dual induction port cylinder head arrangement provides an opportunity for implementing a simple variable geometry manifold, based on the flow area reduction principle, by employing deactivation of one of the ports at lower engine speeds, typically by closing a throttle valve positioned as close as is practicable to the poppet valves.

FIGURE 5d compares the predicted volumetric efficiency for the three manifold configurations shown in FIGURE 5a, each with ram pipe lengths of 0.8 m. The baseline manifold comprises two fully independent ports, each bounded by a single poppet inlet valve. This would typically be the arrangement for a fixed geometry manifold. The port deactivated manifold has the flow through one of the ports isolated and shut-off whilst allowing both valves to breathe through the other port. The port and valve deactivated manifold also has one of the ports deactivated but in this case, since the ports are fully independent, the engine can only breathe through a single inlet valve.

Comparing the respective volumetric efficiency curves displayed in FIGURE 5d, it is apparent that the port and valve deactivation manifold performs worst of all displaying only marginal benefits at the lowest speeds and serious flow restriction in the mid to high speed range. For all practical purposes this manifold design has little to offer. For speeds up to about 3000 RPM the port deactivation manifold displays superior volumetric efficiency and above this speed the twin port baseline manifold performs best. A variable geometry manifold with the given ram pipe dimensions should therefore adopt a strategy of flowing through a single port below about 3000 RPM and through both ports above this speed. FIGURES 5e and 5f show similar sets of curves for ram pipe lengths of 0.5 and 0.2 m respectively. The benefits of port deactivation are still apparent although the engine speed at which the valve should be actuated shifts to progressively higher speeds.

This case study shows that port deactivation offers a compact and mechanically simple form of variable geometry manifold. Simulation analyses permit the engine designer to quantify the potential benefits for a given engine and valve geometries without, in the first instance, extensive prototyping or test bed measurements.

4.3 Exhaust Emissions

Emissions Model - The combustion-emissions model [11] is a quasi 2-D multi-zone one in which various processes of fuel atomisation, droplet evaporation, air mixing, ignition, heat release and emissions formation are included, FIGURE 6. In this model, injection is assumed to be characterised in the shape of intermittent pockets of fuel segments distributed within the spray angle. Combustion zones are formed at the start of atomisation with subsequent air entrainment into each zone. The amount of entrained air is calculated based on jet theory and zonal momentum conservation with allowance for air swirl and spray wall impingement.

Fuel droplet evaporation is calculated from the point of fuel atomisation together with ignition delay calculation. The combustion rate of fuel vapour is modelled by a second order reaction kinetic and is related to the zonal fuel vapour and oxygen concentrations. Heat transfer is modelled at zonal level taking into account both convective and radiative modes.

The NO formation is based on the extended Zeldowich mechanism with fuel bound Nitrogen assumed to convert completely to NO. Dry soot and CO formation and oxidation are assumed to be kinetically controlled and based on respective kinetic reactions. An overall chemical energy balance at zonal levels yields the level of unburnt HC emissions.

FIE Model - The fuel injection process is represented by a fuel line pressure diagram and a nozzle coefficient of discharge. Measured values are normally supplied but provision is made for the generation of fuel line pressure, based on a non-dimensional unified characteristic approach [19]. This option is needed for engine transient operation anyway where quasi-steady models for fuel pump and governor are also implemented.

As a third option, a complete fuel injection system simulation is being integrated into the software to generate fuel line pressure and nozzle coefficient of discharge. FIGURE 7 shows the emissions calculation procedure.

Transient Emissions Prediction - The combustion-emissions model is shown to predict the correct trends with the majority of engine design parameters and operating conditions [11]. It has also been successfully applied to performance-emissions trade-offs studies for a HSDI diesel engine [20]. FIGURE 8 shows an example of the application of the model to transient emissions prediction for a high-speed vehicle type engine, where the effect of turbocharger turbine size on engine transient response and emissions is studied.

As a result of a 20% reduction in turbocharger turbine size, engine and turbocharger transient response are improved with a significant reduction in combustible exhaust emissions including unburnt HC, carbon monoxide and particulates. NOx level increases under steady-state conditions but reduces during transient operation.

5. FUTURE DEVELOPMENTS

The above examples have demonstrated some of the current capabilities of MERLIN. Current areas of model development include: spark ignition combustion, piping network transients, gas turbine simulation, duty cycle emissions, Graphical User Interface and integration with Lloyd's Register's Shafting System Analysis Suite.

Future potential development areas for the model have been identified which include: combustion generated noise, real-time simulation, integration into multi-dimensional models, SI engine emissions, exhaust after treatment systems, hybrid electro-mechanical prime movers with and without regenerative flywheel storage systems.

6. CONCLUSIONS

6.1 Diesel engine thermodynamic simulation and its application to new technical areas are continuing. Of particular interest are areas of flow dynamics and engine breathing characteristics, combustion-emissions including fuel injection system and engine transient response. The case studies presented cover the above areas.

6.2 The user friendliness of such codes has been enhanced through improved user-interface with a general move towards graphical user interface.

6.3 The component oriented EBB approach to modelling ensures modularity and versatility of the developed software. Such an approach saves significantly in maintenance effort and aids the quality of the software.

7. ACKNOWLEDGEMENT

The authors wish to thank the Committee of Lloyd's Register and Ford Motor Company for permission to publish this paper.

REFERENCES

(1) BOMAN, G.L. Mathematical Simulation of Internal Combustion Engine Processes and Performance including Comparisons with Experiment - PhD Thesis, University of Wisconsin, 1964.

(2) MCAULY, K.J., WU, T., CHEN, S.K., BORMAN, G.L., MYERS, P.S. and UYEHARA, O.A. Development and Evaluation of the Simulation of the Compression Ignition Engine - SAE Trans., 74, Paper 650451, 1965.

(3) JANOTA, M.S., HALLAM, A.J., BROCK, E.K. and DEXTER S.G. The Prediction of Diesel Engine Performance and Combustion Chamber Component Temperature Using Digital Computers - Proc IMechE, Vol. 182, Pt. 3L, 1967-68.

(4) STREIT, E.E. Mathematical Simulation of a Large Pulse-Turbocharged Two-Stroke Diesel Engine PhD Thesis , University of Wisconsin , 1970.

(5) WATSON, N. and MARZOUK, M. A Non-Linear Digital Simulation of Turbocharged Diesel Engines Under Transient Conditions - SAE 770123 , 1977.

(6) BANISOLEIMAN, K., SMITH, L.A., BAZARI, Z. and MATHIESON, N. Simulation of Diesel Engine Performance - I Mar E Conference, London, 2nd March 1993.

(7) BENSON, R.S., GARG, R.D. and WOOLLATT, D. A Numerical Solution of Unsteady Flow Problems Journal (Int) of Mechanical Science, Vol 6 No 1, 1964.

(8) BENSON, R.S. A Comprehensive Digital Computer Program to Simulate a Compression Ignition Engine Including Intake and Exhaust System - SAE 710113, 1971.

(9) SAPSFORD, S.M., RICHARDS, V.C.M., AMLEE, D.R., MOREL, T. and CHAPPELL, M.T. Exhaust System Evaluation and Design by Non-Linear Modelling SAE 920686, 1992.

(10) YOSHIZAKI, T., NISHIDA, K. and HIROYASU, H. Approach to Low NOx and Smoke Emission Engines by Using Phenomenological Simulation - SAE 930612, 1993.

(11) BAZARI, Z. A DI Diesel Combustion and Emission Predictive Capability for Use in Cycle Simulation - SAE 920462, 1992.

(12) LOWE, A.S.H. and MOREL, T. A New Generation of Tools for Accurate Thermo-Mechanical Finite Element Analyses of Engine Components - SAE920681, 1992

(13) SMITH, L.A. and BANISOLEIMAN, K. MERLIN-A Modular Approach to Diesel Engine Performance Simulation - CIMAC 20th International Congress on Combustion Engines, London, May, 1993.

(14) LYN, W.T., SAMAGA, B.S. and BOWDEN, C.M. Rate of heat release in High Speed Indirect Injection Diesel Engines. Proc. I.Mech.E. Vol.184 pT.3J 1969/70.

(15) KAMEL, M.M. Thermodynamic Analysis of Indirect Injection Diesel Engine Operation. Ph.D. Thesis, Imperial College, London, 1977.

(16) WATSON, N. and KAMEL, M.M. Thermodynamic efficiency evaluation of an indirect Injection Diesel Engine. SAE 790039, February 1979.

(17) WATSON, N., PILLEY, A.D. and MARZOUK, M. A Combustion Correlation for Diesel Engine Simulation. SAE 800029.

(18) PROSSER, T.G. Induction Ramming a Motored High-Speed Four-Stroke Reciprocating Engine - Influence of Inlet Port Pressure Waves on Volumetric Efficiency - Proc. I Mech E, Vol. 188, 1974.

(19) BAZARI, Z. Diesel Engine Exhaust Emissions Prediction under Transient Conditions - SAE Paper 940666, 1994.

(20) BAZARI, Z. and FRENCH, B.A. Performance and Emissions Trade-offs for a HSDI Diesel Engine - An Optimisation Study - SAE 930592, 1993.

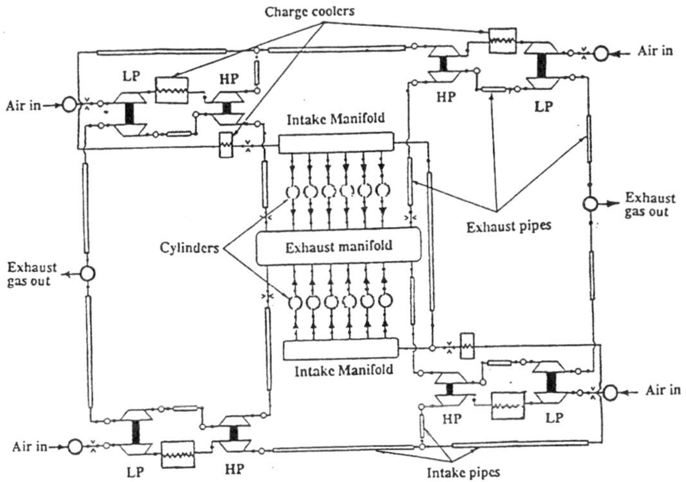

Fig 1 - Schematic of a modern sequentially, two-stage turbocharged engine

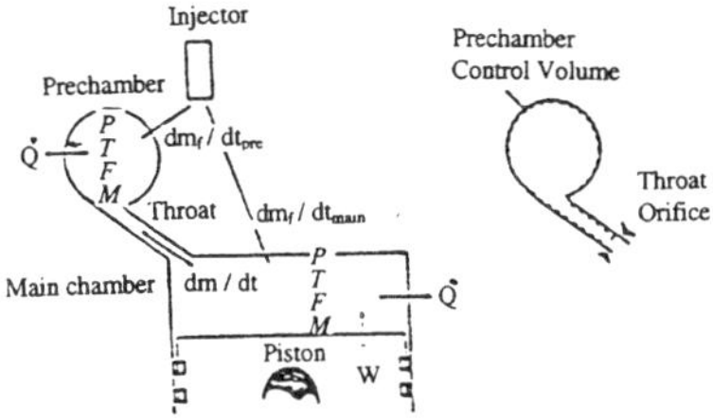

(a) - Control volume model of pre-chamber.

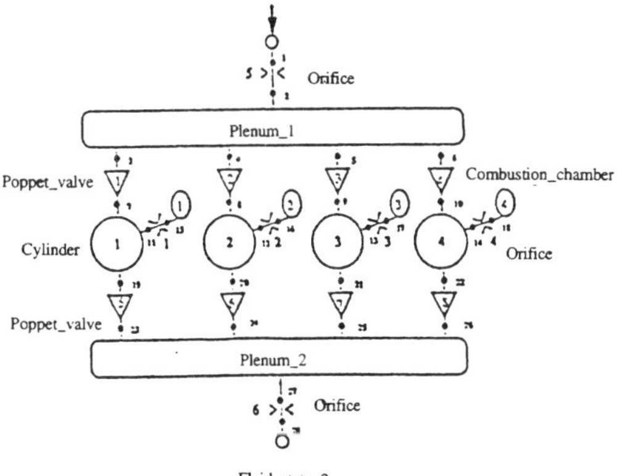

(b) - MERLIN representation of engine system.

Fig 2 - IDI engine model

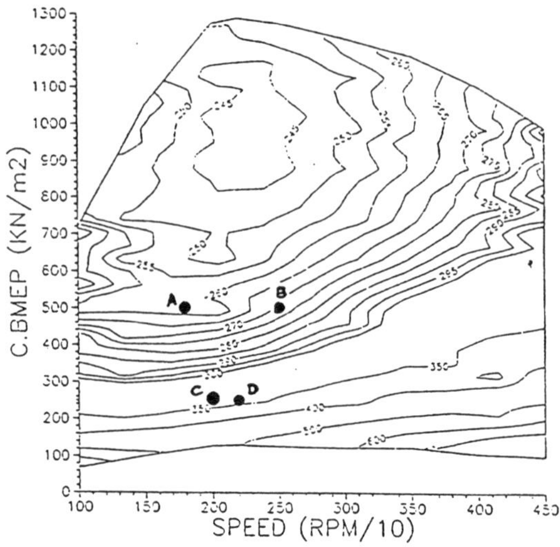

Fig 3 - IDI engine performance map and four part-load measured data points.

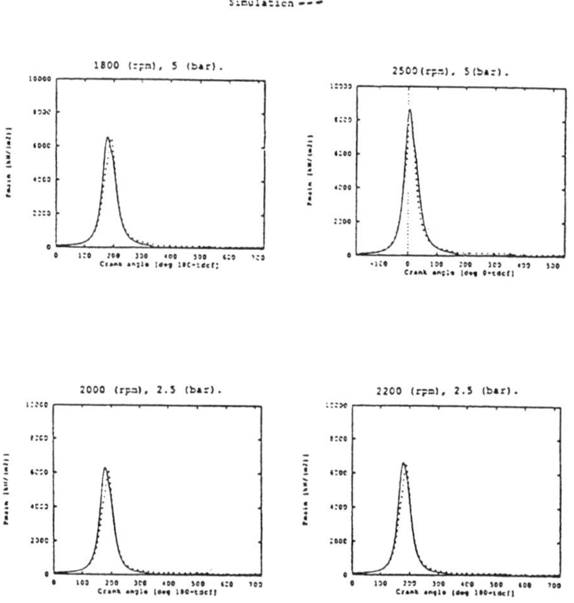

Fig 4 - Comparison of measured and predicted pressures for the main chamber of IDI engine.

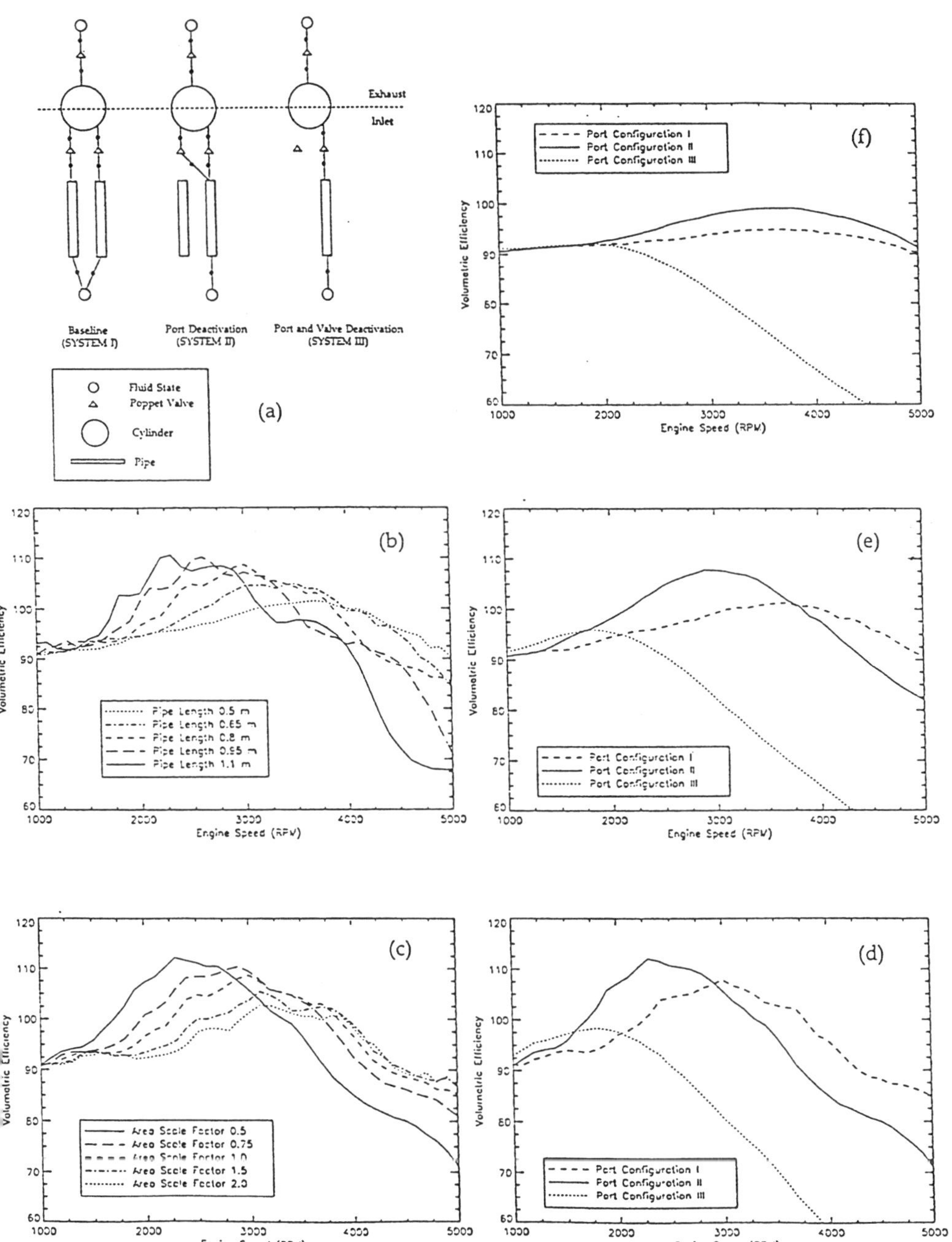

Fig 5 - MERLIN connectivity specification and simulation results for induction manifolds with and without port deactivation.

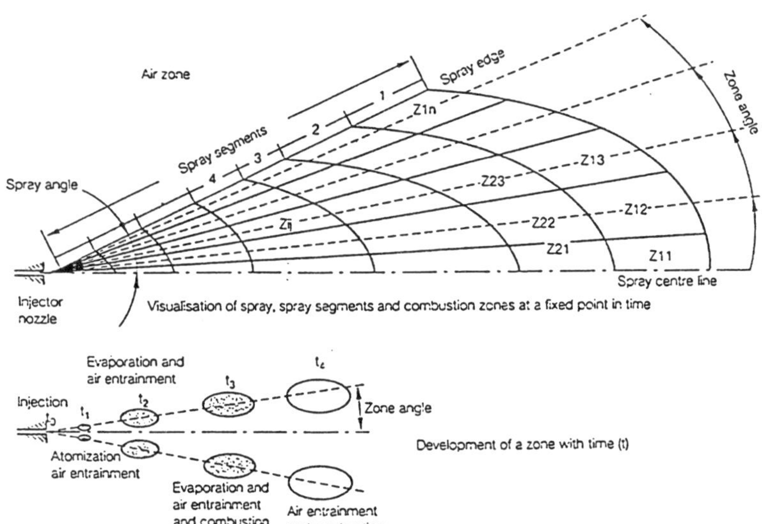

Fig 6 - Visualisation of multi-zone concept of spray and surrounding air.

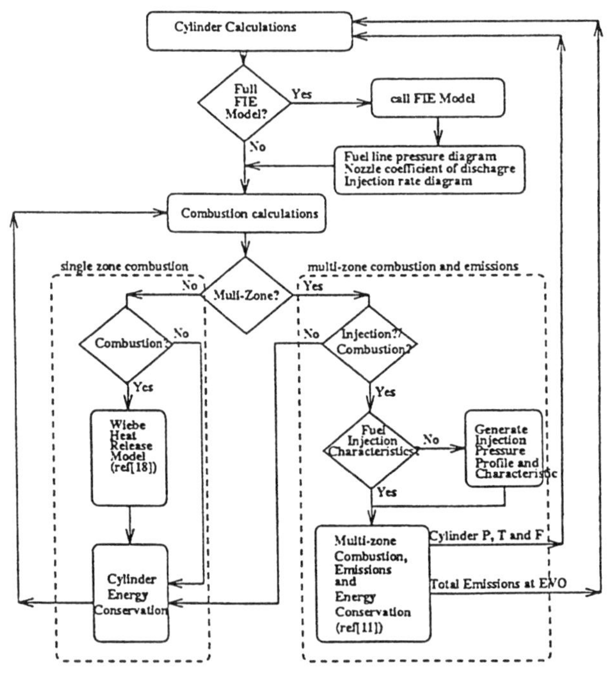

Fig 7 - FIE and combustion-emissions calls procedure within cylinder EBB model.

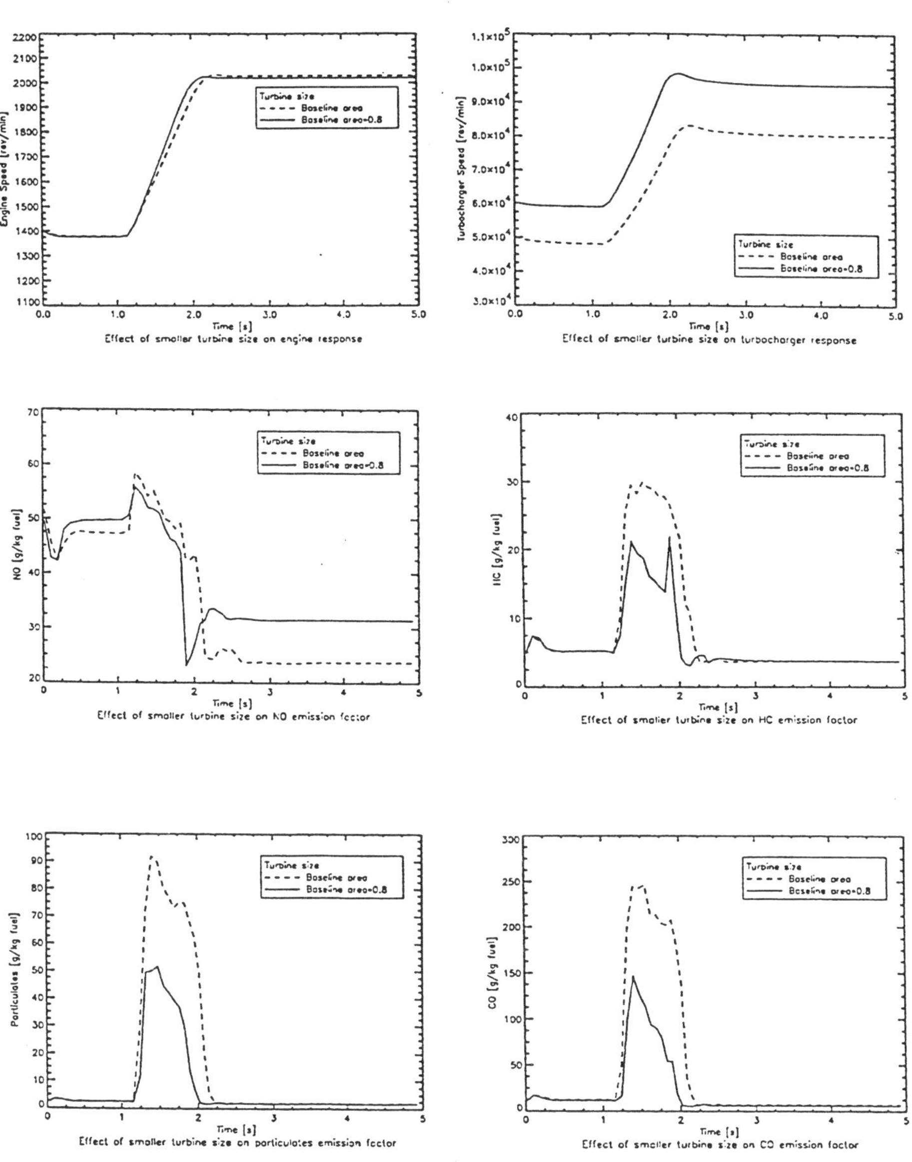

Fig 8 - Predicted engine performance and exhaust emissions under transient operating conditions with alternative turbocharger sizes.

C499/052/96

A white noise approach for rapid gas dynamic modelling of IC engine silencers

A ONORATI MSc, PhD
Department of Energetics, Politecnico di Milano, Italy

SYNOPSIS

A non linear model is described for the prediction of i.c. engine silencer performance in the frequency domain. A white noise pressure signal, numerically generated, is adopted to excite the plane wave motion in the silencer ducts, to achieve a rapid computation of the attenuation curve. The simulation code, based on the Lax-Wendroff and MacCormack shock-capturing methods for the numerical solution of the 1-d conservation equations, has been tested on different reactive and dissipative muffler configurations. In particular, perforations have been modelled, getting a satisfactory agreement between predicted and measured results.

1. INTRODUCTION

The prediction of silencer performance, in terms of noise attenuation in the frequency band, is certainly of practical relevance in design studies concerning not only reciprocating internal combustion engines, but also reciprocating and rotary compressors, fans, gas turbines, and generally any machine which may be regarded as a specific source exciting pressure waves in a duct system. In fact, the optimization of the muffler geometry, in order to achieve a satisfactory reduction of pulse noise levels radiated by the open ends, is a complex and time-consuming operation which cannot be carried out only on the basis of an experimental 'trial and error' procedure. The aid of numerical codes, able to simulate the acoustic and fluid dynamic behaviour of the muffling systems, is fundamental during a preliminary investigation to select a few proper configurations, to be successively refined by measurements and calculations.

Computer models based on the one-dimensional linear acoustic theory (1,2), which assumes small pressure perturbations travelling along the ducts, apply the four pole transfer matrix method to determine the attenuation characteristics of any silencer, reactive or dissipative, with a complex internal structure, in presence of zero and non zero mean flow. The limitations of this approach may be summarized as follows: firstly, the inability of modelling finite amplitude pressure waves, generally propagating along the pipes of the exhaust muffler, which consequently behaves like a non linear system; secondly, the difficulties in simulating noise sources, responsible for the wave motion in the silencing device, due to the experimental parameters needed to characterize the source from an acoustic point of view (3). The advantages of this first group of models are certainly the simplicity, the capability of simulating any type of muffler, the very short computer run-times.

However, the muffler may be optimized to attain its best response only taking account of its interaction with the engine, for different condition of engine speed and load, to point out the influence of muffler characteristics on the engine performances. Therefore it is preferable to make use of more general models, capable of simulating the noise source and its interaction with the intake and exhaust silencers. Following this strategy, the one-dimensional gas dynamic codes can calculate the unsteady flows in engine pipe-systems, on the basis of the solution of the conservation equations by means of shock-capturing numerical techniques, modelling the finite amplitude pressure waves propagation within the ducts and the silencer. In this way, the non-linear response of the muffler may be taken into account, and the prediction of tailpipe noise spectrum carried out. Much research work has been done about this aspect in the last three decades, considering both a rotary valve pulse simulator (4,5) to generate finite amplitude waves in the ducts, and the engine source itself (6,7,8). By means of suitable boundary conditions, today a lot of typical muffling elements may be simulated, ranging from column and Helmholtz resonators to expansion chambers with internal orifices or axial side-branches, from folded straight ducts to flow reversals and perforates (9,10).

The 1-d non linear gas dynamic approach may be successfully employed also for the investigation of silencer performances in the frequency domain, to predict the attenuation curve in dB and outline the resonance frequencies of the filter (11), just like linear acoustic models do. This analysis of the silencer behaviour in the frequency band, considering the system apart from the engine, is fundamental to improving the simulation of complex shape mufflers by means of boundary conditions and acoustically equivalent schemes. It is also important to determine the appropriate corrective lengths for each geometrical discontinuity, to be used in the extensive numerical codes which simulate the pressure wave dynamics in the whole engine intake and exhaust systems, involving the muffling device and the noise source.

The advantages of this second group of models, compared with the linear acoustic ones, are the following: firstly, the validity is not restricted to small pressure waves; secondly, the effects of mean flow, friction and temperature gradients at duct walls may be directly included in the conservation equations; the drawback is generally represented by the very long computer run-times. In fact, the transfer function of the acoustic filter (i.e. the attenuation curve) in the frequency band can be precisely evaluated through the calculation of the steady wave motion in the silencer, arising when a harmonic pressure fluctuation of fixed frequency and amplitude (pure tone) is imposed upstream of the silencer (11). By repeating the calculation for different frequencies, spread over the band examined with a given step, it is possible to obtain the transfer function of the system. This procedure gives accurate predictions of attenuation but is rather time-consuming.

In order to avoid this long computation, a faster technique has been developed. Instead of exciting the wave motion in the ducts by means of a simple pure tone, a 'white noise' pressure perturbation has been imposed upstream of the silencer, that is a particular signal characterized by a discrete spectrum with constant amplitude components, over a specific frequency band. In this way, a FFT spectral analysis of the resulting standing waves upstream and downstream of the muffler allows to determine the transfer function of the system, with very short computer run-times. This approach proves to be advantageous especially in the simulation of perforates, which generally need a notable computational effort, due to the great number of short ducts necessary to model the holes, the perforated pipe and the cavity liner.

This paper describes the major aspects of the computer model, dealing with the generation of the white noise pressure signal and the numerical techniques adopted in the code. Predicted results have been compared with experimental measurements on different silencers, tested in a semi-anechoic room. In particular, mufflers with perforated pipes have been simulated, achieving a satisfactory agreement with experimental data.

2. NUMERICAL MODEL

The non-linear code developed consists of dedicated routines for: (i) the calculation of plane waves in the silencer ducts, (ii) the modelling of the most common muffling elements (such as expansion chambers with internal orifices, extended inlet and outlet, resonators, flow reversals, perforates, etc.), (iii) the analysis of the predicted wave motion to evaluate the attenuation curve in the frequency band, via a fast Fourier transform .

The plane wave motion in the silencing pipe-system, resulting from the excitation imposed upstream, is evaluated through the solution of the conservation laws for one-dimensional, unsteady, compressible flows. By introducing the vectors ***W, F, C*** (the conserved variable, the flux and the source term vectors respectively):

$$W(x,t)=\begin{bmatrix}\rho\\ \rho u\\ \rho e_0\end{bmatrix},\quad F(W)=\begin{bmatrix}\rho u\\ \rho u^2+p\\ \rho u h_0\end{bmatrix},\quad C(W)=\begin{bmatrix}\rho u\\ \rho u^2\\ \rho u h_0\end{bmatrix}\frac{d(\ln F)}{dx}+\begin{bmatrix}0\\ \rho G\\ -\rho q\end{bmatrix} \tag{1}$$

(p, u, ρ, e_0, h_0 are pressure, flow velocity, density, stagnation specific internal energy and enthalpy, respectively; F is the cross-sectional area, $G = 4f_w u|u|/2d$, d is the duct diameter, f_w is the friction factor at the duct-wall, q is the heat transferred per unit mass per unit time), the quasi-linear hyperbolic system of conservation equations (continuity, momentum and energy) may be written in conservative form as follows (12):

$$\frac{\partial W(x,t)}{\partial t}+\frac{\partial F(W)}{\partial x}+C(W)=0. \tag{2}$$

The equations directly include the change of cross-sectional area, friction and heat transfer at duct-walls as source terms. An additional equation concerning the fluid properties must be introduced to solve the problem; a common assumption is to consider a perfect gas with constant specific heats ($p/\rho=RT$, $k=c_p/c_v$=const.).

The simulation code developed can make use of different numerical techniques for the solution of the hyperbolic problem: (i) the traditional mesh-method of characteristics (13), (ii) the two-step Lax-Wendroff method, (iii) the MacCormack predictor-corrector method (8,14). These last two are explicit, symmetric shock-capturing schemes with second order accuracy ($O(\Delta x^2,\Delta t^2)$), capable of handling possible discontinuities in the solution (shock waves and contact discontinuities). Only the third method is briefly described hereafter for conciseness.

The MacCormack method calculates the solution vector W^* in the node i at a provisional time t^*, on the basis of the known vectors W^n, F^n, C^n at the time t^n by a predictor step (forward differences):

$$W_i^*=W_i^n-\frac{\Delta t}{\Delta x}\left(F_{i+1}^n-F_i^n\right)-\Delta t\ C_i^n. \tag{3}$$

W_i^* is used to evaluate the vectors F_i^*, C_i^* at the same time t^*. The final solution vector W^{n+1} at the time t^{n+1} is achieved by a corrector step (backward differences):

$$W_i^{n+1} = \frac{1}{2}\left[W_i^n + W_i^* - \frac{\Delta t}{\Delta x}\left(F_i^* - F_{i-1}^*\right) - \Delta t\ C_i^*\right] \tag{4}$$

To carry out an accurate calculation of a discontinuous solution, the orientation of finite differences is continuously reversed at each time step, that is: backward predictor-forward corrector at time t_k, forward predictor-backward corrector at time t_{k+1}.

The time step Δt is fixed by the Courant-Fredrichs-Lewy criterion, which states:

$$CFL = (a+u)\frac{\Delta t}{\Delta x} \leq 1 \tag{5}$$

for the stability of the calculation.

In order to eliminate the numerical overshoots in the proximity of discontinuities, typical of second order schemes, the Flux Corrected Transport (FCT) technique and the Total Variation Diminishing (TVD) algorithm (14) have been implemented in the model. However, if small pressure waves propagate along the silencer ducts and the gas temperature is uniform (such as during an acoustic excitation of the silencer), these damping techniques are not necessary and may be excluded , since shock waves and significant contact discontinuities do not occur.

2.1 White noise approach and boundary conditions

Different flow boundary regions are generally found in a silencer. For example, the excitation source and the open end at the extremes of the system, often accompanied by abrupt cross-sectional area changes, junctions of pipes, orifices, perforated ducts, axial side-branches, etc.. Suitable boundary conditions have been developed to model these components, on the basis of the classic assumption of quasi-steady flow, involving the steady conservation equations of mass, momentum and energy. The method of characteristics has been used to evaluate the interaction of waves at boundaries, in terms of incident and reflected Riemann variables (13).

The upstream boundary condition representing the excitation source deserves a particular attention. If the imposed upstream perturbation is a 'white noise', that is a pressure signal whose discrete spectral content is uniformly distributed in the frequency band, it is possible to evaluate the gas dynamic response of the muffler to all the concurrent excitation frequencies at once. Hence, having reached the situation of standing waves after the decay of the initial transient motion, pressure and velocity oscillations in the system are characterized by a significant amplitude of their spectral components for each frequency of the white noise introduced in the silencer. Thus a Fourier analysis of the signals, to get the sound pressure level spectra upstream and downstream of the acoustic filter, allows to predict the transfer function, i.e. the attenuation curve. This white noise approach proves to be very rapid, compared with the acoustic excitation by means of single harmonic pressure oscillations, frequency by frequency (10,11).

The numerical generation of a white noise periodic pressure signal may be carried out as follows. Actually, it is possible to build it as the sum of N sinusoidal pressure oscillations with a fixed amplitude Δp and frequency multiple of a fundamental f:

$$p(t) = p_0 + \sum_{n=1}^{N} \Delta p \cdot sin(n \cdot 2\pi f \cdot t + \varphi_n). \tag{6}$$

where p_0 is a constant value representing the mean ambient pressure upstream of the flow duct system. For example, if the frequency band to be examined is 20 to 2000 Hz, the fundamental frequency f is 20 Hz, while the number N of sinusoids is 100. In this way the resulting periodic pressure $p(t)$ has a discrete spectrum with a 20 Hz step in the chosen band.

It is important to take a small value for the amplitude Δp, when the calculation is concerned with the acoustic performance of the silencer, to be compared with experimental measurements carried out under a loudspeaker excitation. In this case, Δp must lie in the range 0.01 to 0.0001 bar, and the amplitude of the resulting signal $p(t)$ must remain within a proper range as well. In fact, during the acoustic excitation by a loudspeaker, the pressure perturbations travelling along the ducts are very small, i.e. acoustic, so that the system behaves linearly (the superposition of effects holds). Similarly, the numerical simulation must involve only small pressure waves, to ensure a linear behavior of the system and achieve comparable results with measured data. To satisfy this request, a suitable choice of the phase φ_n in eq. (6) is essential. A generation of N random numbers k_n in the range 0 to 1, via a dedicated numerical routine, allows to have a random phase $\varphi_n = k_n \cdot 2\pi$ in each sinusoidal component of the sum. This choice yields a pressure signal $p(t)$ with a random character, which falls in the band of prescribed pressure values.

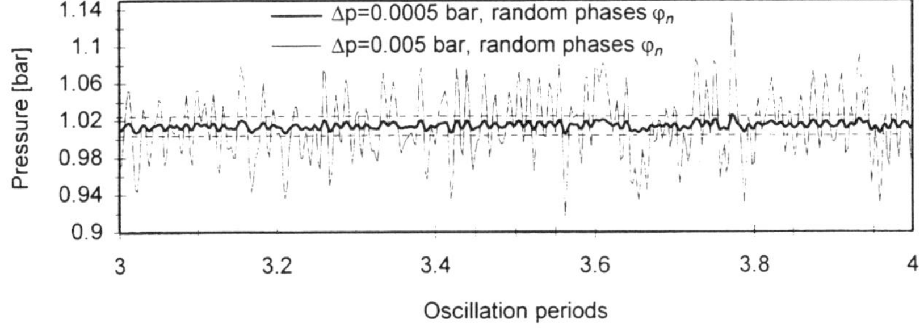

Fig 1 White noise pressure traces $p(t)$ during an oscillation period T (the last of four imposed upstream of the silencer). The signals are periodic, generated by fixing f=20 Hz, N=100 (so the frequency band is 20 to 2000 Hz), random phases φ_n, p_0 = 1.013 bar, and Δp equal to 0.0005 and 0.005 bar respectively.

In Fig. 1 the 'white noise' pressure trace is shown for different choices of the amplitude Δp and random phases φ_n. The signal has been generated to have a significant spectrum in the band 20 to 2000 Hz, taking f=20 Hz and N=100 (T=0.05 s). A good result is achieved by fixing a small Δp, equal to 0.0005 bar. In fact, it may be noticed that the sum signal $p(t)$ remains in the band of acoustic perturbations ($p(t) \leq p_0 \pm 0.01$ bar, with p_0 =1.013 bar, the atmospheric pressure) in the time domain, during an oscillation period T. Conversely, the resulting signal has an excessive, non-acoustic amplitude if Δp is set to 0.005 bar, with random phases φ_n. Furthermore, in Fig. 2 a comparison is reported between an appropriate signal $p(t)$ (again, obtained with Δp=0.0005 bar and random phases φ_n) and another signal $p(t)$ built with the same Δp but phases φ_n equal to zero. It is evident that this second choice provides an excessive, non-acoustic magnitude both at the beginning and in the end of the pressure oscillation. Fig. 3 shows the spectrum of the white noise pressure signal in the band 20 to 2000 Hz, confirming the discrete and uniform quality of the spectral content. Consequently, this 'white noise' periodic pressure trace $p(t)$ may be imposed in the first cross-section of the silencer as a boundary condition, repeating the excitation for several oscillation periods, until a steady wave motion is established in the flow duct system. An expression of $p(t)$, in terms of the incident and reflected

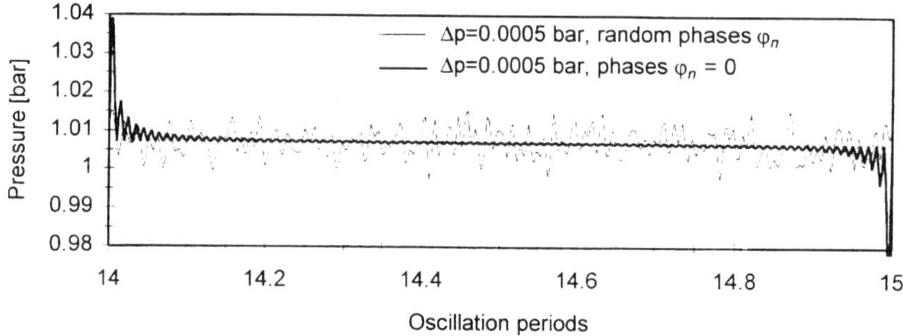

Fig 2 White noise pressure traces *p(t)* during an oscillation period *T* (the last of fifteen imposed upstream of the silencer). In this case f=20 Hz, N=100, p_0 = 1.007 bar, Δp=0.0005 bar in both traces, with different phases φ_n .

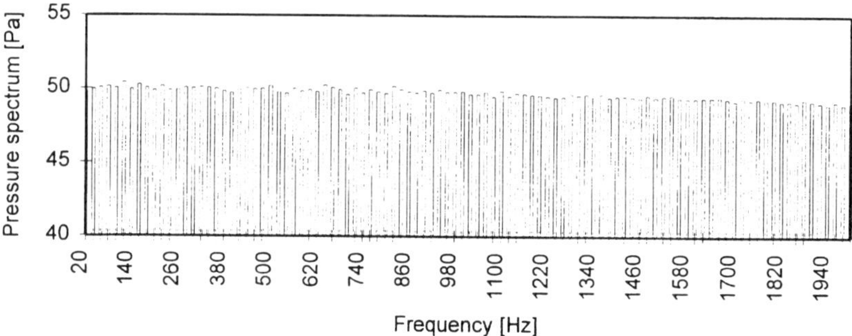

Fig 3 Spectral content of the pressure trace *p(t)*, calculated by a FFT analysis of the periodic signal. The spectrum has a nearly constant amplitude, equal to about 50 Pa (Δp=0.0005 bar) for each frequency component (N=100).

Riemann variables λ_{in} , λ_{out} and the entropy level A_A at the boundary, may be simply derived in different ways (11).

Briefly, with regard to the other boundary conditions, the modelling of the muffler open end has been carried out with the assumption of constant static pressure in the *vena contracta*, equal to the atmospheric pressure. The sudden contractions and sudden enlargements have been treated respectively by the pressure loss and pressure recovery models (13), whereas the junctions of three or more ducts have been treated by the equal total enthalpy (constant pressure) model (15).

2.2 Prediction of the attenuation curve

The simulation program developed applies the following method to determine the transfer function of a typical silencer, i.e. its attenuation characteristic in the frequency domain. The standing wave motion in the acoustic filter is reached after a few oscillation periods of the perturbation, imposed in the first cross-section of the system. Hence, a FFT analysis of the calculated pressure trend allows to determine the sound pressure level spectrum ($SPL_{upstr.}$ [dB] in the 20 to 2000 Hz frequency band) in the upstream cross-section of the silencer. At the same time, the predicted velocity at the open end permits to estimate the radiated sound spectrum at a distance r from the tailpipe outlet, on the basis of the monopole radiation formula (7):

$$p(r,t+\frac{r}{a_0})=\frac{\rho_0 F}{C\pi r}\cdot\frac{d[u(t)]}{dt} \qquad (7)$$

(F: area of the terminal cross-section; ρ_0 , a_0 : density and sound velocity in the surrounding medium; C =*4* for spherical radiation, *2* for hemispherical radiation). The velocity $u(t)$ is known as a set of discrete values versus time instants, so that an interpolation is needed (by cubic splines, for example) to determine the velocity time derivative corresponding to each instant. Then the FFT analysis may be used to process the set of velocity derivative values, computing the spectral content. Thereby, the radiated noise spectrum, i.e. the sound pressure level spectrum ($SPL_{downstr.}$) at a distance r from the outlet, may be evaluated by eq. (7). Hence, the transfer function of the silencer can be calculated directly in the chosen band, as the difference between the two spectra $SPL_{upstr.}$ and $SPL_{downstr}$.

The procedure explained above supplies a fairly good prediction of the attenuation curve, and reduced computer run-times (see Table 1) in comparison with the alternative time-consuming (but more accurate) approach, based on the frequency-by-frequency excitation via a single harmonic perturbation. It is important to choose appropriately the mesh size, to ensure a reliable simulation of high frequency components of the wave motion. In the calculation performed, good results have been achieved by fixing Δx to 1 cm with the MacCormack and Lax-Wendroff methods, and to 0.5 cm with the mesh-method of characteristics. Furthermore, about 10 to 15 oscillation periods of the white noise pressure signal must be imposed upstream of the silencer, to achieve a sufficient stabilization of the fluid dynamic quantities in the system. Hence, the spectral analysis may consider only the last period.

3. RESULTS

Different muffler configurations have been considered to test the validity of the model developed. The experimental attenuation curves have been measured in a semi-anechoic room (11) by means of a spectrum analyzer. The excitation source adopted was a loudspeaker radiating white noise, forcing the acoustic wave motion with zero mean flow. Two microphones (the position is shown in Figs. 4, 5, 6) have been used to measure the sound pressure level spectra, upstream and downstream of the duct system, and get the transfer function of the acoustic filter. In all the experiments the overall sound power level of the excitation source was in the range 65 to 70 dB, while the overall sound pressure level induced by the source in the upstream cross-section lied in the range 115 to 125 dB; it was verified that no modification in the measured spectra and transfer function occurred by varying the source energy level in that range, thus proving the validity of the assumption of linear propagation of sound waves.

The comparisons between the experimental and the simulated attenuation curves are shown in Figs. 4, 5, 6, concerning an expansion chamber with an internal orifice, a column resonator and a perforate respectively. In each figure it may be observed that the curve predicted via the white noise approach is rather similar to the one predicted via the frequency-by-frequency approach(based on sinusoidal perturbations imposed in the first cross-section), in spite of a more irregular aspect in some points. The simulated results have been achieved by the MacCormack numerical technique with both the approaches, showing a notable difference in the computer run-times, as reported in Table 1. A big increase in computation speed is evident (the simulation is generally about 30 to 40 times faster), so that the transfer function may be evaluated in few seconds by a rapid gas dynamic modelling of the silencer. This is particularly useful in the simulation of perforates, since the large number of short ducts, needed to model the numerous holes, the cavity liner and the perforated pipe, makes the calculation rather heavy.

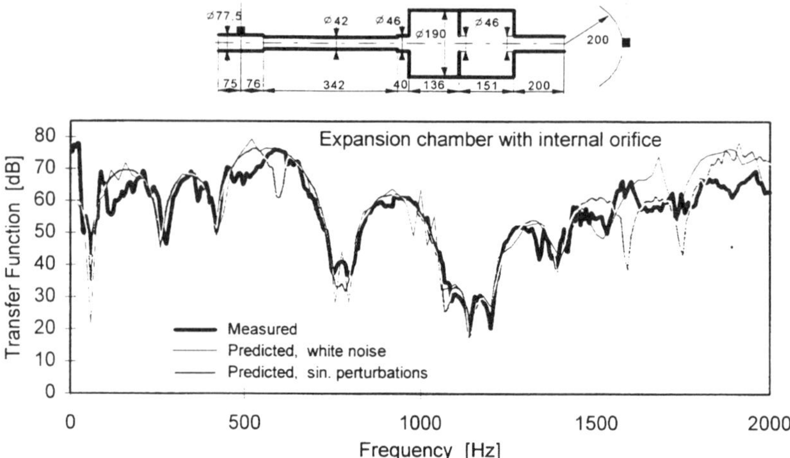

Fig 4 Comparison between the experimental and predicted transfer function of an expansion chamber with an internal orifice.

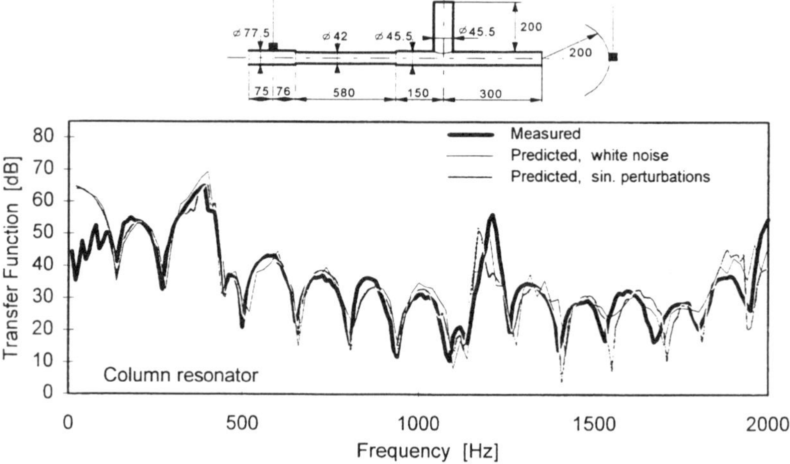

Fig 5 Comparison between the experimental and predicted transfer function of a column resonator.

In this case the white noise approach allows to reduce drastically the computer run-times (15.1 min. versus 10.4 h), rendering practicable the simulation of this type of reactive-dissipative silencers. The one shown in Fig. 6 involves a partially perforated duct (hole diameter 2 mm, hole spacing 3 mm), which can be simulated resorting to the boundary condition for the junction of several pipes (constant pressure model (15)) and to appropriate corrective lengths (10). Briefly, the n holes of each group, distributed on a circle, may be represented by n very short ducts, with the same hole diameter and a length related to the wall thickness. If l is the length of the perforated duct and p the hole spacing, the subdivision of the system into $m=l/p$ axial elements allows to join, on one end, the n holes (short ducts) of each group to 2 adjacent axial elements.

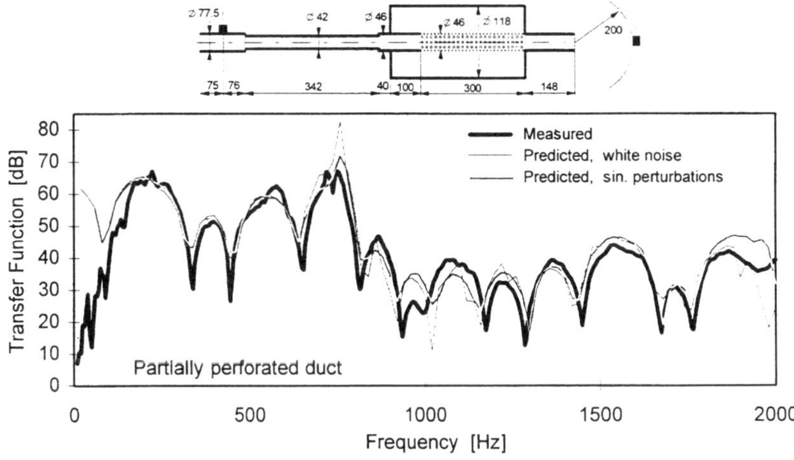

Fig 6 Comparison between the experimental and predicted transfer function of a perforate.

This results in $m+1$ groups of holes and $m+1$ junctions of several pipes ($n+2$ in each junction), which may describe comprehensively the flows in the individual holes. On the other end, the holes must be joined to the cavity liner. The correct way to do that is to adopt an analogous subdivision of the cavity liner into m axial elements and join them to the groups of holes, originating $m+1$ further junctions of $n+2$ ducts each.

Table 1 Computer run-times of the simulations by the white noise approach and the frequency-by-frequency harmonic excitation (Alpha 7600 DIGITAL computer).

Configuration	White noise excitation	Frequ. by frequ. excitation	Ratio
Expansion chamber with internal orifice	11.7 s	6.4 min	32.8
Column resonator	12.1 s	7.6 min	37.7
Partially perforated duct	15.1 min	10.4 h	41.3

The measured and calculated performance of the silencers considered refers to the situation of small (i.e. acoustic) plane pressure waves propagating along the pipes, with zero mean flow. Hence, the attenuation curves versus frequency traced in Figs 4, 5, 6 are representative only of the linear behaviour of the systems. In this situation the comparison between measured and predicted transfer functions is especially useful to test the validity of boundary conditions, acoustically equivalent schemes and corrective lengths adopted and refine the model, in that resonance phenomena (peaks and troughs in the curves) are well recognizable and evident. However, the attenuation features of the same mufflers are rather different when high amplitude pressure waves are involved, so that the behaviour of the systems is non linear. Generally, the resonance frequencies slightly change, and the attenuation peaks and troughs of the curves become less marked (11). To predict the noise abatement of the muffling duct systems when finite amplitude waves propagate, one could simply increase the value of Δp in eq. (6) and resort to a white noise signal of high amplitude. The difference between the upstream and downstream sound pressure level spectra in dB would give an evaluation of the attenuation curve of the silencing component due to its non linear response. Anyway, the best approach is certainly to model directly the actual excitation source, which imposes a series of pulses and a significant mean flow (i.e. the engine, or a rotary valve pulse simulator) (5,6,8), in order to calculate the noise levels radiated by the open termination and the silencer performance in the real working conditions.

CONCLUSIONS

The numerical generation of a white noise pressure signal, to be used as the excitation source of the wave motion in the silencing duct system, has allowed to set up a rapid procedure to simulate the muffler attenuation curve in the frequency domain, by a 1-d gas dynamic code. In this way the performances of the non linear model developed, with regard to the calculation run-times, are comparable with the ones of a linear acoustic model. The white noise approach has proved to give satisfactory predictions of the transfer function of different silencers, ranging from expansion chambers and resonators to perforated liners.

REFERENCES

(1) MUNJAL, M.L. Acoustics of ducts and mufflers, 1987 (John Wiley & Sons, New York).
(2) DAVIES, P. O. A. L. Practical flow duct acoustics, J. Sound Vib., 1988, 124, pp. 91-115.
(3) COULON, J.M., GUERINOT, J.M., GARCIA, P., KUNZ, F. Influence of downpipe dissymmetry on the tailpipe noise of 6 and 8 cyl. engines, ATA Int. Conf. , Florence, Italy, 1994, paper n. 94A1058.
(4) BLAIR, G.P., COATES, S.W. Noise produced by unsteady exhaust efflux from an internal combustion engine, SAE Transactions, 1973, 82, paper n. 730160.
(5) COATES, S.W., BLAIR, G.P. Further studies of noise characteristics of internal combustion engines, 1974, SAE paper n. 740713.
(6) JONES, A. D, BROWN, G. L. Determination of two-stroke engine exhaust noise by the method of characteristics, J. Sound Vib. 1982, 82, pp. 305-327.
(7) ONORATI, A., WINTERBONE, D.E., PEARSON, R.J. A comparison of the Lax-Wendroff technique and the method of characteristics for engine gas dynamic calculations using fast Fourier transform spectral analysis, SAE Int. Congress, Detroit, USA, 1993, paper n. 930428.
(8) FERRARI, G., ONORATI, A. Determination of silencer performances and radiated noise spectrum by 1-d gas dynamic modelling, XXV FISITA Congress, Beijing, 1994, paper 945135.
(9) MOREL, T., MOREL, J., BLASER, D. Fluid dynamic and acoustic modeling of concentric-tube resonators/silencers, SAE Int. Congress, Detroit, USA, 1991, paper n. 910072.
(10) ONORATI, A. Fluid dynamic modelling of internal combustion engine mufflers, submitted to the Noise Control Engineering Journal, 1995.
(11) ONORATI, A. Prediction of acoustical performances of muffling pipe systems by the method of characteristics, J. Sound Vib., 1994, 171, pp. 369-395.
(12) ANDERSON, D. A., TANNEHILL, J.C., PLETCHER, R.H. Computational fluid mechanics and heat transfer , 1989, (Hemisphere Publishing).
(13) BENSON, R.S. The thermodynamics and gas dynamics of internal combustion engines, Vol. I, 1982 (Clarendon Press, Oxford).
(14) PEARSON, R.J., WINTERBONE, D.E. Calculating the effects of variations in composition on wave propagation in gases, Int. J. Mech. Sci., 1993, 35, pp. 517-537.
(15) CORBERAN, J.M. A new constant pressure model for N-branch junctions, Proc.Instn Mech.Engrs, 1992, 206, pp. 117-123.

C499/025/96

Numerical analysis of the flow in the intake manifold of a DI diesel engine

Y TAKENAKA, **H YOKOTA** JSAE, JSME, **H NAKAJIMA**, and **Y AOYAGI**
Hino Motors Limited, Tokyo, Japan

SYNOPSIS

Flow in box type intake manifold is studied using a CFD calculation. The flow pattern is investigated under steady state flow conditions. Three dimensional flow analysis based on the finite volume method is performed. The characteristics of the intake manifold flow are described related to the influence of the flow entering the intake port. The calculated results are compared with experimental results using Tracer Laser Sheet and Laser Doppler Velocimetry techniques.

1 INTRODUCTION

The in-cylinder flow has a great influence on the performance and emission levels of DI diesel engine. Therefore, a variety of research on the flow through an intake port are performed(1-5). It is considered that the in-cylinder flow is also influenced by the flow in the intake manifold. Therefore it is meaningful to investigate the flow pattern in the intake manifold. Several works for a flow in an intake manifold also has been done(6-10). The research for a box type manifold in which a flow tune is very difficult is not enough.

A flow analysis for a flow in a box type manifold is performed in this study. The box type intake manifold is often used in medium heavy duty diesel engines, because of its benefit of weight and space saving. However, basically this type of intake manifold generates a different flow pattern for each cylinder, as the relation between the intake pipe and the intake port location differs for each cylinder. Ideally, the flow pattern at the inlet of the intake port should be same. It seems very important for realizing the same flow pattern to analyze the flow in the intake manifold. The intake pipe part seems to be important for the flow in the intake manifold, so a model with both intake pipe and manifold is used in this study.

Numerical analysis is used mainly in this study to analyze the flow. Experimental results are used to confirm the calculated result and add additional information on the flow. The Laser Sheet technique and Laser Doppler Velocimetry are used to analyze the flow experimentally.

Figure 1 shows the intake system with cross section definition and the experimental apparatus schematically. The intake pipe and manifold have a wall to separate each passage into three cylinders to the front and rear. The separating wall in the intake pipe has a complex shape and is twisted to keep the same passage length for each set of three cylinders. Transparent glass is installed in the intake manifold to visualize the flow inside the manifold. The area available to observe is limited. Only the front part of the intake manifold is visualized, because it is difficult to visualize the rear part due to the intake pipe location. As the connection of the intake pipe and manifold is located in front of the third cylinder, the visualized area near the third cylinder is also limited.

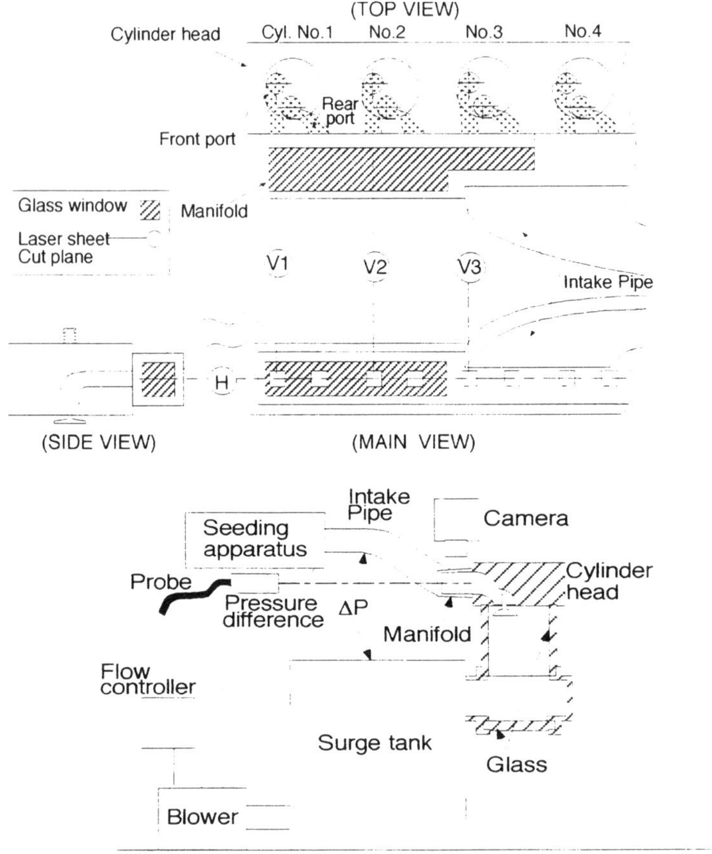

Fig 1 Sketch of intake system and experimental apparatus

Basically, two different measurements are performed. The flow patterns are obtained using the Tracer Laser Sheet(TLS). The laser used in this experiment has a power at 3 to 3.5 W. The thickness of the laser sheet is 5 mm. The pictures at each cross section in Figure 1 are taken to investigate the flow in an intake manifold.

The Laser Doppler Velocimetry (LDV) is also used to measure the flow velocity by the backward scattering method. The same laser as the TLS is used and adjusted to have a power of 1W. These two experimental methods have different characteristics. A

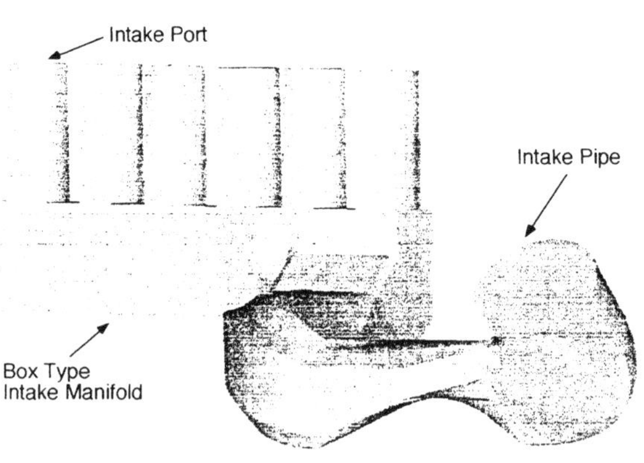

Fig 2 View of the model geometory

global flow pattern is obtained easily by TLS, while an exact flow velocity is obtained by LDV.

2 NUMERICAL ANALYSIS

The flow in the intake pipe and manifold is calculated using CFD code which is based on the Finite Volume Method. A brick type cell is used to descretize a flow region. A flow boundary is treated using a technique of VOF which does not use a deformed cell. Generally, the accuracy of a calculation decreases due to cell deformation in a usual finite volume formulation. The accuracy decrease does not occur in this type of boundary treatment. As a boundary location is automatically calculated, a mesh generation process is not required. This capability is quite important for applications to actual engine development.

Figure 2 shows the solid model for the calculation. The geometry is created using IDEAS and transferred to the CFD code VECTIS through an STL format data. The intake pipe, manifold and part of the intake port are modeled considering the influence of each part and CPU resources. The intake pipe part connected to the manifold directly seems to have a strong influence on the flow in the intake manifold. A flow pattern in the intake port is not essential for the purpose of knowing the difference in the flow entering an intake port. A calculation for the full modeling of the intake pipe, manifold and port is CPU intensive. Therefore the modeling of the intake port is limited to a straight part of the port and simplified to an extent which seems to be enough to investigate the flow pattern to the intake port.

Mesh size is different for each analysis case to optimize the mesh concentration. A typical mesh size is around 150000 cells. However this type of structure grid formulation has a disadvantage of having dead cells. As the boundary of the intake pipe is very complex, an option to adaptive mesh for boundary is used. Therefore the mesh is concentrated effectively in the live cell region.

Basically, the calculations are performed for transient conditions. Sometimes a case with flow separation does not reach a steady solution, because a vortex generation occurs. Time averaged solutions are used in this study.

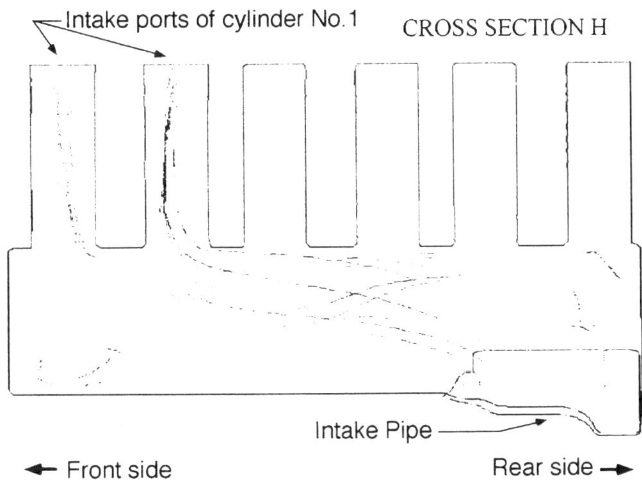

Fig 3 Particle trace of flow entering cylinder 1

3 CALCULATED RESULTS

Figure 3 shows particle trace for the calculated result in the case of flow entering the first cylinder. The flow characteristics obtained from this figure are as follows. The flow starting from the rear side at the inlet of the manifold impinges on the cylinder head wall and changes its direction. This flow finally enters into the front side intake port. The flow starting from the front side changes its direction immediately and enters into the rear intake port. The trace lines on the left side of the starting location seem to indicate that the flow separation occurs at the corner part of the manifold inlet. The trace lines on the right side near the starting location show that the flow pattern is distorted. This seems to be caused by the distorted intake pipe shape. Basically, the flow goes from the rear side to the front side but is rotating weakly and

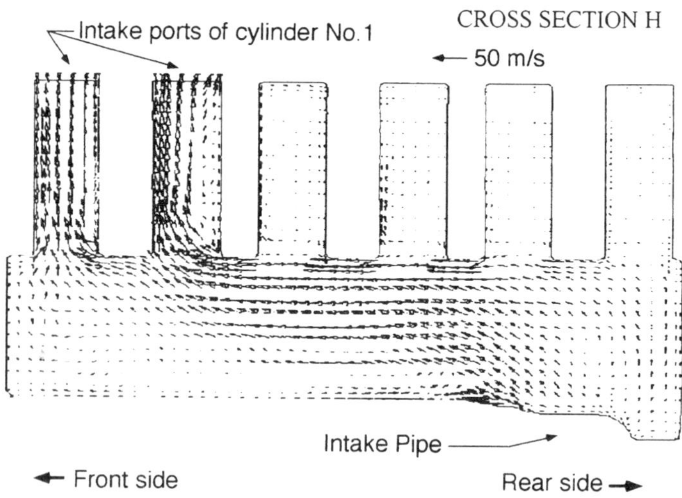

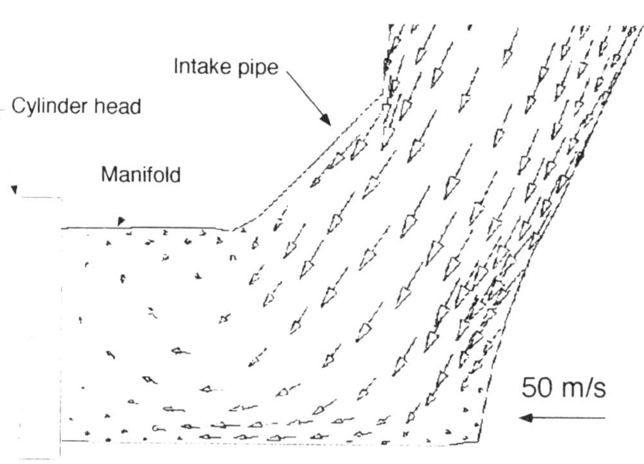

Fig 4 Velocity vectors for flow entering cylinder 1

twisting. It seems that the flow has a larger velocity component parallel to the cylinder head surface in front of the intake port of the second cylinder, as the separation occurs near the intake pipe connection.

The particle trace has an advantage to be able to obtain a global flow pattern easily but only gives information of the location where particles exist. On the other hand, a velocity distribution is more convenient to investigate the local flow pattern in detail at any location.

The velocity distribution on the cross section V3 and H are shown in Figure 4. It is clear from the cross section V3 that the flow has a rotating component on the cross section normal to the main flow direction. It is considered that the origin of flow rotation is the relation between the intake pipe and manifold. The intake pipe and manifold are connected with an installation angle. This angle seems to cause the rotating flow in the intake manifold.

This velocity distribution shows that the flow is complex at the connection of the intake pipe and manifold as shown in the particle trace figure. A separation of the flow is

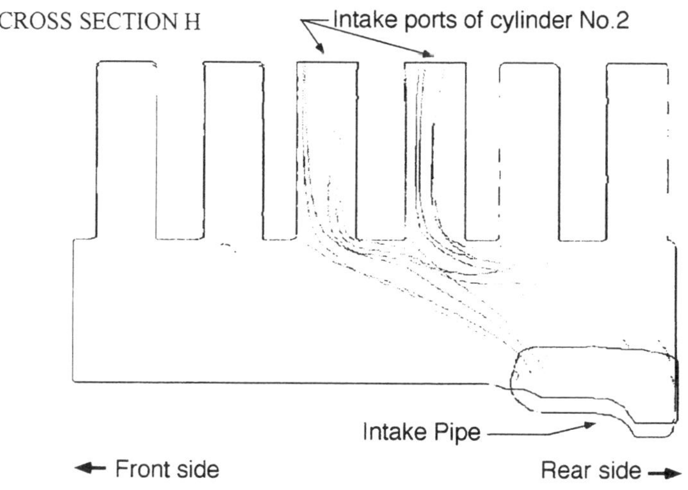

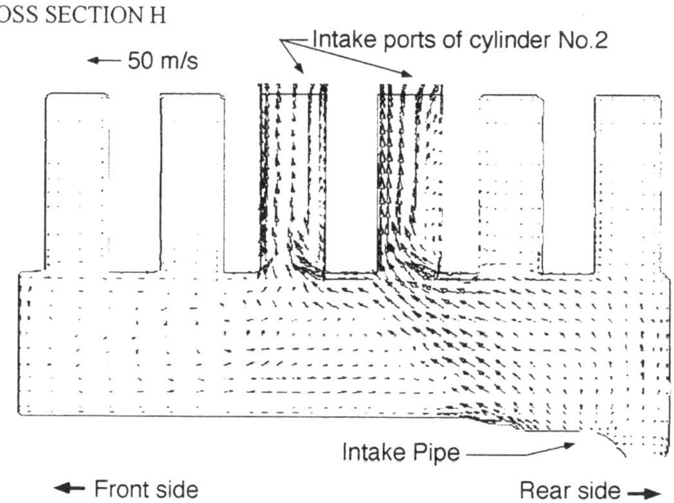

Fig 5 Particle trace and velocity vectors for flow entering cylinder 2

found near the connection. This is caused by the edge where the flow has to change its direction rapidly. A recirculation zone is formed after this flow separation. The recirculation zone reduces the effective flow area of the cross sections which intersect the recirculation zone. Therefore, the flow which passes by this area has a larger flow velocity compared to flows in other regions. The separation influences the velocity distribution of the region in front of the first cylinder. The flow velocity in this figure is averaged. However, it was observed from the flow velocity distribution at each time step that the flow is changing and includes a travelling vortex which is generated in the separation area. Unfortunately it is difficult to observe such a phenomena from the averaged flow velocity field.

The flow pattern at the inlet of the intake port has a small separation and is slightly different for each port. The separation is larger at the rear port and small at the front port. The reason for this difference is considered to be the existence of the wall. The flow velocity component parallel to the cylinder head wall is larger in front of the rear port than the front port. This velocity component is reduced because of flow direction change before the front side wall of the intake manifold.

The case of the flow entering the second cylinder was also calculated. The results of the particle trace and velocity distribution on the cross section H for the calculated flow are shown in Figure 5. The flow pattern differs from the case of the flow entering the first cylinder, as the distance between the inlet of the intake manifold and the intake port is closer. A region of high velocity exists between the inlet of the manifold and second intake ports. A separation exists at the corner of the manifold inlet. This separation reduces the effective flow area, therefore the flow velocity component parallel to the cylinder head is accelerated by the existence of this separation.

The global features of the flow seem to be similar to the case of flow entering the first cylinder. The flow entering the manifold impinges on the cylinder head wall and changes its direction rapidly. It is clear that the high velocity area is limited in the region between the intake pipe outlet and second cylinder's intake ports and the flow to the right hand side of the intake pipe outlet is disturbed. The flow velocity distribution on cross section V3 in the case of flow entering the second cylinder is almost the same as in figure 4, therefore the figure is omitted. This means that this flow also has a rotating velocity component. As the flow path is shorter than the previous case, the flow seems to rotate only half that of the previous case.

It is apparent that the flow is complex in the region near the inlet of the intake manifold. The high velocity region can be seen between the inlet part of the manifold and the intake port of second cylinder. It is found from both figures that a large recirculation zone exists in front of the first cylinder. The flow has a small velocity and is rotating slowly in this region. The flow velocity distribution in this region seems to have an insignificant influence on the flow in the intake port.

4 COMPARISON BETWEEN CALCULATION AND EXPERIMENT

The flow in the intake manifold is measured experimentally to confirm the accuracy of the flow calculation and obtain more information on the flow. The TLS and LDV techniques are used in the experiment. First of all, a flow pattern is compared between the calculated and experimental result. Figure 6 shows the results of TLS, LDV and CFD.

The global flow pattern can be obtained from the result using TLS. Strictly, the TLS is two dimensional, while the particle trace is three dimensional. As both results provide a global flow pattern, they are compared. It is found from these result that the recirculation zone near the inlet of the manifold is quite similar between the calculation and experiment. The flow directions in both cases show good agreement except in the region in front of the first cylinder. This seems to be caused by a velocity fluctuation. The results of TLS are considered to express instantaneous flow velocities, while CFD results are averaged. It is observed from the TLS result that the flow fluctuation is larger than the other region. As an other possibility, CFD might over estimate the recirculation zone slightly.

This kind of prediction for flow separation has a close relationship with the accuracy of the numerical scheme. A separation is predicted very weakly using a low order scheme or sometimes does not occur. A higher order schema, for example QUICK etc, is desirable. A

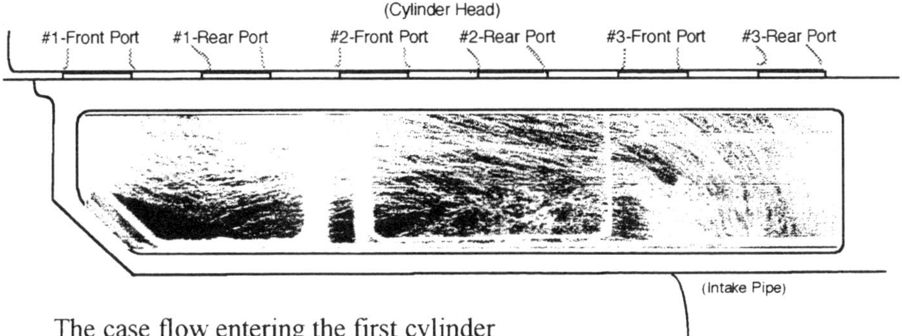

The case flow entering the first cylinder

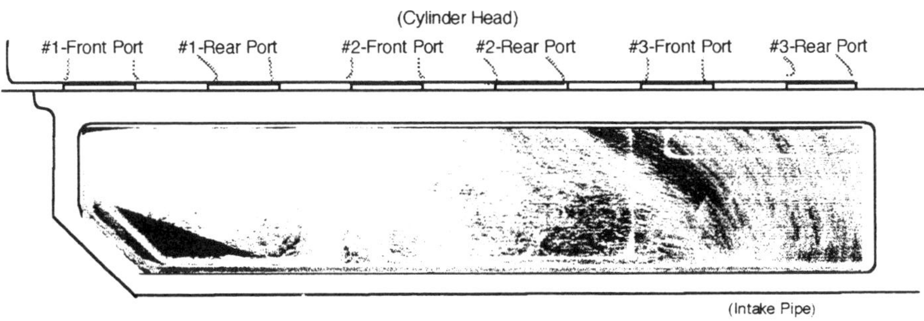

The case flow entering the second cylinder

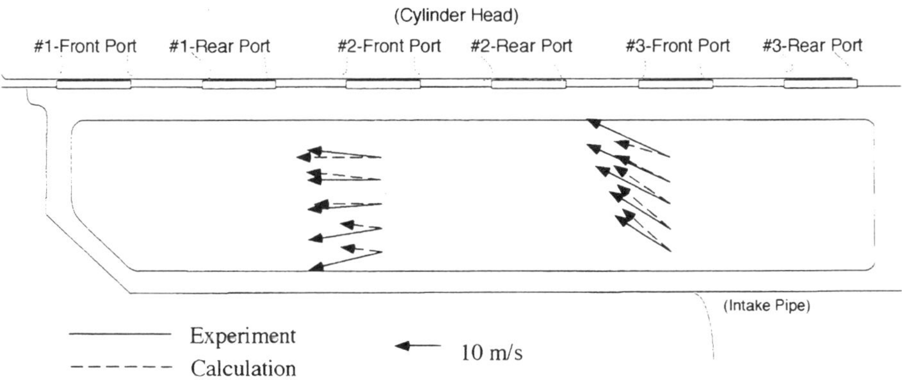

The case flow entering the first cylinder

Fig 6 Experimental results using TLS and LDV

second order scheme is used in this study. As another factor, the deformation of cells decrease the accuracy in the finite volume method. However, as the code used in this study is adopting the boundary treatment similar to VOF, the accuracy is maintained at second order. Therefore the separation size seems to be predicted well but the fluctuation by a vortex generation from the separation area makes it difficult to compare the calculation and experimental result directly.

So far the qualitative comparisons has been made for the cross section H between CFD and TLS. A quantitative comparison is performed between CFD and LDV. The bottom figure of Figure 6 shows the compared results of velocity at 10 locations which are on the cross sections V2 and V3. The velocity vectors are the projection onto the cross section H. The calculated flow velocity vectors on cross section V3 are smaller than those of LDV. For these points, the magnitude of the flow velocities agree well but the calculated flow vectors direct downward more than the calculated ones. Therefore the velocity component parallel to the cross section H has a smaller value. The intake pipe has a very complex surface which may not be expressed well in the model geometry. This seems to cause the velocity vector difference on the cross section V3.

The velocity vectors on the cross section V2 agree well except for 2 points in the lower part of the figure. These 2 points are located in the recirculation zone in the calculation. The measured higher velocity vectors means that the flow separation re-attaches to the wall. The size of the recirculation zone of the calculation seems to be over predicted. However the prediction for the flow velocity field in a dead water region is quite difficult subject.

It is concluded from the comparison of the flow fields that the CFD results agree well except at the tip of the recirculation zone. Various idea on the manifold design may be available but difficult to evaluate soon experimentally. At the beginning of the design, a quick and rough evaluation is required and can be performed by CFD effectively. Because a slight difference between the calculation and experiment is not important in such a design stage. However the effort to get more accuracy of CFD in the recirculation zone is required.

5. CONCLUSION

The flow in a typical box type intake manifold is analyzed using a CFD calculation and compared with the results obtained using the TLS and LDV techniques. The cases of the flow entering the first and second cylinder are investigated. A flow separation occurs near the intake pipe outlet and influences much of the flow pattern in the box type intake manifold. The relation between the intake manifold and pipe also has a significant influence on the flow in the intake manifold. Therefore the design of the connection part of an intake manifold and pipe is important to optimize the flow entering each intake port.

A flow analysis is effective for a design of an intake manifold. It has been proved that the CFD analysis has enough capability for the flow field prediction except for complex area like the recirculation zone. The experimental methods by the TLS and LDV techniques also are quite effective to enhance the CFD analysis. In particular, the TLS has an advantage in obtaining a global flow pattern easily but has difficulty in obtaining quantitative information. The opposite can be said of the LDV method. The combination of CFD, TLS, and LDV is powerful tool. Especially, the CFD seems to be popular in future by a progress of numerical schema and hardware.

(1) Ishida S. et al, "Development Status of a Small, Direct Injection Diesel Engine at Isuzu" SAE 8500068, 1985.

(2) Y. Takenaka, M. Yabe, Y. Aoyagi and T. Shiozaki,"Three Dimensional Computation of In-Cylinder Flow with Intake Port in DI Diesel Engine" COMODIA, 1990, p.425.

(3) S. Aita, A. Tabbal, G. Munck, N. Montmayeur, Y. Takenaka, Y. Aoyagi and S. Obana,"Numerical Simulation of Swirling Port-Valve-Cylinder Flow in Diesel Engines" SAE 910263, 1991.

(4) C.Arcoumanis and S.Tanabe,"Swirl Generation by Helical Ports" SAE 890790, 1989.

(5) D. Haworth et al,"Multidimensional Port and Cylinder Flow Calculations for Two- and Four-Valve -Per-Cylinder Engines: Influence of Intake Configuration on Flow

Structure", SAE 900257, 1990.
(6) R. Margary, E. Nino and C. Vafidis,"The Effect of Intake Duct Length on the in-cylinder Air Motion in a Motored Diesel Engine", SAE 900057,1990.
(7) C. Arnold,"Time Dependent Fluid Dynamic Simulations for More Accurate Time Averaged Results", SAE 921090,1992.
(8) M. Chapman, "Two Dimensional Numerical Simulation of Inlet Manifold in a Four Cylinder Internal Combustion Engine", SAE790244, 1979.
(9) C. Arcoumanis et al,"Flow in the Inlet Manifold of a Production Diesel Engine", Proc. Instn. Mech. Engrs., Part C, 1989, 203, p.39-49.
(10) Y. Zhao and D. E. Winterbone,"Numerical Simulation of Multi-Dimensional Flow and Pressure Dynamics in Engine Intake Manifolds", Proc. Instn. Mech. Engrs., C430/039, 1991, p.47-56.

C499/055/96

Three-dimensional simulation of the flow through a twin-intake port engine

A CHEN MSc, PhD, **Z MAHMOOD** BEng, MSc, MRAeS, and **M YIANNESKIS** BSc, MSc, DIC, PhD, CEng, FIMechE
Mechanical Engineering Department, King's College London, UK
G GANTI MS, PhD
Ford Motor Company Limited, Laindon, UK

SYNOPSIS

The three-dimensional flow structure in the intake ports and cylinder of a Ford 'Zetec' dual-intake valve engine operating under steady flow conditions has been modelled using computational fluid dynamics. A detailed 3D numerical grid incorporating all the features of the engine inlet ports, valves and cylinder head was produced using CAD surface data. Grid sensitivity tests showed that the CFD predictions were grid-independent. Laser-sheet flow visualisation and laser-Doppler anemometry were also employed to investigate the flows and to assess the predictions. The results show that the mean velocity characteristics have been well-predicted in all planes.

NOTATION

C_1, C_2, C_μ	Constants of k-ε model.
k	Turbulent kinetic energy, m^2/s^2.
P_k	Production rate of turbulence energy, $J/s.m^3$.
U, V, W	Time-mean velocities in x, y and z directions, m/s.
U_i	Time-mean velocity in i direction, m/s.
u_i	Fluctuating velocity in i direction, m/s.
x, y, z	Cartesian co-ordinates, m.
δ_{ij}	Kronecker delta.
ε	Dissipation rate of turbulence kinetic energy, m^2/s^3.
μ	Molecular dynamic viscosity, $N.s/m^2$.
μ_t	Turbulent dynamic viscosity, $N.s/m^2$.
ν	Molecular kinematic viscosity, m^2/s.
σ_ε	Turbulent Prandtl number for ε.
σ_k	Turbulent Prandtl number for k.

ABBREVIATIONS

CAD	Computer-aided design.
CFD	Computational fluid dynamics.
LDA	Laser-Doppler anemometry.
r.m.s.	Root mean square.

1. INTRODUCTION

In recent years, motor manufacturers throughout the world have placed great emphasis on understanding and controlling the intake flow processes of internal combustion engines. The desire to achieve improvements in engine fuel economy and increasingly stringent vehicle exhaust emissions regulations have together helped to devise many innovative strategies for the control of induction flows. Modern designs of internal combustion (I.C.) engines incorporating dual-intake valve cylinder heads offer the designer not only the advantage of greater volumetric efficiency, and thus more engine power output than their single intake port counterparts, but also allow greater scope for manipulation of the in-cylinder flow field for enhancing turbulence and combustion parameters. Recent developments in lean-burn four-valve engine design, such as charge stratification and variable valve timing, have served to improve engine performance whilst allowing national and international emissions regulations to be observed.

Computational fluid dynamics has become an increasingly important tool in engine design since the mid-1980s due to the availability of powerful computer hardware and developments in CFD codes. The complex shape of production engine intake ports and cylinder heads means that often great effort is required to accurately model all the features present in an actual cylinder head. The difficulty of the task of modelling is highlighted further when the computational grid is required to alter with time to account for the movement of the piston and of the valve assembly in a reciprocating engine. Although increases in the complexity of a CFD model will lead to correspondingly higher processing times, accurate representation of an engine configuration is essential if reliable results are to be achieved from a simulation.

A large number of studies involving simulation of multi-valve engine flows has been reported as part of current engine research programmes with a view to identifying strategies which offer optimum levels of turbulence near the top-dead-centre of the compression stroke. The study of Henriot et al (1) compared computational results with those acquired using LDA in a reciprocating engine. The shape of the cylinder head was modelled in detail and boundary conditions were applied at orifices representing the inlets, as the intake ports were not modelled. Flow calculations were performed using the KIVA computer code and the standard k-ε turbulence model. The study reported fairly good agreement between numerical and experimental results for configurations in which one inlet valve was kept closed, but no satisfactory agreement was acquired for the configuration where both valves were open. Naitoh et al (2) computed the 3D flow in the ports and cylinder of a multi-valve engine using a third-order upwind scheme. The results revealed complex flow phenomena in the engine cylinder under unsteady flow conditions, and good agreement between computed and experimental results was reported. Fujii et al (3) computed the in-cylinder flow in an engine, without modelling the ports. Comparisons of computational and experimental results obtained with LDA revealed relatively good agreement between axial and tangential flow velocities. A study by Haworth et al (4) compared the computed flow structure in a reciprocating two-valve-per-cylinder engine with that in a four-valve cylinder head configuration. The k-ε turbulence model was used to model the turbulent transport of momentum, and the flow was computed using the PISO (Pressure-Implicit-Split-Operator) algorithm (5) for pressure-velocity coupling. The results showed that swirl was the dominant motion in the single intake valve case whilst high levels of tumble and turbulence were acquired with the dual-intake port configuration.

This paper presents a simulation of steady incompressible flow through the intake ports and cylinder of the Ford 'Zetec' engine with both inlet valve lifts set at 10 mm. This flow simulation was carried out to assess the 3D numerical simulation before adapting the CFD code to calculate transient flows in the engine. The mean velocity predictions have been validated against experimental data obtained using laser-Doppler anemometry to measure the flow through a Perspex replica of the Zetec engine. Details of the experimental system are reported in Mahmood and Yianneskis (6, 7). As the experimental technique involved the turbulent flow of a liquid instead of air under steady flow conditions, both computed and measured velocities are lower than would be achieved with air flow at the same Reynolds number. The similarity of air and liquid flows through ports has been previously demonstrated (8). At the same time, CFD predictions performed with flow of both air and liquid at the same Reynolds number revealed that identical flow structures are obtained in the ports and cylinder in both cases. Extensive testing of the effect of liquid mass flow rate on the flow structure was also carried out and both computational and experimental results showed no difference in the flow field with flow rate variation.

2. NUMERICAL MODEL

2.1 Mean flow equations

The conservation of mass and momentum for steady, turbulent and incompressible flow in Cartesian co-ordinates can be described by the following equations:

$$\frac{\partial}{\partial x_i}(U_i) = 0$$

$$\frac{\partial}{\partial x_j}\left(\rho U_i U_j\right) = -\frac{\partial p}{\partial x_i} + \frac{\partial}{\partial x_j}\left[\mu\left(\frac{\partial U_i}{\partial x_j} + \frac{\partial U_j}{\partial x_i}\right) - \rho\overline{u_i u_j}\right]$$

2.2 Turbulence model

The standard k-ε model of turbulence was used to model the Reynolds stress terms $-\rho\overline{u_i u_j}$ (9). The transport equations for the turbulence energy and its dissipation rate are:

$$\frac{\partial}{\partial x_j}\left(\rho U_j k\right) = \frac{\partial}{\partial x_j}\left[\left(\mu + \frac{\mu_t}{\sigma_k}\right)\frac{\partial k}{\partial x_j}\right] + P_k - \rho\varepsilon$$

$$\frac{\partial}{\partial x_j}\left(\rho U_j \varepsilon\right) = \frac{\partial}{\partial x_j}\left[\left(\mu + \frac{\mu_t}{\sigma_\varepsilon}\right)\frac{\partial \varepsilon}{\partial x_j}\right] + C_1\frac{\varepsilon}{k}P_k - C_2\rho\frac{\varepsilon^2}{k}$$

where k, the turbulence energy, is given by

$$k = \frac{1}{2}\overline{u_i u_i}$$

ε is the dissipation rate of k

$$\varepsilon = \nu\overline{\frac{\partial u_i}{\partial x_j}\frac{\partial u_i}{\partial x_j}}$$

and P_k is the production of turbulence energy

$$P_k = -\rho\overline{u_i u_j}\frac{\partial U_i}{\partial x_j}$$

Like their viscous counterparts, the Reynolds stresses $-\rho\overline{u_i u_j}$ can be related to the rate of strain through the Boussinesq hypothesis:

$$-\rho\overline{u_i u_j} = \mu_t\left(\frac{\partial U_i}{\partial x_j} + \frac{\partial U_j}{\partial x_i}\right) - \frac{2}{3}\rho k \delta_{ij}$$

where μ_t, the turbulent viscosity, is acquired from

$$\mu_t = C_\mu \rho \frac{k^2}{\varepsilon}$$

and is assumed to be isotropic.

The empirical constants in the equations of the standard k-ε turbulence model are:

$C_\mu = 0.09$ $C_1 = 1.44$ $C_2 = 1.92$
$\sigma_k = 1.0$ $\sigma_\varepsilon = 1.20$

2.3 Numerical Model and Boundary Conditions

The CFD predictions of the flow were performed using the STAR-CD (10) program. A "law of the wall" formulation was applied at all solid surfaces. Computation of the flow was accomplished by employing the SIMPISO solution algorithm (10) with an upwind differencing scheme. With the inlet valve lift set at 10 mm, the numerical grid contained 68822 cells. Figure 1(a) gives a view of the three-dimensional grid for the inlet ports and cylinder together with the co-ordinate system definition, while Figure 1(b) shows the cell structure in a cross-section of the grid through the centre of one inlet valve. The concentration of the grid cells was increased in the vicinity of the inlet valves, as steep velocity gradients are encountered there amidst the intake jet flows. The cylinder bore of the Zetec engine modelled was 80.6 mm and a cylinder length of 134.3 mm was modelled.

An inlet boundary condition was specified at the entry plane of the inlet ports and the outlet boundary condition was defined at the cylinder exit plane. The mass flow rate of the working fluid was ascribed as 1.54 kg/s and its kinematic viscosity as 1.71 x 10^{-6} m^2/s to match with the properties of the liquid used in the experimental investigation. At the inlet plane, the velocity distribution was specified as a uniform mean velocity profile based on the measured mass flow rate. The outlet boundary condition was prescribed at the exit of the cylinder, and defined such that the flow at all locations at the exit would be outward directed. Estimation of the outlet conditions was carried out in two stages. Initially, the distribution of variables at the outlet plane were computed by extrapolation from upstream by assuming a zero gradient along the mesh lines intersecting the planar outlet surface. The velocities were then adjusted to produce the mass flowrate of 1.54 kg/s.

Highly-concentrated grid lines were located in regions of steep gradients, and the grid structure was based on experience gained from earlier CFD studies of generic ports. Grid sensitivity tests were performed with the model to determine the grid-independence of the solutions. The total number of cells across the flow field was increased by 26%, with the additional cells either distributed across the entire field or concentrated in regions around the valve stem. In both cases, the variations in the mean flow and turbulence results with grid size were found to be smaller than the convergence criteria specified for the predictions. Consequently, it was established that the solutions are independent of grid size.

3. RESULTS AND DISCUSSION

The main features of the flow in the vertical plane x = -17.6 mm through the centre of one inlet valve are indicated by the velocity vector plot on Figure 2. Computed flow velocities are high inside the inlet port, with magnitudes of around 1.2 m/s at entry into the port and approximately 2.5 m/s at regions closer to the valve stem. To the right hand side of the valve stem, as shown on Figure 2, there is a recirculation in the upper region of the port, where the velocities are low, at approximately 0.5 m/s. Velocity magnitudes increase as the working fluid travels towards the valve lip. The highest velocities, at around 2.75 m/s, were predicted in the valve gap, within the intake jet to the right of the inlet valve. A large recirculation region centred on the valve seat is formed on the left side of the valve. There is no separation, however, on the right of the valve.

The flow structure inside the cylinder shows the presence of two large scale vortices. A weak tumble-like vortex is created by the motion of the fluid within the cylinder. The flow towards the right hand side of Figure 2 travels downwards at velocities of approximately 1.0 m/s. Lower velocity magnitudes, between 0.18 m/s and 0.83 m/s are noted for the flow in the central region of the cylinder. Intake fluid motion behind the inlet valve, shown on the left hand side of Figure 2, produces an elongated elliptically-shaped vortex underneath the inlet valve which, together with the main tumble vortex, forms a toroidal ring vortex. Computed velocities are low throughout the left side of the ring vortex. This is related to the proximity of the valve to the cylinder wall in this region and the restriction of the flow area by the recirculation on the right of the valve.

Figure 3 shows the measured velocity vectors in the same plane (x = -17.6 mm). The experimental results show that the flow in the ports is almost uniform, with ensemble-averaged velocities of around 1.75 m/s. Detailed comparison of the CFD and LDA results are made later in this section. On the whole, however, in-cylinder flow structures in both the CFD predictions and the experiment are very similar in this plane.

Flow in the vertical plane x = 0 mm is shown on Figure 4. This is effectively the plane of symmetry of the cylinder geometry but it should be noted that no symmetry conditions were specified in the model. The intake ports generate a strong jet of fluid directed towards the right hand side of Figure 4, which travels downwards after it collides against the cylinder wall. Velocity magnitudes are higher for flow close to the cylinder head, with values of up to 1.76 m/s. Fluid is also directed towards the left of the engine cylinder in Figure 4, and velocities are lower there, at between 0.62 m/s and 0.9 m/s. In the lower part of the cylinder, velocities are very low in the central region of the plane. No large scale flow structure can be identified in this plane, and the majority of the fluid is travelling downwards.

Vector plots of predicted flow motion in horizontal x-y planes are shown on Figures 5(a) and 5(b). The velocities in the z = -10 mm plane are plotted on Figure 5(a) and indicate that flow inside the cylinder is symmetric about the vertical x = 0 mm plane. Swirling motions in opposite directions can be identified along the cylinder wall, and velocity magnitudes are around 0.75 m/s in these regions. In the x = 0 mm plane the flow towards the right hand side of Figure 5(a) is strong due to the combined effect of the intake jets from the two valves in this region. The highest velocities for flow in this plane are approximately 1.39 m/s. The flow in this horizontal plane comprises two counter-rotating swirling vortices of equal strength which can be seen at both the top and bottom of Figure 5(a), in the regions underneath the inlet valves. Velocity magnitudes in these vortices are low, between 0.2 and 0.4 m/s. In general, the flow in this plane is directed away from the inlet valves, and towards the right of the cylinder.

Figure 5(b) presents the velocity vectors in the horizontal plane z = -30 mm. Fluid motion is more pronounced on the right side of the plane, i.e. on the right of the inlet valves. In the y = 0 mm plane, the flow is directed toward the right of Figure 5(b), and velocities of up to 0.5 m/s were computed in this region. The predictions showed that the axial component is considerably larger than the other two at these locations, indicating strong downward flow. Velocity magnitudes in the left half of the figure (i.e. below the intake valves) are much smaller, at around 0.1 m/s to 0.2 m/s. The two counter-rotating vortices at the top and bottom of this figure are more well-defined than at the z = -10 mm plane, although velocity magnitudes are again similar, approximately 0.35 m/s.

Comparisons of the individual flow velocity components have shown very good matching of both numerical and experimental results. In Figure 6(a) the measured and predicted profiles of axial W mean velocity in the x = -17.6 mm plane are compared. The velocity profiles inside the port reveal only small differences between the LDA and CFD W-velocity magnitudes. The computational results exhibit steeper velocity gradients in the z = 25 mm and z = 35 mm profiles than the corresponding experimental data. Inside the cylinder, both the calculated and measured W component velocities shown in Figure 6(a) are very similar, except at some locations close to the cylinder wall. The shape of the velocity profiles indicates that flow features in this plane are well predicted, and the LDA and CFD velocity magnitudes at most locations in the central region of the cylinder are very similar.

Calculated and measured V-component velocities in the horizontal z = -10 mm plane are compared in Figure 6(b). Again, both experimental and CFD results are very alike at most locations in the plane. The LDA results revealed a flow asymmetry towards the front end of the cylinder, at the bottom of Figure 6(b), which has not been simulated by the CFD program. This asymmetry is thought to be caused by uneven splitting of the flow between the two intake ports in the experimental configuration, so that both intake ports may not have equal mass flowrates of fluid even though the valve lifts are identical. CFD predictions with 45% of the mass flowing through one port and 55% through the other resulted in a similar profile shape at x = 35 mm. As the test section and port were geometrically symmetrical about the y = 0 mm plane, this uneven flow split is not expected to stem from asymmetries in the test section and deserves further investigation.

An assessment of the turbulence kinetic energy predictions was made by comparing the measured r.m.s. velocities with those calculated from the k predictions using $u' = v' = w' = (2k/3)^{0.5}$. Measured and calculated profiles of w' are shown in Figure 6 (c). It can be seen that the w' levels are in general under-predicted, especially in regions of high shear at the edges of the jet entering the cylinder. This underprediction may be expected to stem partly from the implicit assumption of isotropic turbulence in the k-ε model; as the measurements of the r.m.s. velocities showed, this assumption is not appropriate for all regions of the flow. In addition, the measured r.m.s. velocities in the vicinity of the intake jets have been reported to be affected by the 'flapping' of the jet issuing from the valve, an effect which is not modelled by the CFD code (11). A more appropriate comparison may be made if time-resolved velocity recordings are obtained to identify and remove non-random contributions to the r.m.s. values through FFT and inverse FFT techniques and such work is already in progress.

The present investigation has shown that qualitatively all main features of the steady flows have been accurately predicted and quantitative agreement between experiment and calculation is very good in most parts of the flow. The work has provided useful information of the mean flow and turbulence structure in the ports and cylinder and allowed the spatial resolution requirements of the CFD method to be determined without the additional complications associated with transient calculations. Measurements and predictions of the flow field under motored conditions are in progress in order to determine the temporal variation of the flow during the engine cycle and will be reported shortly.

4 CONCLUSIONS

The flow motion in the ports and cylinder of the Ford Zetec dual-intake valve engine has been simulated under steady flow conditions, using a CFD code incorporating the k-ε turbulence model. The results have been compared with experimental data obtained by laser-Doppler anemometry.

The design of the inlet ports and cylinder head generates a highly three-dimensional flow structure inside the engine cylinder. At steady flow conditions, a tumble-like vortex is created in vertical planes.

Comparison of the computational and experimental results has shown good quantitative agreement for both radial and axial mean velocity components at most locations in the ports and cylinder. Differences in velocity magnitude are seen at locations close to the cylinder wall. The study has shown that the CFD model can provide very accurate predictions of the steady flow structure in the ports and cylinder, and is currently being adapted to evaluate transient flows inside a reciprocating engine.

5 ACKNOWLEDGEMENTS

The authors gratefully acknowledge financial support provided by the Engineering and Physical Sciences Research Council under grant GR/J65693 and Ford Motor Company Limited for this work.

REFERENCES

(1) HENRIOT, S., LE COZ, J.F. and PINCHON, P. Three-dimensional modelling of flow and turbulence in a four-valve spark-ignition engine - comparison with LDV measurements. SAE Paper 890843, 1989.

(2) NAITOH, K., FUJII, H., URUSHIHARA, T., TAKAGI, Y. and KUWAHARA, K. Numerical simulation of the detailed flow in engine inlet ports and cylinders. SAE Paper 900256, 1990.

(3) FUJII, H., TAKAGI, Y. and URUSHIHARA, T. A study of in-cylinder gas flow in 4-valve engine based on 3-dimensional numerical simulations and LDV measurements. JSAE Review, Vol. 10, No. 4, October 1989.

(4) HAWORTH, D.C., EL TAHRY, S.H., HUEBLER, M.S. and CHANG, S. Multidimensional port and in-cylinder flow calculations for two- and four-valve-per-cylinder engines: influence of intake configuration on flow structure. SAE Paper 900257, 1990.

(5) ISSA, R.I. Solution of the implicitly discretized fluid flow equations by splitting of operators. J. Comp. Phys. Vol. 62., pp. 40-65, 1986.

(6) MAHMOOD, Z. and YIANNESKIS, M. Laser-sheet visualisation of the flow processes in a Ford 'Zeta' engine cylinder under steady flow conditions. Internal Report No. EM/94/07, Mech. Eng. Dept., King's College London, 1994.

(7) MAHMOOD, Z. and YIANNESKIS, M. Velocity and turbulence characteristics of steady flow processes in the ports and cylinder of a Ford 'Zetec' engine. Internal Report No. EM/95/02, Mech. Eng. Dept., King's College London, 1995.

(8) CHEUNG, R.S.W., NADARAJAH, S., TINDAL, M.J. and YIANNESKIS, M. An experimental study of velocity and Reynolds stress distributions in a production engine inlet port under steady flow conditions. SAE Paper 900058, 1990.

(9) LAUNDER, B.E. and SPALDING, D.B. The numerical computation of turbulent flows. Computer Methods in Applied Mechanics and Engineering, 1974, North-Holland Publishing Company.

(10) COMPUTATIONAL DYNAMICS LIMITED, STAR-CD Version 2.112, 1991.

(11) CHEN, A., LEE, K.C., YIANNESKIS, M. and GANTI, G. Velocity characteristics of steady flow through a straight generic inlet port. International Journal for Numerical Methods in Fluids, 1995 (In Press).

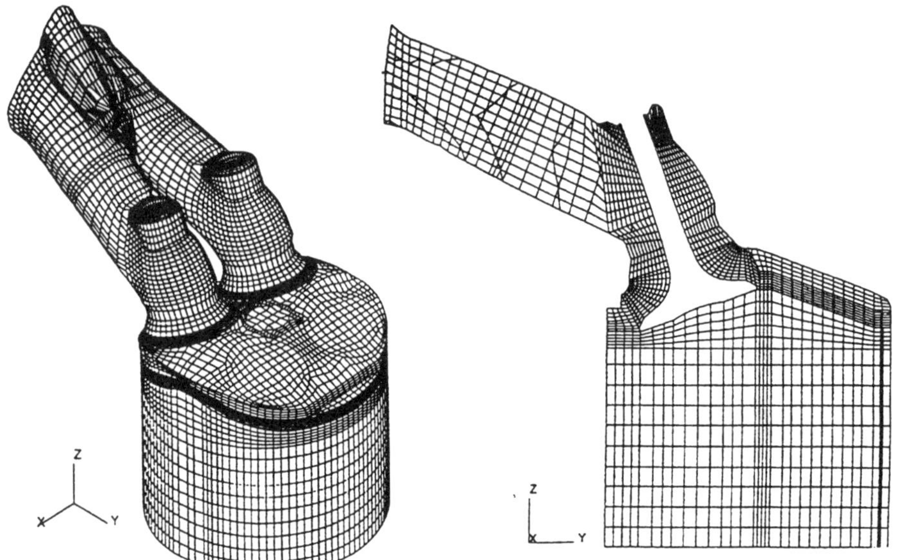

Figure 1(a) Three dimensional grid for the inlet ports and cylinder of the Ford Zetec engine

Figure 1(b) Cell structure within the vertical plane x = -17.6 mm at a valve lift of 10 mm

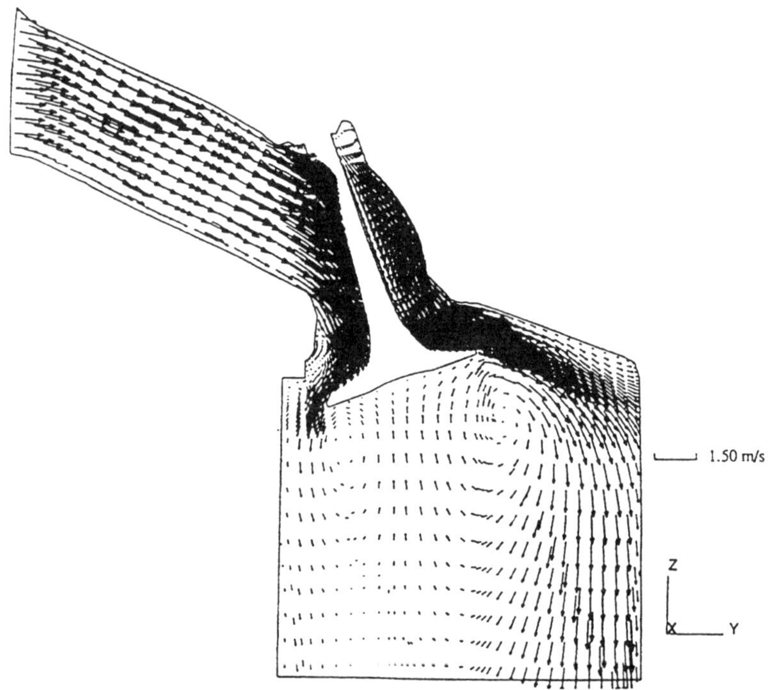

Figure 2 Computed mean velocity vectors in vertical plane x = -17.6 mm

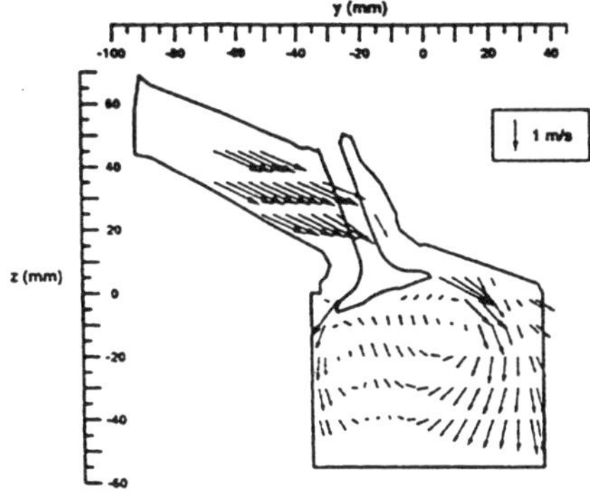

Figure 3 Mean velocity vectors in vertical plane x = -17.6 mm with both valve lifts set at 10 mm

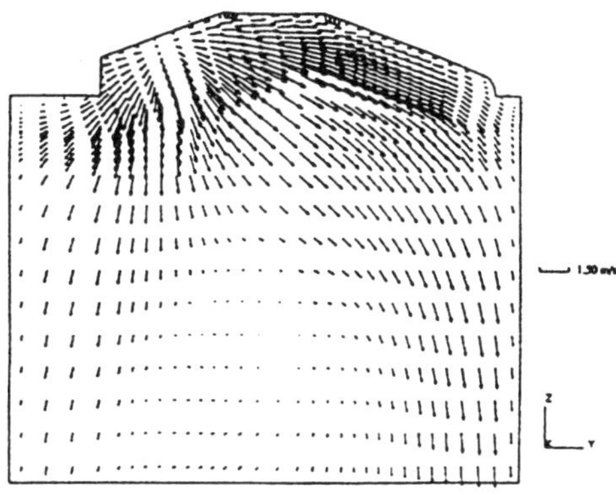

Figure 4 Computed mean velocity vectors in vertical plane x = 0 mm

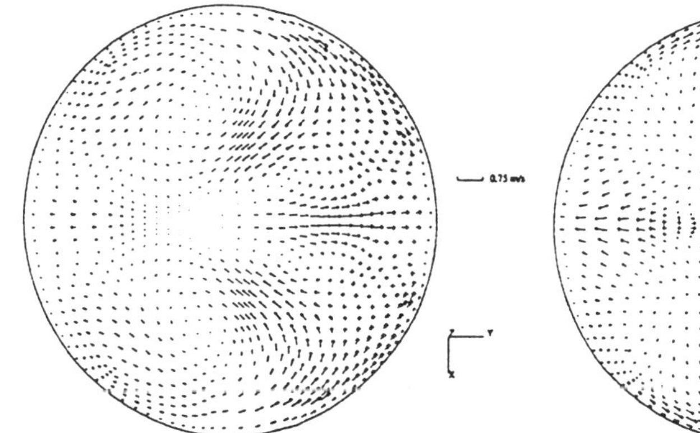

Figure 5(a) Mean velocity vectors in horizontal plane z = -10 mm

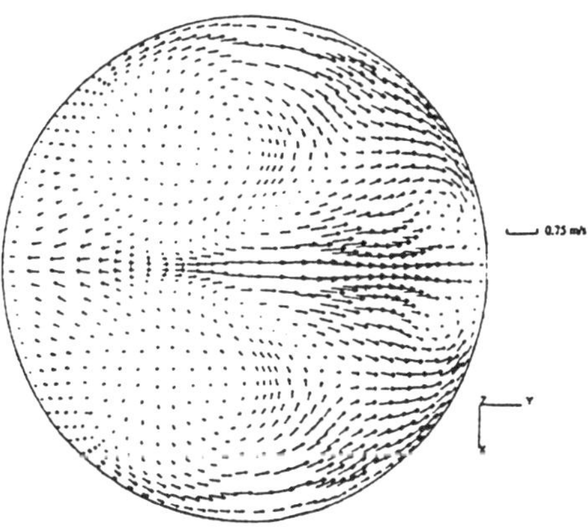

Figure 5(b) Mean velocity vectors in horizontal plane z = -30 mm

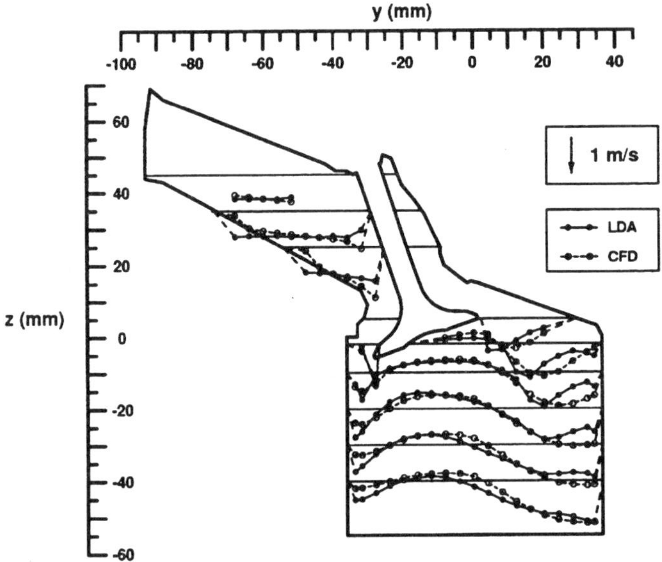

Figure 6(a) Comparison of measured and calculated W component velocities in vertical plane x = -17.6 mm at 10 mm valve lift

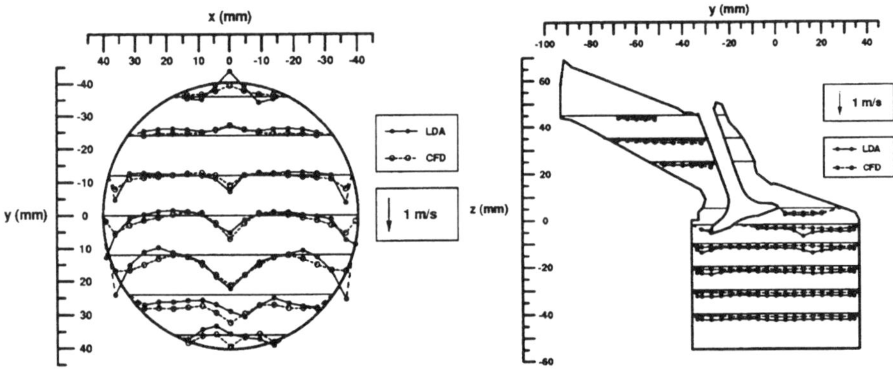

Figure 6(b) Comparison of measured and calculated V velocities in plane z = -10 mm with both valve lifts set at 10 mm

Figure 6(c) Comparison of measured and calculated r.m.s. w' velocities in vertical plane x = -17.6 mm at 10 mm valve lift

C499/057/96

CFD optimization of powertrain components

H P BENSLER and **R OPPERMANN**
Volkswagen AG, Wolfsburg, Germany

Abstract

Due to increasing market demands for better engine performance and the need for engine development time and cost reduction, engine powertrain components need to be optimized before costly investments are made for prototype tooling, hardware acquisition, and testing. In the VOLKSWAGEN AG Powertrain Development department, powertrain components such as intake manifolds, intake ports, and exhaust manifolds are optimized using CFD analysis based on 3D CAD models. A combination of 1D and 3D CFD analysis is used for this optimization process. As it is well known, one of the major hindrances for 3D simulations is the time and effort required for the mesh generation of complex geometries. This problem has been recently solved due to a major advancement based on automatic mesh generation (AMG). To illustrate the CFD optimization utilized, the process relating to intake manifolds, coolant flows, and intake ports will be discussed. Descriptions pertaining to the 1D and 3D CFD analysis as well as details concerning the AMG will be presented. This paper shows that through the use of AMG, CFD optimization of powertrain components has finally reached the point of being fast and cost effective.

1. Introduction

Traditionally, powertrain component design and optimization has been based on experimental methods. This method is not only time intensive but requires costly investments for prototype tooling and manufacturing. In order to accelerate the design process (i.e. reduce the number of design iterations) and to reduce costs, computer methods were developed. In the field of computational fluid dynamics (CFD), computer codes were written which could model everything from a complete 1D engine simulation to a full 3D analysis of complex individual components.

In the VOLKSWAGEN AG (VW) Powertrain Development Department, the two CFD codes utilized for engine and powertrain component optimization are PROMO for 1D engine simulations and VECTIS for 3D cold-flow analysis. These codes are used as stand alone programs or in combination with one another. PROMO, due to its one dimensional nature, requires relatively little setup and CPU time to perform a simulation. Contrary to other conventional 3D codes, VECTIS has drastically reduced the time, manpower and cost of 3D CFD analysis through a major advancement in automatic mesh generation (AMG). Powertrain components such as intake manifolds, intake ports, in-cylinder flows, exhaust manifolds, coolant flows, and catalytic converters are modelled with speed and ease all in house.

It is well known that one of the major problems of performing 3D CFD analysis is the mesh generation. The long and painstaking process of manual mesh generation often made CFD results obsolete due to late calculation results. Engine design time scales have become so compressed, that manual semi-automatic meshing does not "fit" into the design process. This paper will show how the problem of mesh generation for 3D analysis was overcome through AMG. In the following sections brief descriptions of PROMO and VECTIS will be given along with descriptions of how the programs are used as individual packages and in combination with each other. For the 3D analysis, the results for stationary and instationary calculations will be presented.

2. CFD Tools

2.1 1D Analysis: PROMO

PROMO is a powerfull software-system to calculate the working process of S.I. and Diesel engines including turbocharging. The strong point of PROMO, which was developed by the Bochum Ruhr University on behalf of the German Research Organization "Forschungsvereinigung Verbrennungskraftmaschinen (FVV)" is the calculation of charge-exchange. PROMO calculates pressure, temperature and mass-flow rate as a function of time and position in the engine-system. The main results are volumetric efficiency and residual gases over a range of engine speeds. Fig. 1a shows an example of these results in addition to the predicted torque.

PROMO is based on the equations of gas dynamics. The method used to solve the equations of instationary gas dynamics for flow through pipes is a two-step finite difference method primarily developed from Lax and Wendroff [1]. To calculate the flow through boundaries (cylinder, volumes, junctions, orifices, abrupt changes of area, turbocharger, etc.), PROMO uses empirical flow coefficients which can be either measured on a test stand or calculated using a 3D CFD code like VECTIS.

To increase the accuracy, speed, safety and handling of PROMO, VW optimized the program by adding a graphical pre- and postprocessor. An example of a typical PROMO model can be seen in Fig. 1b. In addition to the basic PROMO features, the current version can calculate EGR, two-stroke engines (various scavenging models) and changes of pipe cross-sectional area correctly. At VW, a PROMO job runs parallel on a cluster of SGI workstations; this ensures a very short run time necessary to calculate an entire job (typically a half hour for 20 operation points at full load).

PROMO is a very useful in optimizing the operational process of an engine in time and quality. The main optimization parts are:

- intake and exhaust manifold layout
- camshaft (fixed and variable valve profiles and timing)
- turbocharger
- intake/ exhaust port and valve geometry

Another feature of PROMO is the interface to the 3-D CFD code VECTIS. At any position of the engine model, PROMO can calculate the mass flow rate, pressure and temperature and write it into a file for use as time-dependent boundary-conditions in VECTIS (Fig. 1c).

Regarding the accuracy of PROMO, Fig. 1d shows a comparison of the measured and calculated intake manifold pipe pressure for a 4 cylinder gasoline engine as a function of crank angle at engine speeds of 4250 and 6000 rpm. As can be seen, the calculated pressure correlates well with the measurements; both the pressure peaks and the pressure fluctuations coinside within more than acceptable limits. It should be noted that the calculations and measurements were conducted independently of one another.

2.2 3D Analysis: VECTIS

VECTIS is a 3D CFD code developed by Ricardo Consulting Engineers. For the last 3 years, Ricardo and VW have been working together as members of a European funded EUREKA project called CARP (Computer Aided Rapid Prototyping) [2]. The goal of the CARP project was to develop software tools and a design environment which would enhance and accelerate prototype development. One of the aspects of the CARP project was the improvement of 3D CFD analysis through the application of AMG for the fast analysis of complex component geometries.

VECTIS, a finite-volume based general-purpose code for solving fluid flow problems, gives particular attention to engine and automotive applications. Its primary features are:

- interactive pre- and postprocessor
- the importing of 3D geometries from CAD systems or finite element preprocessors
- AMG for arbitrary complex geometries using an orthogonal structured cartesian mesh with wall cell volume and face area adaptation
- local mesh refinement for the resolution of small scale flows near boundaries
- CFD solver for the 3D time-dependent, compressible or incompressible solution of the Navier-Stokes, continuity, and energy equations
- second-order differencing and Stone and multigrid solvers for accurate and efficient numerical treatment
- k-ε turbulence model
- Magnussen combustion model
- discrete-droplet spray model for the analysis of fuel injection phenomenon
- one-dimensional flow for specified parts of the 3D flow domain (used for catalytic converter simulations)

In order to better understand the AMG utilized, a brief description of the mesh generation process will now be discussed. The basic methodology used by VECTIS for the generation of structured cartesian orthogonal (hexahedral) meshes is the following:

1) A set of mesh lines are defined around the surface model being considered (Fig. 2a). The resulting cells around and in the surface model are refered to as global cells.

2) The mesh generator then determines which global cells have any volume inside the model and eliminates all the cells which lie completely outside of the surface model (Fig. 2b).

3) In order to produce an accurate fit to the surface model, the mesh generator then subdivides (refines) the cells on or near the boundary (Fig. 2c).

4) Triangular surface patches are then superimposed onto the boundary cells to conform to the original shape of the model. These patches modify the cell face areas and volumes and provide the required surface area for heat transfer, friction, etc.. (Fig. 2d). In this manner, an accurate description of the wall boundaries is ensured.

A noted advantage of this type of hexahedral cell automatic meshing compared to mesh generators based on tetrahedral cells, is that the critical wall areas are represented by a finer mesh, while less critical locations "inside" the flow can be more coarsely meshed. Thus a considerably smaller number of computational cells are needed for the flow domain. For more information concerning automatic meshing see [3].

The successful use of VECTIS depends primarily upon the existence of complete unambiguious surface data of the geometry to be analyzed. A completely closed trimmed surface model of the geometry is required. Once the task of attaining such surface data is accomplished, the usage of VECTIS is quite simple and fast. The basic procedure for using VECTIS is the following:

1) read in the surface model
2) graphically define the inlets, outlets, and walls on the surface model
3) define a global mesh around the geomerty including the areas where local mesh refinement is desired
4) start the automatic mesh generator and let VECTIS generate the mesh
5) utilize the CFD solver of the code either on a workstation or on a Cray supercomputer to obtain the fluid flow results
6) visualize the results using the postprocessor

3. CFD Optimization Process

3.1 The CARP Development Environment

The CFD optimization process is dependent on the integration of numerous Computer Aided Engineering (CAE) and Computer Aided Manufacturing (CAM) software and hardware tools. This can be visualized through the schematic of the CARP design envirnoment [4] found in Figure 3. As can be seen, the fundamental building block for the optimization of a specific powertrain component is a 3D CAD model. The vital geometric data supplied by the CAD model provides the needed input for all the different aspects of the optimization process. The CAD model is coupled to software codes performing 1D, 3D CFD,and FE analysis which allows the optimization of the model. The hardware for manufactuing the resulting optimized prototype is either a classical CNC machine or a Fast-Free-Form-Fabrication (FFFF) machine. It should be noted that one of the most recent significant technological developments in the field of prototype manufacturing has been the emergence of FFFF machines. These systems enable prototype models to be generated layer-by-layer using such technologies as stereolithography, solid ground curing, laminated object manufacturing (LOM) or selective laser sintering [5].

In the VW Powertrain development division, powertrain components are now being modelled using the parametric 3D solid modeller, PRO-Engineer (Pro-E) from Parametrics Technology Corporation. Once the design engineer has modelled a 3D geometry with Pro-E, the component is ready to be read either into VECTIS for a 3D CFD analysis or directly into a FFFF machine for the rapid manufacturing of the prototype. As an example of how the CARP design process works, let us consider the recent design of a new intake manifold. The intake manifold was modelled in Pro-E (Fig. 4a); then the surface geometry was read into VECTIS and a stationary 3D CFD analysis was performed; all four pipes were individually analyzed. Thereafter, a LOM model was made for fit/form studies (Fig. 4b).

In order to illustrate how powertrain components are optimized using the CARP development environment, details concerning 1D and 3D CFD analysis will now be presented.

3.2 Instationary 1D Analysis

In practice, before a CAD model is created, a 1D CFD analysis with PROMO is performed. A 1D calcualtion has the advantage of being extremely fast compared to a 3D calculation; results can be obtained within less than an hour. The basic concept geometric information of the intake manifold is used by a PROMO user to build the 1D PROMO model (Fig. 1b). Before a calculation can be performed, assumptions concerning the flow coefficients (α) (effective area loss due to friction effects) for the various orifices, pipe bends, etc.. must be inputted into PROMO. In the past, these α values have been either estimated from previous experience or been determined from measurements. For various T-junctions, a library of measued α coefficients is available in PROMO. One cannot however measure the α coefficients for all the various geometric combinations possible in PROMO. This is especially true for intake manifolds. An alternative to this problem is to calculate the α coefficients using a CFD code like VECTIS. The next and better step would then be to simultaneously perform a PROMO and VECTIS calculation having both of the codes exchanging information after each time step.

After the α coefficients have been defined, a 1D calculation is then performed. The case of an intake manifold optimization, PROMO results tell the design engineer how long and wide to make the individual pipes. This information is then used when constructing the CAD model of the manifold. The PROMO model also provides the instationary boundary conditions needed to an instationary 3D CFD calculation with VECTIS. More details concerning this process will be presented in Section 3.4.

3.3 Stationary 3D Analysis

3.3.1 Coolant Flows

There are very few situations in an operating engine where one has a stationary flow. The various working fluids are continuously flowing through nozzles, ducts, combustion chambers, etc, with ever fluctuating velocities. One exception is the coolant liquid flowing through the engine. Here the coolant liquid flows at a more or less constant mass flow rate (at a constant engine speed) through the engine block around the cylinders and then through the cylinder head cooling the combustion chamber, etc.. The flow between the block and the head is regulated by a gasket perforated with different dimensioned orifices to help regulate the proper cooling of the engine. For S.I. engines, it is important that local hot spots on the combustion chamber and cylinder walls be avoided in order to prevent the engine from knocking and from having other thermal problems.

Taking the above into account, the coolant flow through an engine is perfectly suited for a stationary 3D CFD analysis. There are two major hindering blocks to performing a coolant flow calculation: 1) obtaining a CAD model or some other surface model of the complete water jacket, and 2) the extreme effort required for the manual meshing of the inside of the water jacket. Using conventional CFD codes, the mesh generation could take up to several months of tedious labor. Due to these difficulties, coolant flow calculations were in the past, given by default to CFD Service Bureaus where the coolant jacket surface model was first created from 2D drawings and then the CFD mesh was generated by hand. Needless to say, this entire process was very expensive and time intensive.

The use of AMG does not solve the problem of obtaining a surface model of the coolant jacket; AMG does however, reduce the time involved for the mesh generation from about a month to a single day. And with regards to the costs, a man month of hand meshing is replaced by the costs to run a computer for less than a day. Given one has a 3D model of the coolant jacket and by using the procedure for VECTIS presented in Section 2.2, the CFD user using a workstation can have the results of a coolant flow calculation within a week . Fig. 5a shows the geometry of an engine water jacket which had a mesh size of approximately 450000 internal cells; an example of the flow around the cylinders in the block region can be seen in Fig. 5b.

3.3.2 Intake Manifolds

Another example of how stationary flow 3D CDF analysis can help the design engineer is in the simulation of intake manifolds. When intake manifolds are being designed, it is desired that each of the individual manifold pipes provide a minimal of friction losses and an equal performance (i.e., the same pressure drop at a given mass flow rate). This insures that the manifold provides the maximum and equal amount of air to each of the engine cylinders. These results are obtained by performing a stationary flow calculation through each of the manifold pipes; for the simulation, the inlet manifold and just one of the pipe outlets are left open. This is repeated for all of the pipes; for a 4 cylinder engine, 4 stationary calculations are performed. The resulting pressure drops (or alternatively the α coefficients) across the inlet and the 4 pipe outlets are graphically summarized in a bar chart. Changes are then made to the intake manifold model until a satisfactory uniform pressure drop is achieved for all of the manifold pipes. The VECTIS mesh generation time for a typical intake manifold of approximately 200000 internal cells is about 5 hours on a workstation ; the steady state solution is obtained in 2 to 3 days on a single processor workstation. The same mesh can be used for all of the different stationary pipe simulations.

Stationary CFD modelling can only provide limited flow structure infomation in certain areas of the manifold (primarily at the inlet). It cannot show how the air is really pulsating through the manifold during an engine operating cycle. In order to overcome this short coming, an instationary 3D CFD analysis can be performed which then provides detailed infomation of the flow structure in the intake manifold over an entire engine cycle. Moreover, the same mesh used for the stationary calculation can also be used for an instationary calculation. The advantages and more details concerning instationary intake manifold calculations will be described in Section 3.4.

3.3.3 Intake Ports

The optimization of intake ports which feed the combustion chamber with air can also be performed using stationary 3D CFD analysis. A stationary CFD calcualtion can provide the following information to the design engineer: loss coefficients and swirl numbers for a given port geometry at a constant valve gap and mass flowrate, and a detailed look at the flow in the port and the combustion chamber. The advantage of using VECTIS for port simulations is that it simply meshes all types of port geometries at any valve gap without problems. Everything from simple 2 valve ports to complex 5 valve ports are meshed with speed and ease. One no longer needs to think up mesh topology strategies that are required for conventional hand meshing CFD codes. Figure 6 shows a VECTIS mesh for a 5 valve intake port simulation. This particular mesh had a total of 445000 internal cells and took 12 hours to mesh on a workstation. Local mesh refinement provided the necessary cell density required to represent the high speed flow found in the critical area of the valve curtain. The calculated loss coefficients for this 5 valve engine were found to be within 6 % of loss coefficients measured on a test rig using the same boundary conditions. These results are quite acceptable when taking into account the experimental uncertainties that are present in all measurements [6].

As for the intake manifold, stationary 3D CFD simulations of intake ports however do not really show the design engineer what is occurring in the combustion chamber for an operating engine. Moreover, the only similarity between the flow results of stationary and instationary moving valve/piston (MVP) simulations are found at large valve gaps when the piston is at BDC and the valve has its maximum lift. It is only then that the two types of calculations can provide similar results. Thus a MVP simulation is required for a clear picture of what is happening to the air flow in the combustion chamber. In addition, if one wants to model fuel spray injection either in the port or directly in the combustion chamber, one absolutely needs to perform a MVP calculation. This type of calculation is easily done using VECTIS; however, details concerning spray and MVP calculations will not be presented in this paper.

3.4 Coupling of 1D and 3D CFD

As previously stated, the real advantage of 3D CFD compared to test rig results is that it can provide detailed knowledge of exactly what is taking place with the flow inside the powertrain component. Whether it be an intake manifold or a combustion chamber, the design engineer is now given the chance to understand the actual physics of what is occurring. Given such detailed flow information, geometry changes can be done to the CAD model, and a new CFD calculation can be performed. The calculation results are then interpreted and the optimization process is repeated until the desired results are obtained.

To illustrate this process, let us take an instationary intake manifold simulation as an example. As was seen, stationary 3D CFD can help to develop a CAD model so that it is finely tuned with regards to a uniform pressure drop across the intake manifold pipes. This is surely one step in the right direction; this however, still only applies to the case of steady flows which never really occur in an operating engine intake manifold. Due to piston motion, pressure wave propagation, and other dynamic effects, fluctuating turbulent mixing of the flow in the intake manifold results. The way to visualize and better understand this fluctuating flow, is to perform an instationary 3D CFD calculation using instationary boundary conditions arising from a 1D CFD simulation. The procedure is the following:

1) PROMO performs a 1D CFD calculation and writes out the instationary boundary conditions in a data file. These boundary conditions include the mass flow rate, pressure and temperature as a function of the crank angle.

2) VECTIS performs a 3D CFD calculation using these PROMO boundary conditions for each time step.

3) The 3D CFD calcualtion is done over at least two complete 720 degress cycles to insure cyclic independence. The post processing results are written out only for the last 720 ° cycle calculated.

4) An on-screen video is made from the post processing file and is visualized together with the design engineer.

As an example of what can be seen in such a video, consider Fig. 7. At a crank angle of 130 ° after TDC, a large recirculation builds up behind the inlet of the intake manifold; this recirculation effectively cuts off the flow of the fresh air supply to the collector. Design changes were then incorporated into the 3D CAD model to alleviate this problem. Typical calculation times for such an instationary calculation is about 2 weeks on a single processor workstation. The first results can already be visualized after the first week (i.e. after the first 720 ° cycle has been calculated). This is an established procedure at VW and provides design engineeers with vital information needed to intelligently optimize intake manifolds.

4. Conclusions

It has been shown in this paper that through the use of a CAD based design environment coupled to 1D CFD and 3D CFD analysis based on automatic mesh generation, tremendous time and cost savings have been achieved in the optimization of complex powertrain components. This has been shown for the case of stationary coolant, intake manifold, and intake port flows as well as for instationary intake manifold simulations. With intelligent changes to powertrain component CAD models resulting from 1D and 3D CFD simulations, reduction in the amount of testing is achieved along with quicker more cost effective optimization. With such tools available, CFD optimization of powertrain components has finally become fast and cost effective.

REFERENCES

[1] Lax, P.D., & Wendroff, B., "Difference Schemes with High Order of Accuracy for Solving Hyperbolic Equations", Comm. Pure Appl. Math., Vol. 17, pp. 31, 1964.

[2] Raynor, M.J., "CARP, The Integration of CAD, CAE Tools and Fast Free From Fabrication", Proceedings of Autotech 93, No. C462/215, Birmingham, England, November, 1993.

[3] George, P.L., Automatic Mesh Generation. Application to Finite Element Methods, Wiley, New York, 1991.

[4] CARP (Computer Aided Rapid Prototyping) Newsletter, "Project Introduction", Issue 1, Jan. 1994, Raynor, M. Ed.

[5] Juster, N.P. & Childs T.H.C., "A Comparison of Rapid Prototyping Processes", Proceedings of the 3rd European ´Conference on Rapid Prototyping, Nottingham, England, pp. 35-52, 6-7 July 1994.

[6] Bensler, H.P., Delhaye, J.M. & Favreau, C., "Uncertainty analysis of the volumetric interfacial area and Sauter mean diameter in bubbly flows determined by the ultrasonic transmission and photographic techniques", Proceedings of the 6th Workshop on Two-Phase Flow Predictions, Erlangen, Germany, March 30 - April 2, 1992, (Sommerfeld,M., Ed.)

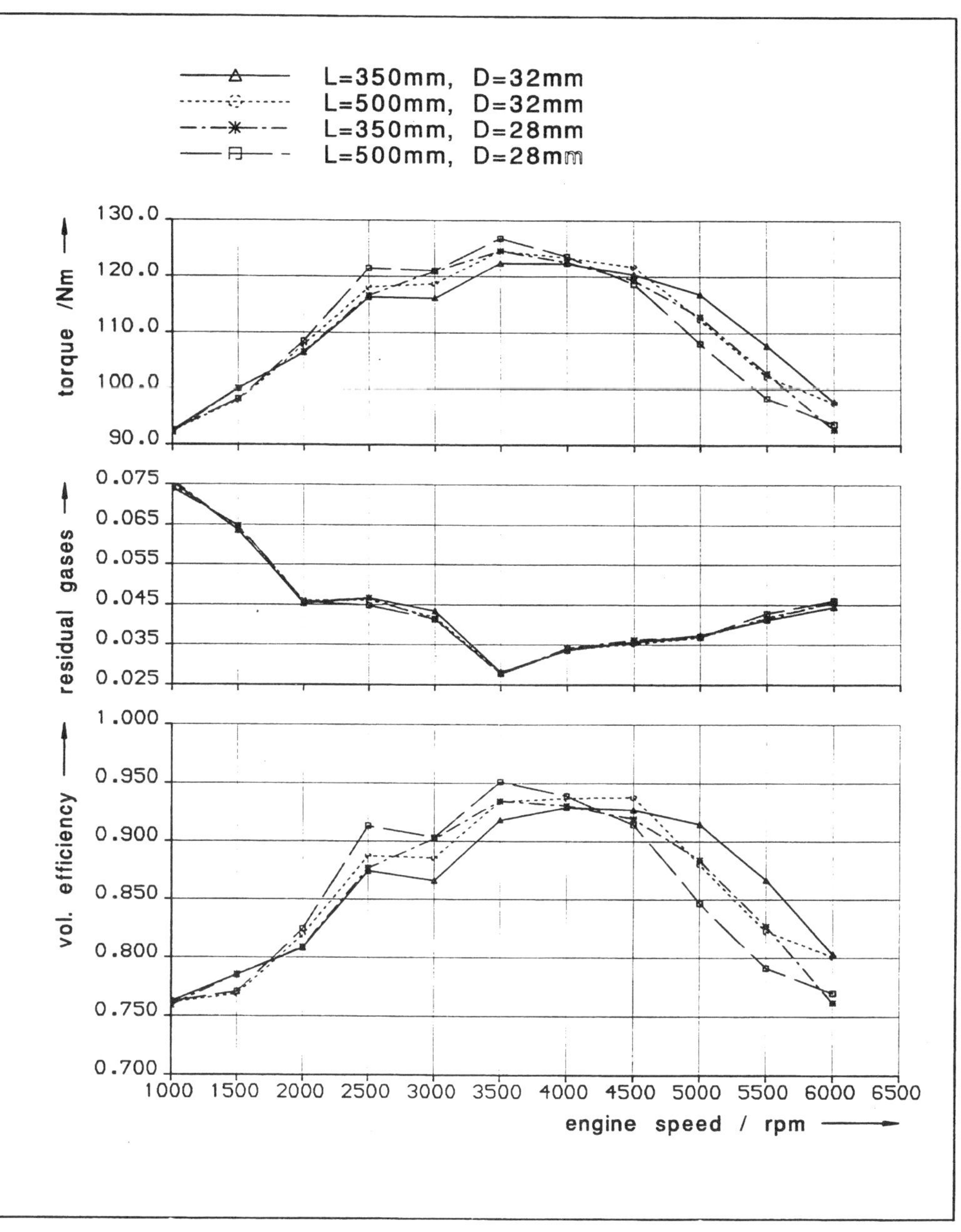

Fig. 1a **PROMO 1D CFD calculation results showing predicted torque, residual gases,& volumetric efficiency for varying intake manifold pipe length and diameter.**

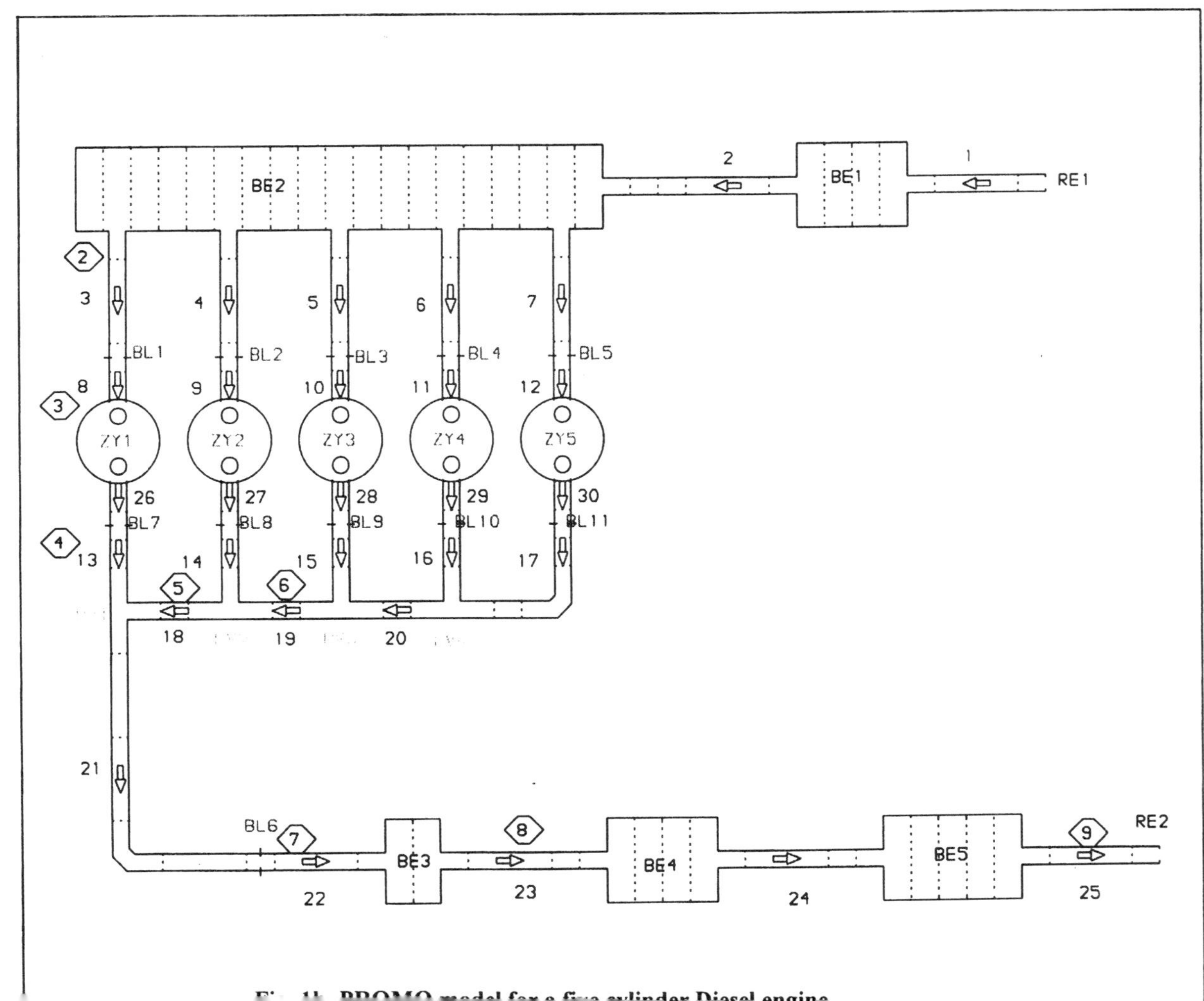

Fig. 11 PROMO model for a five cylinder Diesel engine

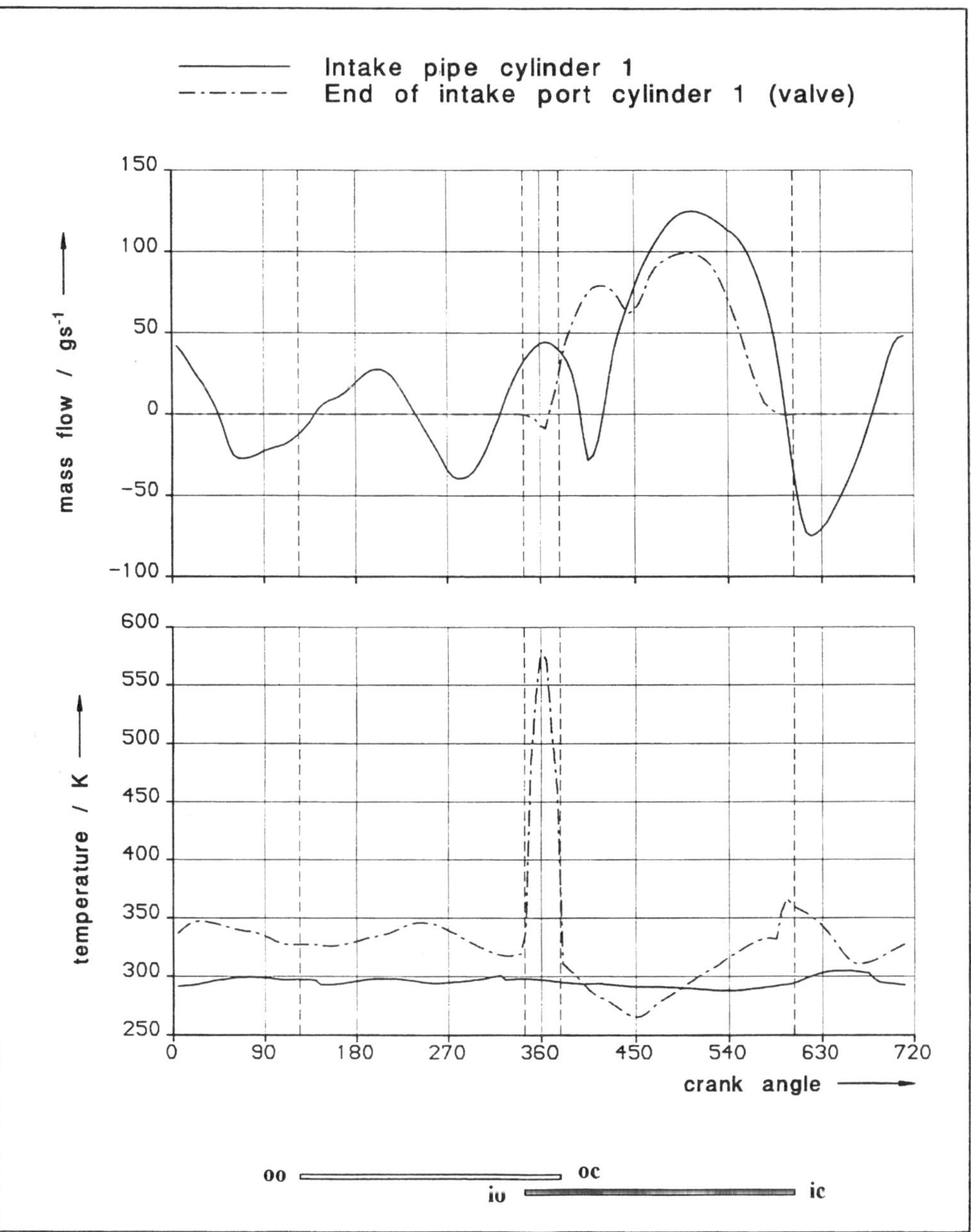

Fig. 1c **PROMO boundary conditions for a 3D CFD calculation (engine speed: 5000 rpm, oo=outlet valves open, oc=outlet valves closed, io=inlet valves open, ic=inlet valves closed).**

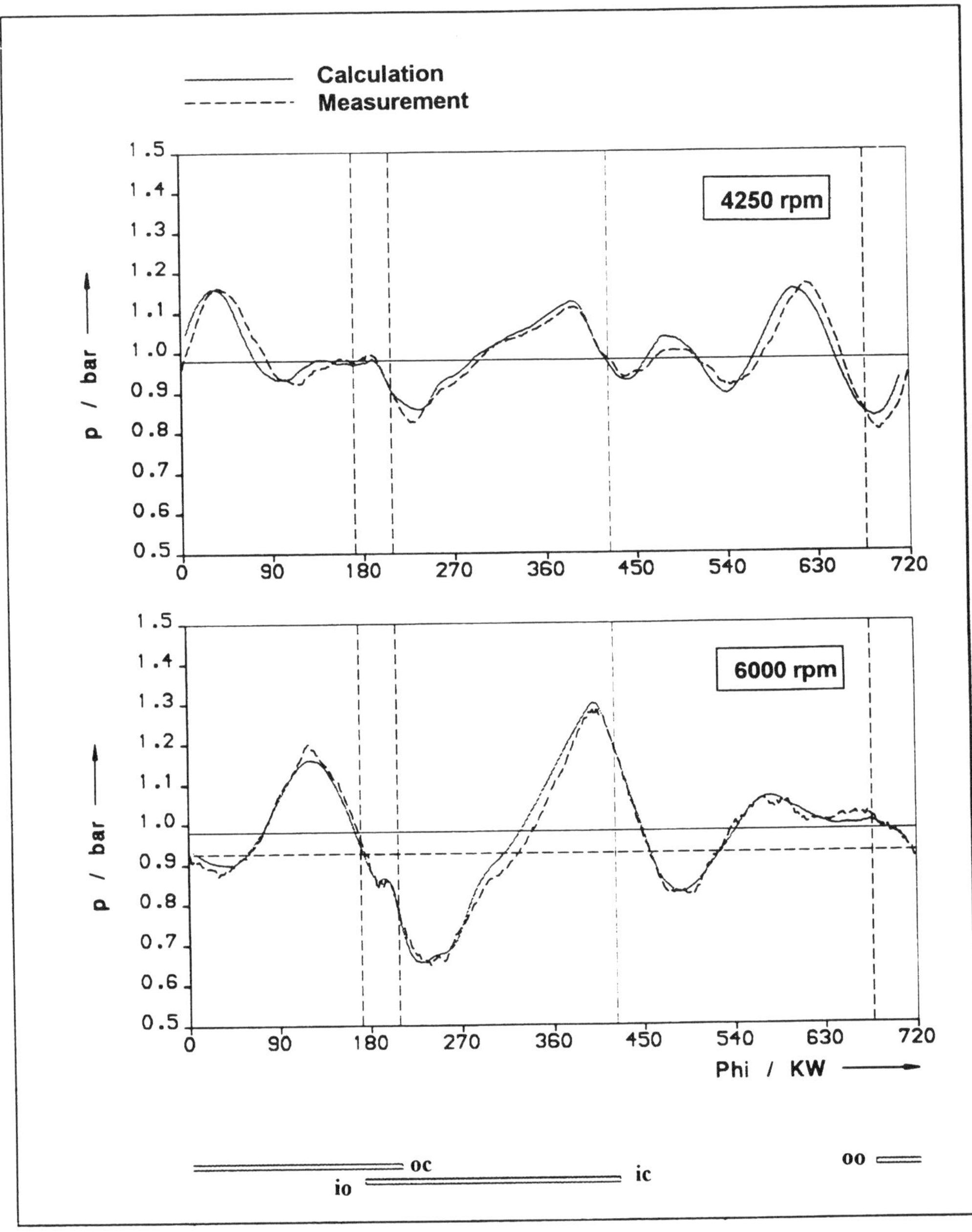

Fig. 1d PROMO calculated intake manifold pipe pressure compared to engine measurements (cylinder 2, engine speed = 4250 and 6000 rpm, oo=outlet valves open, oc=outlet valves closed, io=inlet valves open, ic=inlet valves closed).

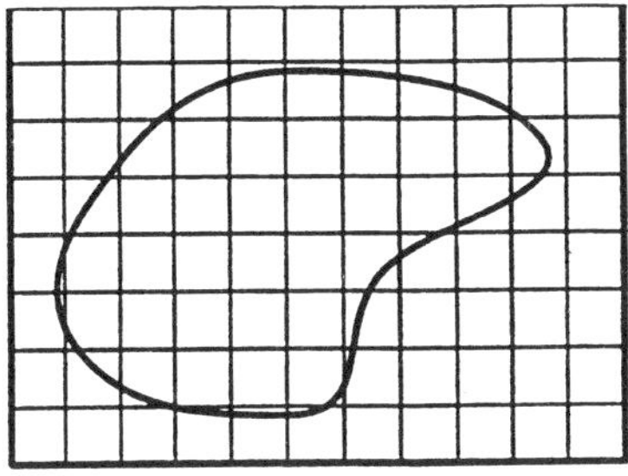

a. Definition of global cells around surface model.

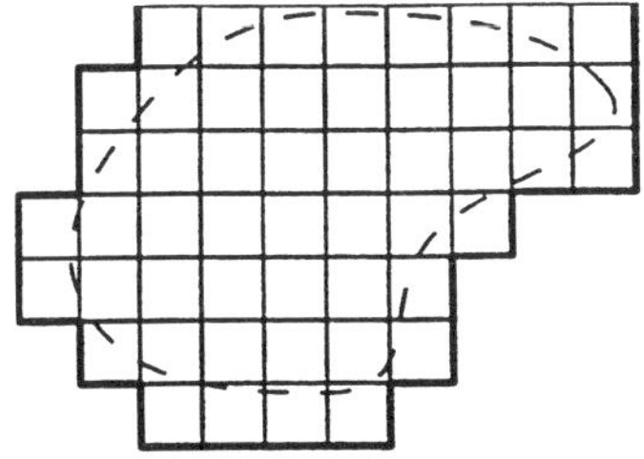

b. Elimination of global cells outside of surface model.

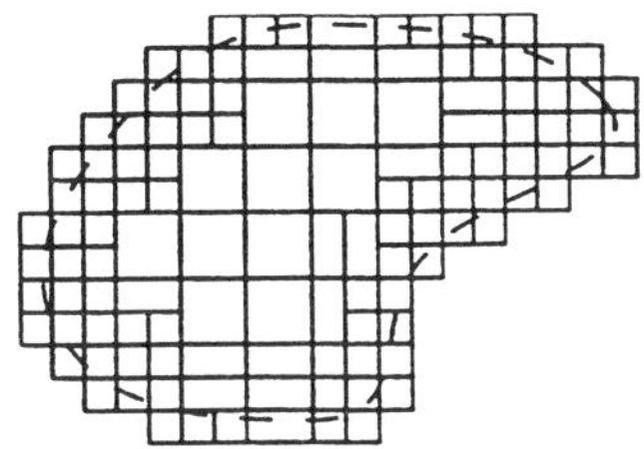

c. Refinement of cells near surface model boundary.

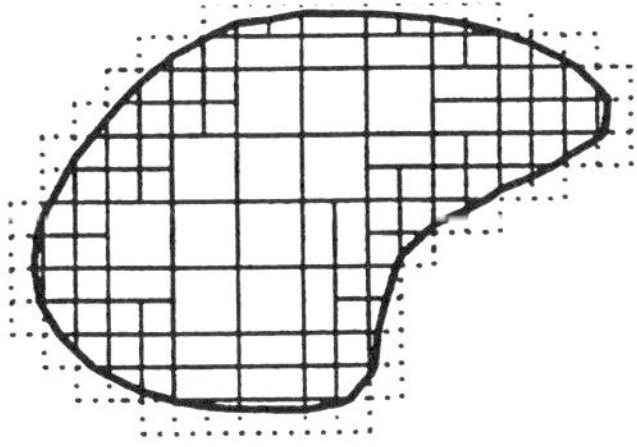

d. Addition of surface patches onto boundary cells.

Fig. 2 Description of VECTIS automatic mesh generation methodology.

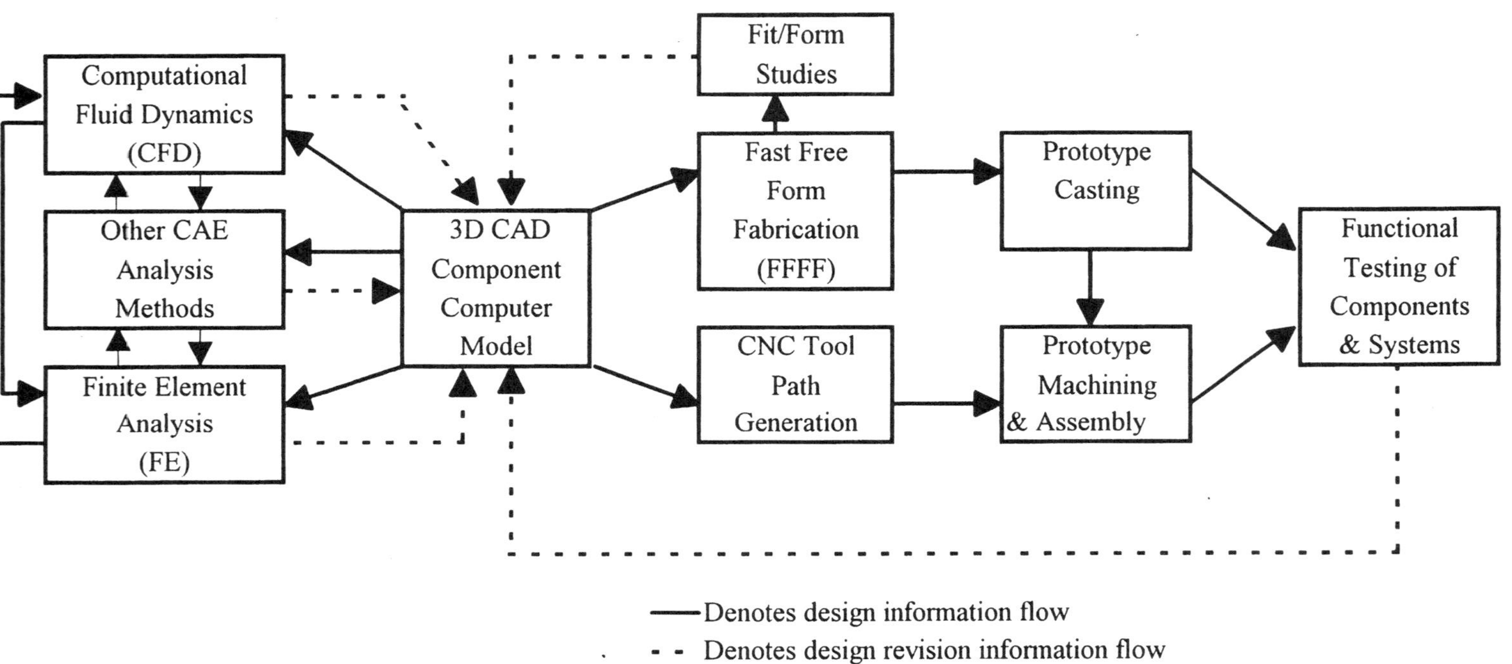

Fig. 3 Schematic Representation of CARP (Computer Aided Rapid Prototyping) Development Environment.

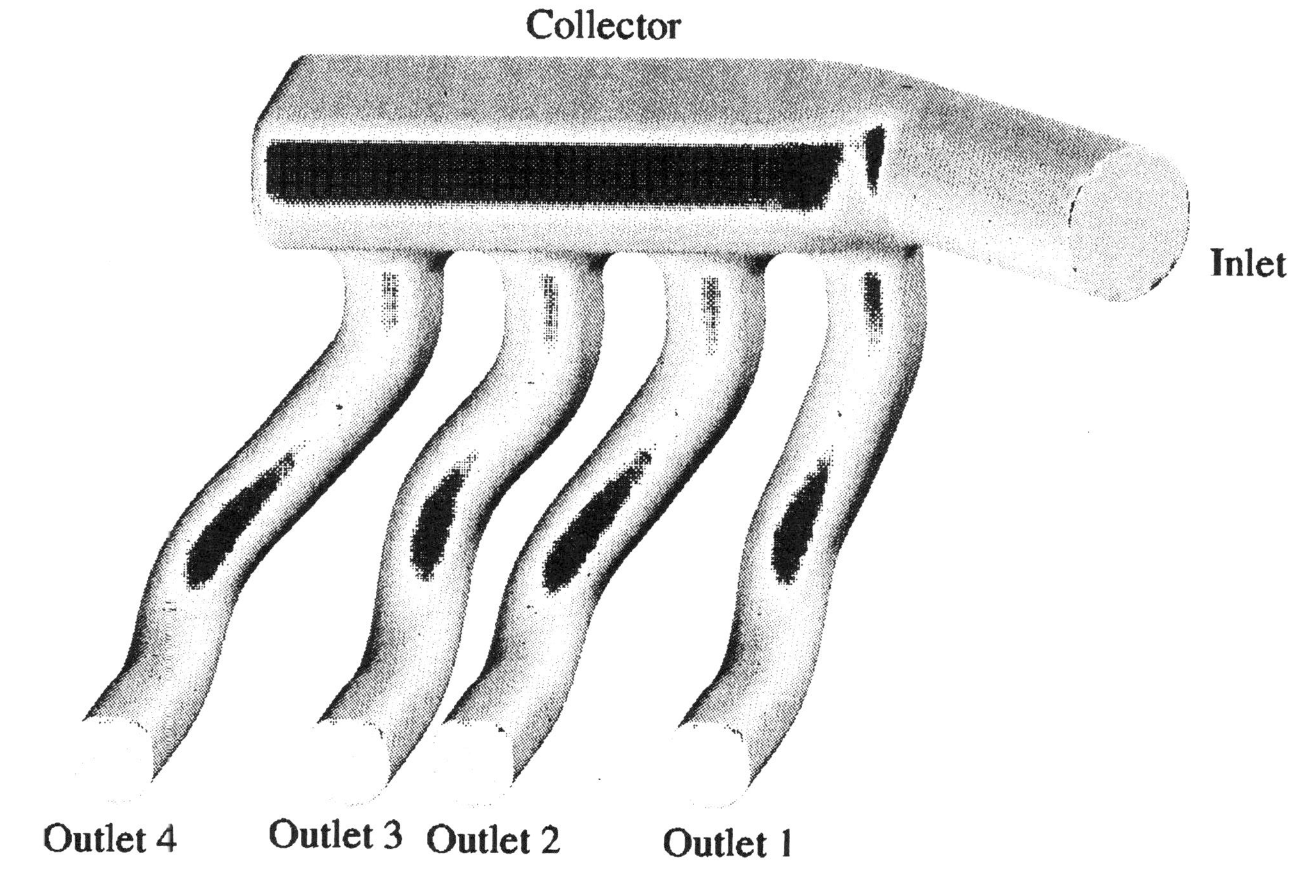

Fig. 4a CAD intake manifold geometry

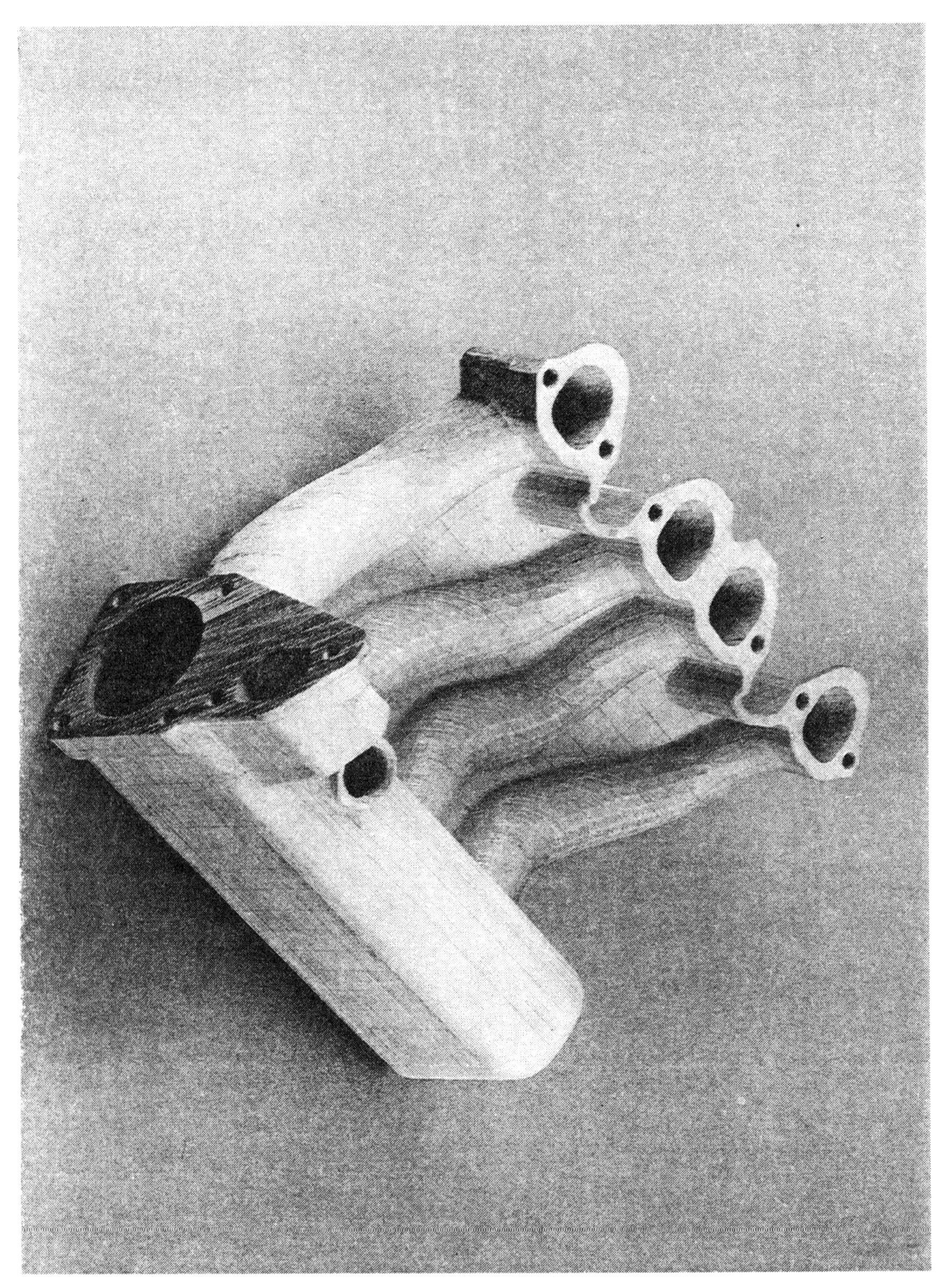

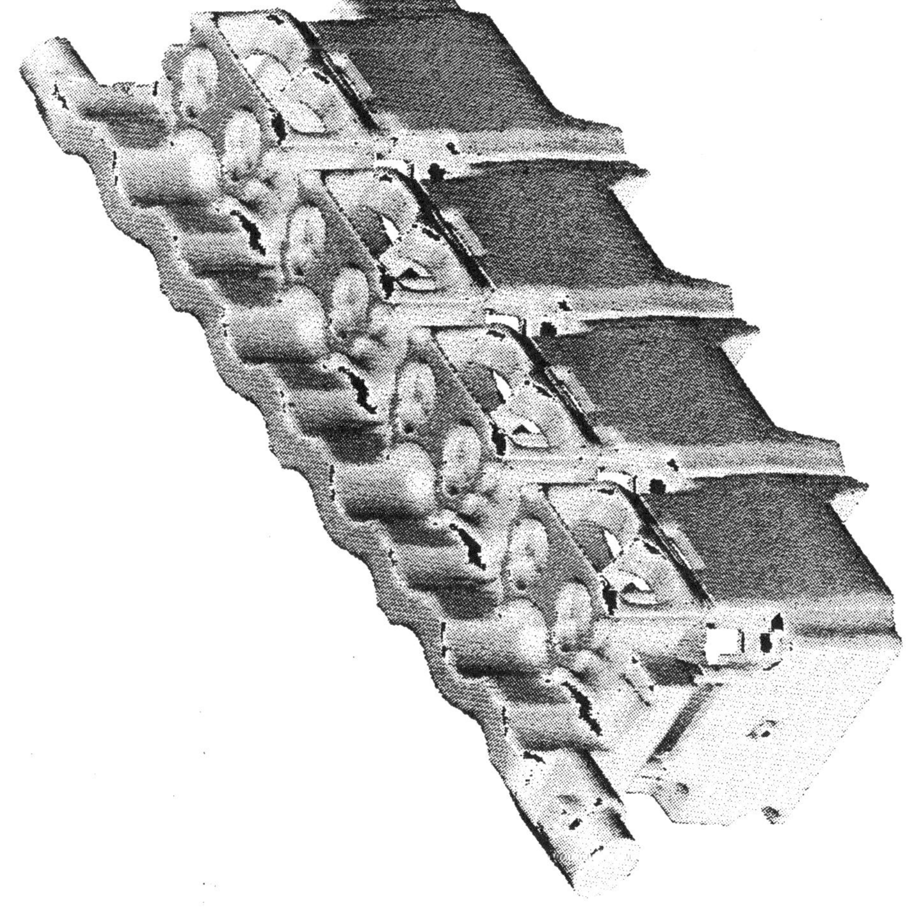

Fig. 5a Geometry of an engine water jacket.

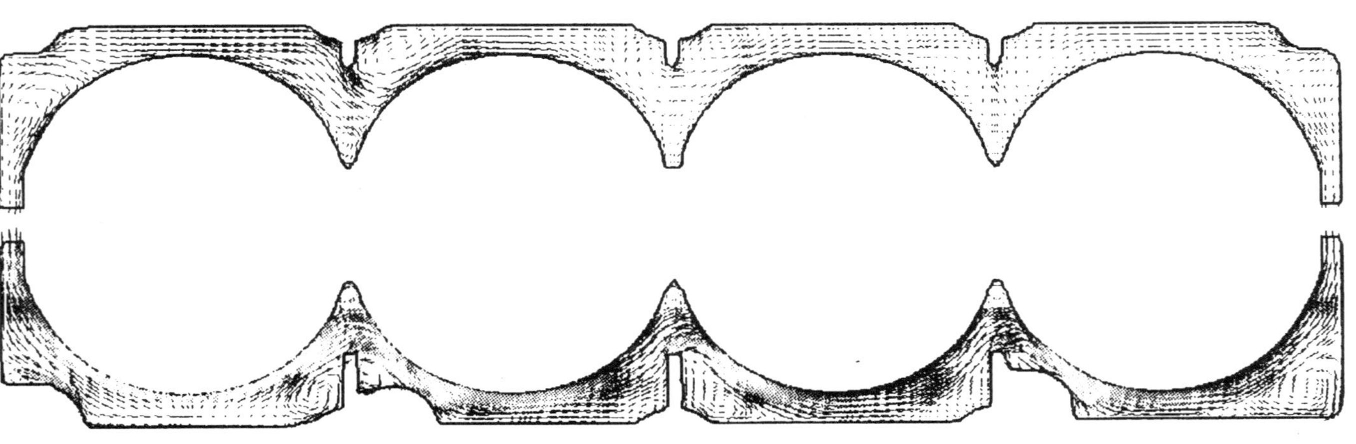

Fig. 5b Stationary 3D CFD results of the flow (velocity vectors) in the cylinder block region of the water jacket.

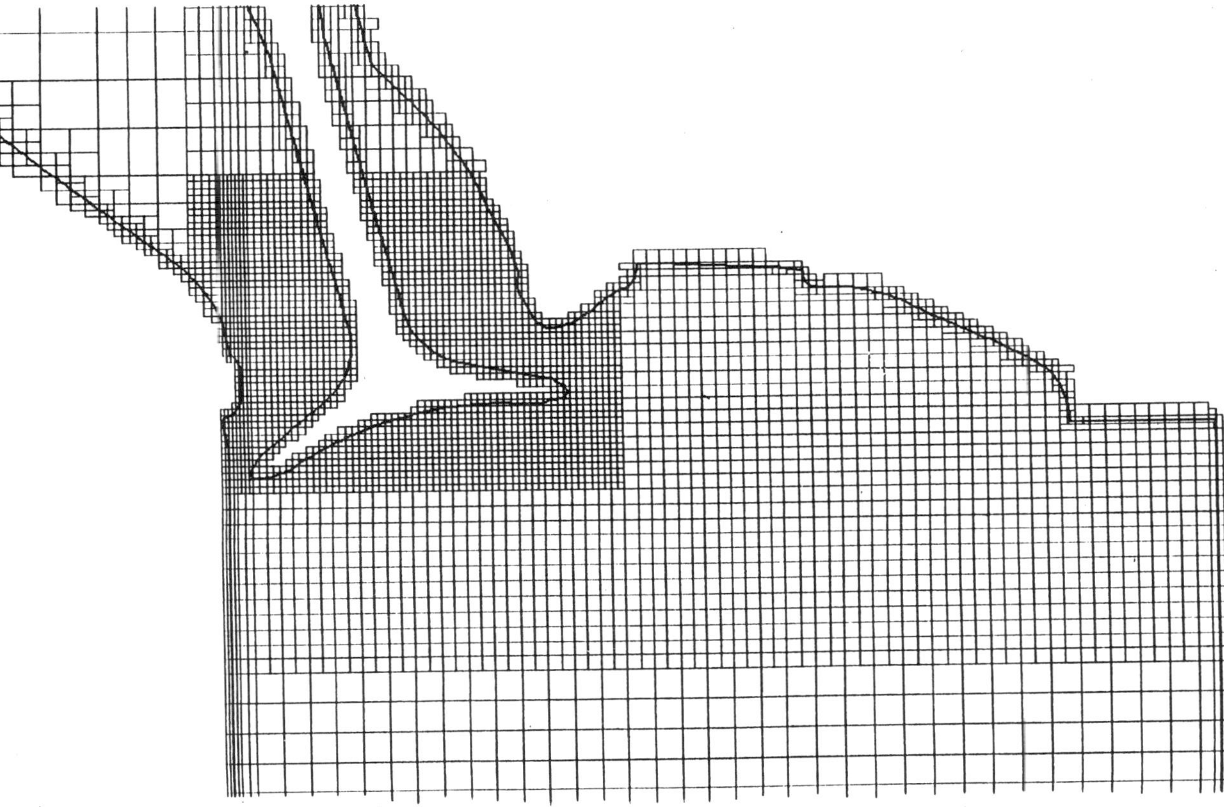

Fig. 6 VECTIS orthogonal structured mesh for a 5 valve intake port simulation.

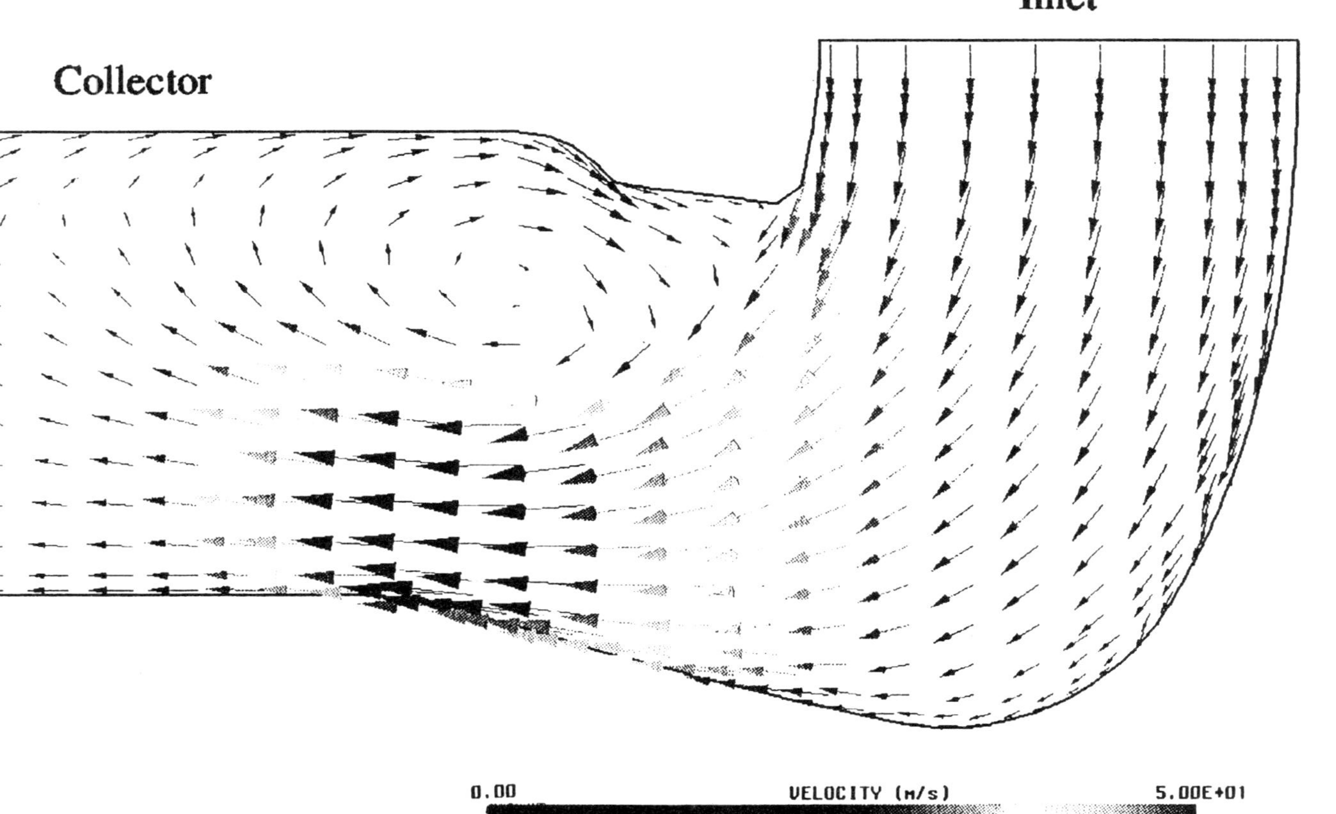

Fig. 7 Flow at the inlet of an intake manifold resulting from an instationary 3D CFD calculation (crank angle = 130 degrees after TDC).

C499/056/96

A numerical investigation of the effect of inlet charge direction on the scavenging behaviour of a two-stroke engine

K NG BEng, MSc, DIC, **M YIANNESKIS** BSc, MSc, DIC, PhD, CEng, FIMechE, and **D P E FOSTER** BEng, MSc, DIC
Mechanical Engineering Department, King's College London, UK
G GANTI MS, PhD
Ford Motor Company Limited, Laindon, UK

SYNOPSIS

In a two-stroke engine, the transfer ports play an important role in ensuring that the incoming jets of fresh charge to the cylinder are directed at optimum angles. In this study CFD predictions were made to investigate the influence of the inlet jet flow angles on the flow characteristics in a generic cylinder. Experiments were performed with an engine motored at 1500 rpm to obtain transient pressure data. The mass flux calculated from the pressure data was used to provide the boundary conditions used for the calculations. CFD simulations with the inlet jets directed at various horizontal and vertical angles were made and the effects are compared.

NOTATION

M	Mass flow rate to the cylinder, kg/s.
C_d	Discharge coefficient.
A	Total port opening area, m^2.
R	Gas constant, J/kg K.
T	Gas temperature, K.
p, P	Pressure, bar.
γ	Specific heats ratio.
x	Distance between the cylinder centre and the projection of the intersection point of the centre-lines of the jets from the inlets, mm.
r	Cylinder radius, m.
α	Vertical inclination angle of an inlet jet, degrees.
k	Turbulence kinetic energy, m^2/s^2.
v	Volume of individual cells, m^3.
$\bar{k}$	Volume-averaged turbulence kinetic energy, m^2/s^2.
V	Total cylinder volume, m^3.

Subscripts

c	conditions in the cylinder.
o	conditions in the crankcase.

1 INTRODUCTION

In a two-stroke engine, understanding of the interaction of the fresh charge and exhaust gas during the scavenging period is important in order to improve its design. Knowledge of the scavenging mechanism is fundamental to the understanding of the operation of the engine as a whole. The charge streams entering through the transfer ports to the cylinder must interact so as to displace the residual gases through the exhaust effectively. For such an engine, the transfer ports play an important role in ensuring that the inlet jets are guided in optimum directions in order to maximise fuel economy and minimise emissions.

Numerical simulation of two-stroke engines has been the subject of a number of previous investigations, some of which have been concerned with the detailed modelling of the transient flow field of the cylinder and the transfer ports (1,2). Much effort has also been expended on experiments to examine the scavenging flow pattern, including LDA measurements (3,4). Ahmadi-Befrui et al (5) investigated the in-cylinder gas exchange process and the effect of the variation of the instantaneous mass flow rate, which was calculated with an unsteady gas dynamics method, on the mixing processes and the flow structure. Sweeney et al (6) compared the results of CFD calculations with the performance characteristics of a firing engine and found that the performance ranking of the cylinders obtained from the engine tests agreed well with the corresponding scavenging efficiencies predicted. Kenny et al (7) investigated the steady flow field of model cylinders and compared CFD predictions with LDA measurements. Different cylinder designs were considered with different horizontal port entry angles and different port locations. For the calculations the port jets were directed horizontally and a uniform velocity profile was assumed at the port exits. General agreement was found with the measured velocity field but not with the turbulence kinetic energy distribution. They also noted that a better prediction could be achieved if the boundary velocity profile was correctly defined. Generally, it has been confirmed that the scavenging efficiency can be predicted better if the inlet flow directions, under steady or motoring conditions, are defined according to experimental observations (8, 9).

As the accuracy of CFD techniques for simulating engine in-cylinder flows is improving, they are becoming an important component in the process of engine design and development. In the present work a numerical model of a generic two-stroke engine has been developed to study the influence of the inlet jet directions on the scavenging flow behaviour in the cylinder.

The geometry of the engine cylinder studied in the present work is of a generic nature and it consists of a flat cylinder head and five inlets whose location and size are typical of two-stroke engine designs. The flow inlet boundaries along the cylinder wall represent the exits of the transfer ports where the directions of the efflux, or the inlet jets, are specified. In this way, the inlet flow directions can be defined precisely and the effects of the jets issuing from individual inlets at various angles can be investigated in a systematic and controlled manner.

The engine geometry investigated was selected as, in order to gain a more fundamental understanding of the general aspects of two-stroke engine in-cylinder flows, there is a need to study the influence of only a few parameters on the flow characteristics. A full simulation of an engine with all the transfer ports and exhaust port included can provide detailed information on the flow behaviour, but it is relevant mostly to the particular engine concerned. Since all the different components have interacting effects, it is often very difficult to attribute the results observed to a single design feature.

This paper reports a study of the effects of the scavenge air jet directions on the in-cylinder flow characteristics. Rather than modelling the entire flow field of one specific engine design, the current approach utilises a generic cylinder geometry, in order to observe and establish any relevant trends in the flow features as a result of systematic changes in the inlet boundary conditions. This approach offers the advantage of minimising the number of interacting factors so that their influence on engine operation can be better understood.

2 THE MOTORED ENGINE CONFIGURATION

A motored two-stroke engine was used to obtain crankcase and cylinder pressure data. The engine used was a five-ported, loop scavenge type engine: its main operating parameters are listed in Table 1 below.

Table 1 Specifications of the motored engine used in the experiment.

Engine compression ratio	10.5
Bore, mm	84
Stroke, mm	72
Exhaust port opens, °ATDC	90
Transfer ports open, °ATDC	120
Transfer ports close, °ATDC	240
Exhaust port closes, °ATDC	270
Motoring speed, rpm	1500

Pressure transducers were placed at the crankcase and the cylinder head to record the transient pressure characteristics of the engine motored at the speed of 1500 rpm. The data were ensemble-averaged over 50 cycles and the corresponding traces are presented in Figure 1. From the pressure data the instantaneous mass flow rate into the cylinder, M, was calculated with one-dimensional compressible flow equations, thus:

For sub-critical flow,

$$M = \frac{Cd \cdot A \cdot p_o}{\sqrt{RT_o}} \left(\frac{p_c}{p_o} \right)^{\frac{1}{\gamma}} \left\{ \frac{2\gamma}{\gamma - 1} \left[1 - \left(\frac{p_c}{p_o} \right)^{\frac{\gamma-1}{\gamma}} \right] \right\}^{\frac{1}{2}} \tag{1a}$$

For choked flow,

$$M = \frac{Cd \cdot A \cdot p_o}{\sqrt{RT_o}} \gamma^{\frac{1}{2}} \left(\frac{2}{\gamma + 1} \right)^{\frac{\gamma+1}{2(\gamma-1)}} \tag{1b}$$

where A is the total port opening area; the ratio of the specific heats, γ, used was 1.4; the gas constant, R = 287 J/kg K; subscripts o and c designate conditions in the engine crankcase and cylinder respectively.

The mass flow rate was then used to compute the velocity components at the cylinder inlet boundaries. To account for the frictional losses across the transfer port passages a discharge coefficient, C_d, of 0.65 was assumed. Typically the value of C_d varies from about 0.65 to 1.0, therefore the lower value used here would provide a conservative estimate for the air flow into the cylinder.

It was observed from the pressure traces shown in Figure 1, and subsequently from the calculated values of the instantaneous mass flow rate, that the air flux oscillates significantly during the engine cycle, and considerable reverse flow from the cylinder to the transfer ports was found, particularly after BDC. In a two-stroke engine, back flows are commonly observed from the cylinder to the transfer ports, or from the exhaust to the cylinder (5); the boundary conditions used for the CFD calculations had to take into account this mass flow variation, which is described in the following section.

3 THE COMPUTATIONAL MODEL

A generic CFD model was created to simulate isothermal air flow within the cylinder of a two-stroke engine, in order to investigate the flow and the mixing of the inlet charge and its interaction with the cylinder geometry. The computer code STAR-CD was used for the predictions. It is a general purpose numerical simulation software which solves the equations for fluid momentum, energy and the turbulence parameters by means of a finite volume method. In transient calculations such as those carried out here, the code employs the PISO algorithm for solving the algebraic finite-volume equations by iterative means (10).

The computational mesh consisted of 23040 cells and only the cylinder volume was modelled. The size and location of all the inlets and the exhaust have been accurately represented in the model to match the experimental geometry as closely as possible.

The k-ε turbulence model was used in the calculation. The model is recognised to be less than perfect largely because of the implicit assumption employed in the formulation of the model that the turbulence is isotropic. In particular for jets interacting in an engine cylinder, the level of turbulence kinetic energy has been found to be generally under predicted (9). Therefore the turbulence kinetic energy results should only be taken as indicative, and their interpretation is best confined to comparison of the different cases calculated to establish trends.

The cases studied in the present investigation are listed in Table 2, together with the horizontal and vertical angles specified at the inlets. The parameter x is defined as the distance between the cylinder centre and the projection of the intersection point of the centre-lines of the jets from the different inlets. The vertical inclination, α, of the inlet jets emerging from the main, auxiliary and boost ports is also listed. A schematic diagram of the cylinder geometry is shown in Figure 2 where x, r and α are defined.

Table 2 Inlet jet boundary parameters.

CASE	x (mm)	α^{o} (main)	α^{o} (auxiliary)	α^{o} (boost)
1	0	10	40	60
2	0	30	30	30
3	0.5 r	10	40	60
4	0.75 r	10	40	60
5	0.75 r	30	30	30

3.1 Boundary conditions

The boundary conditions were set by means of specifying directly the three velocity components at the inlets. Five different cases were investigated where the inlet jets were directed at various horizontal and vertical angles. The mass flow rate into and out of the cylinder for all the cases studied was identical and thus dynamic similarity was preserved. When a back flow is present, the 'inlet' boundary velocities are specified in the reverse direction, that is, the fluid flows out of the cylinder at the same angles as the inflow.

The velocities at each inlet were specified with a uniform, or plug flow, profile. For the purpose of the present work, the variation of flow angles due to the specific port curvatures and the shape of the port opening was not modelled, as this not only simplifies the problem specification but also allows a more fundamental approach to be adopted. It has been observed (8) that the jets issuing from the ports do vary across the port both horizontally and vertically with time, but at any one instance they are relatively uniform across the port area. This finding provides some support for the use of the plug flow assumption employed.

At the exhaust port a pressure type boundary was prescribed with the local pressure set equal to atmospheric. At the start of the calculation the fluid in the cylinder was stationary at TDC, and an initial pressure of 13.1 bar, taken from the motored engine measurements, was applied.

The calculation commenced with the piston at TDC and was performed for one complete cycle. During the calculation the cylinder volume expands and contracts by means of stretching and compressing the cell volumes, as well as by adding and deleting cell layers. The mesh movement simulating the piston motion corresponded to an engine speed of 1500 rpm. A scalar was defined at the inlets to represent the fresh charge, to allow the composition of the fluid in the cylinder to be expressed as a percentage of the fresh charge at any instant.

4 RESULTS AND DISCUSSION

Of the five cases considered, case 1 can be regarded as a reference against which the results of the other cases can be compared. Figures 3 and 4 show the velocity vectors and fresh charge distributions within the cylinder at four stages of the cycle, namely at crank angles of 140°, 180°, 240° and 300°, for cases 1 and 3 respectively. Corresponding results are not shown for the other cases due to space limitations. In both figures, the piston head is shown at the appropriate position for the crank angle concerned. The boost port is located on the left side of each figure and the exhaust port on the right. Figures 3 (a) and 4 (a) show the velocity vectors at horizontal planes, Figures 3 (b) and 4 (b) show the vectors at the symmetry plane and at the top of the cylinder. The contours of the fresh charge concentration at the symmetry plane and at the top of the cylinder are shown in Figures 3 (c) and 4 (c).

4.1 Velocity and concentration fields

Figures 3 (a) - (c) present the results for calculation case 1. The jets are directed to meet at the symmetry plane, but they are inclined vertically at different angles so that those issuing from different ports do not actually meet at the same location.

The velocity vectors show that shortly after the inlets are opened, at 140°, the magnitude of the flow velocities near the cylinder head is relatively small in comparison with those near the piston. At BDC (180°) a 'loop' vortex is already formed; the boost port jet is attached to the back wall, and when it reaches the cylinder head it is deflected and is subsequently directed toward the exhaust. At 240° when the inlets are closed a large 'loop' vortex is the dominant feature occupying a large part of the cylinder volume. From the mass flow rate data it is observed that the reverse flow from the cylinder to the 'inlets' occurs from approximately 190° to 240°. A back flow from the exhaust into the cylinder is also induced and this has the effect of enhancing the development of the vortex in the cylinder. The loop persists for a while before it is gradually broken up by the ascending piston. The fresh charge concentration contours can reveal the amount of fresh air escaping to the outlet. From Figure

3 (c) the formation of the scavenge loop is evident. It can be observed that in the symmetry plane near the exhaust, the fresh charge concentration reaches values of up to 50 % at BDC.

Case 2 differs from case 1 only in the vertical angle of the inlet jets, as they are all inclined at $\alpha = 30^o$ to the vertical. The flow field (not shown here) was broadly similar to that in the previous case. However, as a result of a smaller inclination of the boost port jet, the flow has detached from the back wall during the initial part of the port opening period. The jets from the main and auxiliary ports are not sufficiently strong to force the boost port jet to flow vertically upward. Instead much of the boost port jet is directed towards the opposite wall and the exhaust. The size of the loop vortex at the symmetry plane is reduced and a smaller loop is formed. Nevertheless at the symmetry plane the loop structure is evident in the cylinder for a longer period of time than for case 1. At BDC, shortly before reverse flow occurs, the maximum fresh charge concentration at the exhaust is quite high at about 90 %.

Case 3 (Figures 4 (a) - (c)) is different from case 1 because the projection of the intersection of the jet centre-lines is located nearer the back wall, at a distance of half a cylinder radius from the centre. In this case the upward flow at the back wall (Figure 4 (b)) is much stronger than the two previous cases. Since the inlet jets are all inclined at different vertical angles they do not interact as much as they would if the inclinations are the same. The general flow direction towards the back wall initially results in the formation of a well-defined vortex much earlier: the loop can be observed soon after the inlet ports have opened (at 140°).

The enhancement of the early creation of the scavenge loop is also evident in the results of case 4 (again, figures are not shown here for economy of presentation). The vortex is better defined at BDC compared with case 3. But generally the flow pattern and the concentration distribution for cases 3 and 4 are quite similar.

For case 5 all the inlet jets are directed at 30° to the vertical axis. The reduced inclination of the auxiliary and boost port, compared with case 4, results in a reduction of the fluid momentum in the vertical direction, particularly noticeable in the flow above the boost port. A weaker scavenge loop is generated even though the jets enter at the same horizontal angles as in case 4 with identical x. The loop is better-defined than those in cases 1 and 2. There is little deflection of the boost port jet above the piston surface and a significant portion of this jet travels in the direction of the exhaust.

At the occurrence of reverse flow soon after BDC, the flow patterns established during the early half of the cycle are quickly altered, for all the cases modelled. More significantly the reverse flow from the exhaust, rather than the reverse flow to the transfer port, is the dominant factor affecting the subsequent evolution of the flow. As the piston continues to move upward after the exhaust port has closed the in-cylinder flow continues to evolve, but the main features of the established flow pattern are generally preserved. During the rest of the cycle little mixing of the jets occurs and the fresh charge remains non-homogeneously distributed up to TDC. This could have implications for the combustion process and indicates that different cylinder head shapes should be investigated to determine whether it is possible to enhance mixing after the exhaust port closes and thus improve the conditions for combustion.

4.2 Fresh charge concentration at the exhaust

The manner in which the fresh charge is directed by the transfer ports has an important influence on the development of the in-cylinder flow, and the extent of flow short circuiting to the exhaust is also affected. Monitoring the arrival of the fresh charge at the exhaust is a convenient means of gauging the engine's scavenging effectiveness.

The average concentration of the fresh charge at the exhaust is shown in Figure 5 for the five cases studied. The first two cases might be considered to have the poorest scavenging characteristics, as in the earlier part of the cycle the fresh charge concentration at the exhaust is up to three times higher than the other cylinder configurations. For case 2 where the inlet jets are all aimed at the same vertical location at the centre of the cylinder, the fresh charge is quickly dispersed and a considerable amount is lost to the exhaust; this may be attributed to the fact that the scavenge loop is not well formed.

Cases 4 and 5 show better scavenging characteristics. This could be related to the air jets being directed closer to the back wall; as seen from the flow patterns in Figures 3 and 4, the air flow path in the form of a loop might be expected to have the effect of delaying the fresh charge reaching the exhaust. When reverse flow occurs at about 190° in Figure 5, the exhaust is rapidly occupied by the exhaust fluid and therefore the profiles show a steep decline. After 240° a higher concentration of fresh charge at the exhaust is observed for case 4; this is probably due to the fact that in this case the jets are directed at higher angles which reduce the length of the path required for the fluid to reach the exhaust.

4.3 Volume-averaged turbulence kinetic energy

The variations of the volume-averaged turbulence kinetic energy, $\bar{k}$, with crankangle for the five cases are plotted in Figure 6. $\bar{k}$ is defined by the following equation:

$$\bar{k} = \sum \frac{kv}{V} \tag{2}$$

where k and v are the turbulence kinetic energy and the volume of a cell respectively, and V is the total cylinder volume.

In all cases a sharp rise in $\bar{k}$ is observed, most noticeably just after the inlets are open. This is likely to be caused by the associated high shear occurring during the initial rush of air to the cylinder. Subsequently the levels fall gradually and rise again just after BDC when the reverse flow starts to occur. As the ports start to close $\bar{k}$ continues to rise, reaching a maximum just before the transfer ports are closed. Thereafter as the piston moves upward the vortex is broken up and $\bar{k}$ decreases continuously to the TDC. The suppression of the levels of $\bar{k}$ from around 150 m^2/s^2 at BDC to 7.5 m^2/s^2 at TDC is not dissimilar to that observed in four-stroke engines.

5 CONCLUDING REMARKS

A CFD model of a generic two-stroke engine cylinder has been developed to study the influence of inlet charge direction on the scavenging flow characteristics of two-stroke engines. Experimental data obtained in an engine motored at 1500 rpm were used to calculate the transient mass flow rate to the cylinder.

In all the cases studied, the main feature of the in-cylinder flow is the formation of the scavenge loop. However, the flow pattern can be markedly affected by the way in which the inlet charge is guided into the cylinder. The occurrence of reverse flow considerably alters the in-cylinder flow. Particularly, back flow from the exhaust has a large influence on the subsequent flow development.

The time taken by the inlet charge to reach the exhaust port varies partly as a result of the differences in the flow entry angle. The distribution of the fresh charge in the cylinder remains non-uniform throughout the cycle.

The predictions enabled the effect of the inlet jet directions on the mean flow patterns and turbulence levels in the cylinder to be determined and could prove useful for the optimisation of the design of transfer ports. Work is in progress to simulate the flows inside the inlet and exhaust ports as well as in the cylinder. In addition, LDA data of the in-cylinder flow are being obtained in the motored engine to help assess these predictions.

6 ACKNOWLEDGEMENTS

The authors wish to acknowledge the support for this work provided by the Engineering and Physical Sciences Research Council (Grant GR/J65693) and Ford Motor Company Limited.

REFERENCES

(1) EPSTEIN, P.H, REITZ, R.D and FOSTER, D.E. Computations of a Two-Stroke Engine Cylinder and Port Scavenging Flows, SAE Paper No. 910672, 1991.

(2) AMSDEN, A.A, O'ROURKE, P.J, BUTLER, T.D, MEINTJES, K and FANSLER, T.D. Comparisons of Computed and Measured Three-Dimensional Velocity Fields in a Motored Two-Stroke Engine, SAE Paper No. 920418, 1992.

(3) IKEDA, Y, OHHIRA, T, TAKAHASHI, T and NAKAJIMA, T. Flow Vector Measurements at the Scavenging Ports in a Fired Two-Stroke Engine, SAE Paper No. 920420, 1992.

(4) McKINLEY, N.R, FLECK, R and KENNY, R.G. LDV Measurement of Transfer Port Efflux Velocities in a Motored Two-Stroke Cycle Engine, SAE Paper No. 921694, 1992.

(5) AHMADI-BEFRUI, B, BRANDSTATTER, W and KRATOCHWILL, H. Multidimensional Calculation of the Flow Processes in a Loop-Scavenged Two-Stroke Cycle Engine, SAE Paper No. 890841, 1989.

(6) SWEENEY, M.E.G, SWANN, G.B.G, KENNY, R. G and BLAIR, G.P. Computational Fluid Dynamics Applied to Two-Stroke Engine Scavenging, SAE Paper No. 851519, 1985.

(7) KENNY, R.G, McKINLEY, N.R and RAGHUNATHAN, B.D. Experimental and Theoretical Studies of Two-Stroke Loop Scavenging, IMechE Seminar - The validation of computational techniques in vehicle design - Stage one: Computational Fluid Dynamics, UK. April 1994.

(8) SMYTH, J.G, KENNY, R.G and BLAIR, G.P. Motored and Steady Flow Boundary Conditions Applied to the Prediction of Scavenging Flow in a Loop Scavenged Two-Stroke Cycle Engine, SAE Paper No. 900800, 1990.

(9) McKINLEY, N.R, KENNY, R.G and FLECK, R. CFD Prediction of a Two-Stroke, In-Cylinder Steady Flow Field An Experimental Validation, SAE Paper No. 940399, 1994.

(10) STAR-CD 2.21 Manuals. Computational Dynamics Ltd., London, 1994.

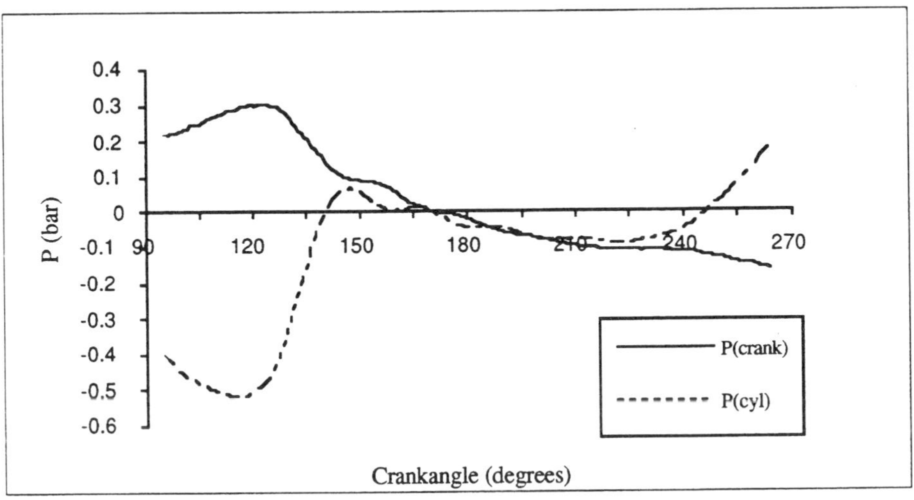

Figure 1 Measured variation of the crankcase and cylinder pressures with crankangle.

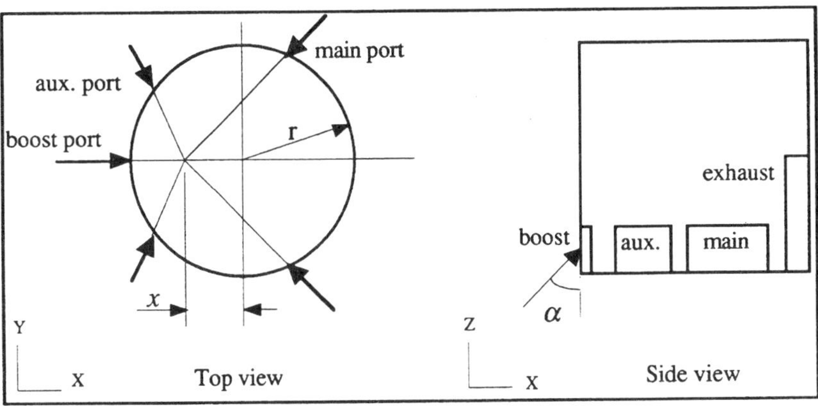

Figure 2 Schematic diagram of the cylinder geometry and definition of inlet flow direction parameters x and α.

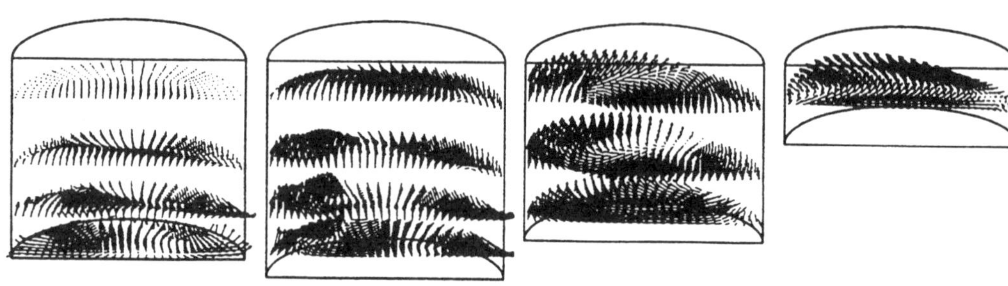

Figure 3 (a) Velocity vectors at different horizontal planes.

→ = 240 m/s   → = 80 m/s   → = 65 m/s   → = 19 m/s

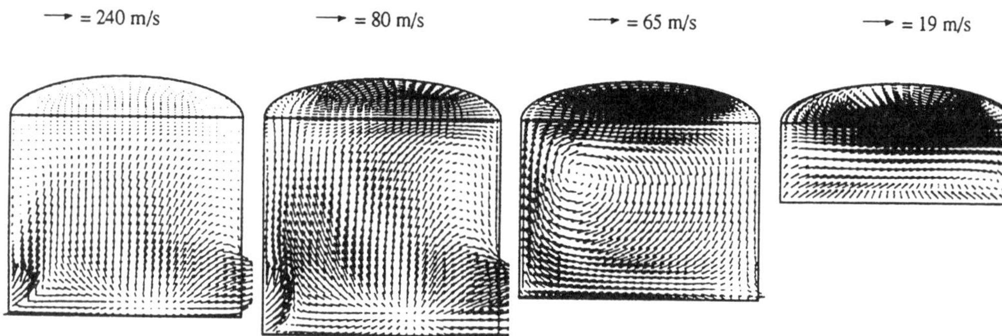

Figure 3 (b) Velocity vectors at the cylinder head and the symmetry plane.

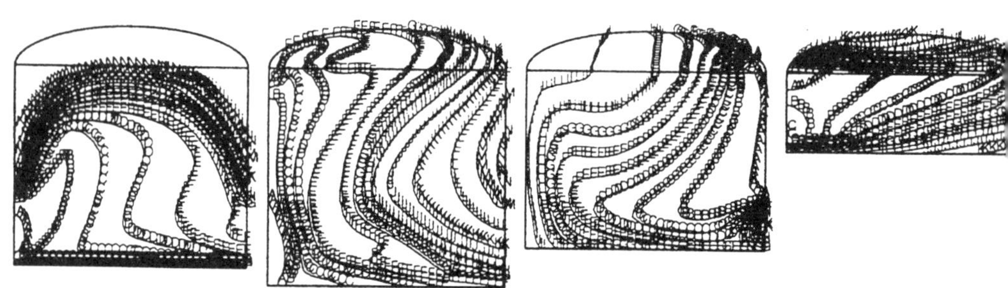

contour values

Figure 3 (c) Contours of fresh charge concentration (%).

Figure 3. Case 1: In-cylinder flow patterns (at 140°, 180°, 240° and 300° ATDC).

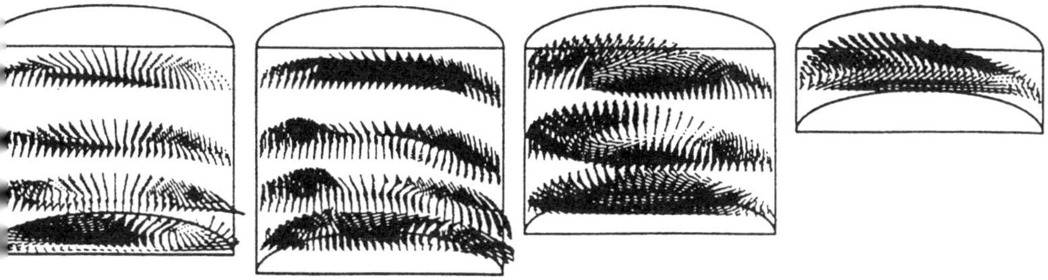

Figure 4 (a) Velocity vectors at different horizontal planes.

→ = 180 m/s → = 80 m/s → = 50 m/s → = 17 m/s

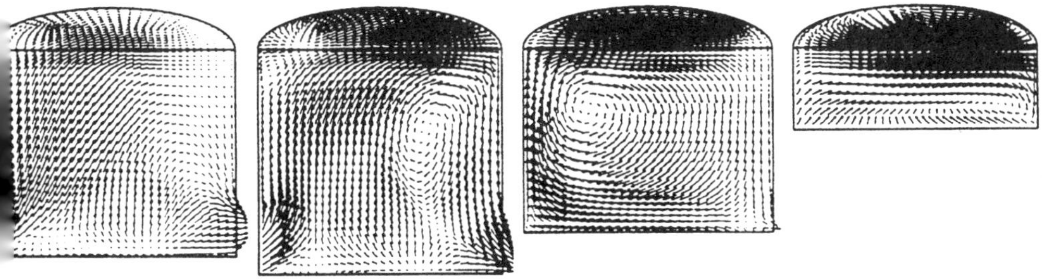

Figure 4 (b) Velocity vectors at the cylinder head and the symmetry plane.

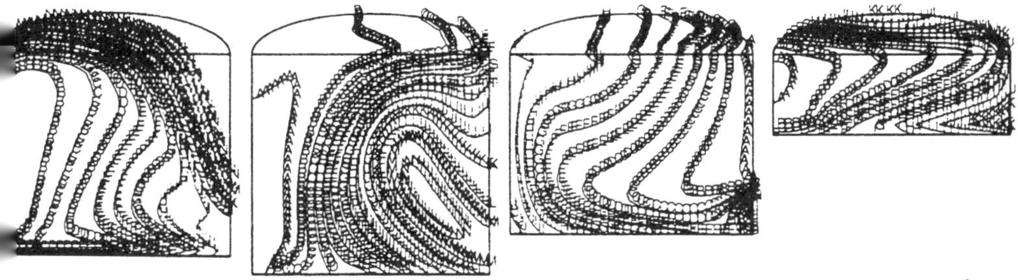

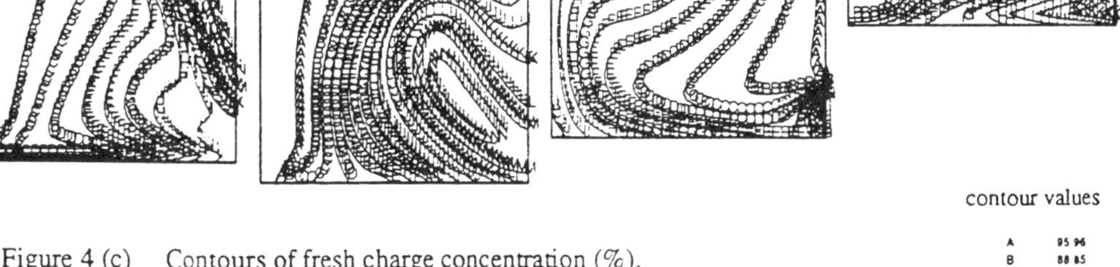

contour values

A	95.96
B	88.85
C	81.74
D	74.63
E	67.53
F	60.42
G	53.31
H	46.20
I	39.09
J	31.99
K	24.88
L	17.77
M	10.66
N	3.554

Figure 4 (c) Contours of fresh charge concentration (%).

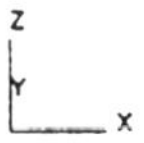

Figure 4. Case 3: In-cylinder flow patterns (at 140°, 180°, 240° and 300° ATDC).

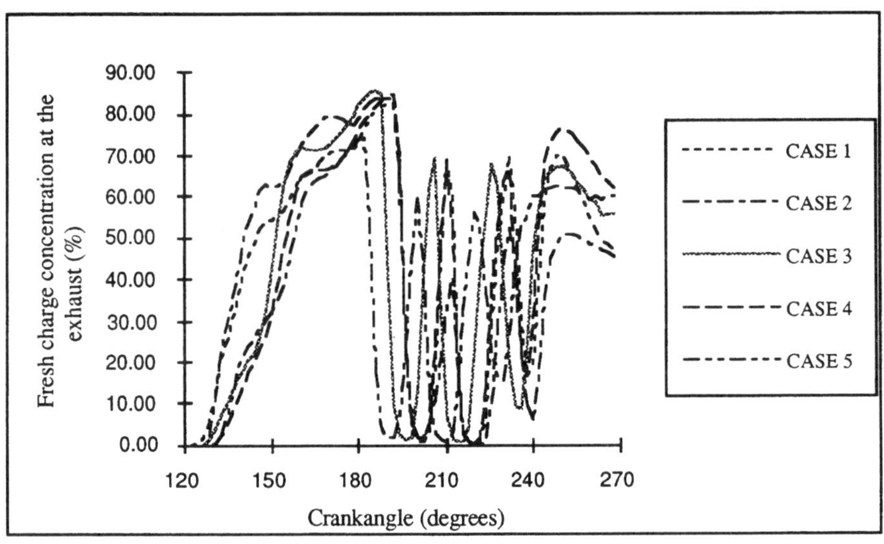

Figure 5 Variation of fresh charge concentration at the exhaust with crankangle.

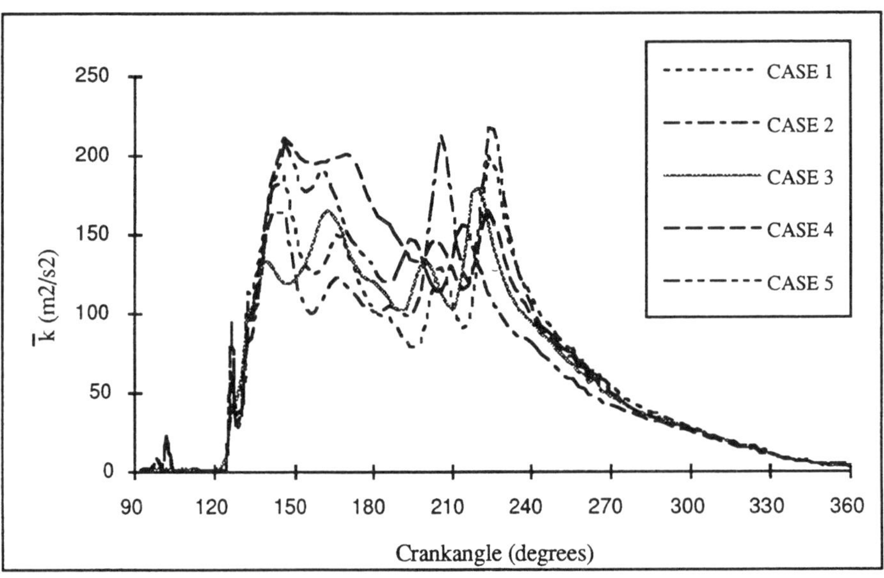

Figure 6 Variation of volume-averaged turbulence kinetic energy with crankangle.

C499/005/96

Turbulent flow simulations in model reciprocating engines with a differential stress model

A P WATKINS BA, MSc, DIC, PhD, MSAE, **T BO** BSc, MSc, PhD, and **C J LEA** BSc, MSc, PhD, AIMechE
Mechanical Engineering Department, UMIST, UK

SYNOPSIS

A differential stress model (DSM) of turbulence is applied to flows in a number of model reciprocating engines, with and without swirl, and results compared to those from k-ϵ models and experimental LDA data. During the intake stroke there is relatively little difference between the turbulence models at the mean flow level for swirling or non-swirling cases. However, the k-ϵ model apparently shows better agreement with experimental turbulence intensity data during intake than the DSM, but the possible effects of cycle-to-cycle flow variations lead to some uncertainty remaining. During the compression stroke, for non-swirling flows, the DSM returns vastly superior results to the k-ϵ model at both the mean flow and turbulence intensity levels. This is because the DSM suppresses the main vortical structure, which spuriously persists with the k-ϵ model. Much smaller differences are found for swirling flows, if the k-ϵ model incorporates a Richardson number modification to suppress turbulence intensity levels. However, the DSM still gives better agreement overall.

NOTATION

k	Turbulence kinetic energy
L	Valve lift
R	Valve radius
r	Radial coordinate
u	Axial turbulence intensity
U	Mean axial velocity
V	Mean radial velocity
V_p	Piston speed
$\bar{V}_p$	Mean piston speed
W	Mean swirl velocity
z	Axial coordinate

Greek symbols

ϵ	Dissipation rate of k

1 INTRODUCTION

Flows in reciprocating internal combustion engine chambers are invariably unsteady, three-dimensional and turbulent. The combination of these features makes computer calculations lengthy and expensive. The nature of the flow changes dramatically during the different strokes of the engine cycle. Thus the computer code must be capable of predicting high velocity, possibly swirling, flow during intake, with its concomitant large generation of highly structured turbulence, followed by relaxation of the flow during the early stages of compression. Towards the end of compression the turbulent structure may be subject to 'spin-up', as the volume available to the flow reduces, with an accompanying increase in density. The structure may be further modified if the engine is equipped with geometric features, such as piston bowl or cylinder head cavities, into which the fluid is forced, generating more turbulence.

The ability to accurately predict the conditions at the end of the compression stroke is crucial to the usefulness of a computer code for engine applications. For it is here that combustion takes place. The combustion rate is strongly affected by the turbulence of the flow, and is considerably higher than an equivalent laminar flame. However sophisticated a combustion model is employed, one cannot hope to accurately predict the flame spreading rates and such fine details as the amounts of unburned hydrocarbons and other emissions, if the turbulent flow structure is not correct.

For this reason considerable attention has been focused on deriving the optimum turbulence model for reciprocating engine flows. In the main, these have been derivatives of the standard k-ϵ model. The main modification is designed to account for the bulk compression/expansion undergone by the gas during the compression and expansion strokes of a reciprocating engine. The magnitude of this effect is represented by the divergence of the mean velocity vector.

For reasons of computational economy, reduction of numerical errors by using finer grid spacings, and because of the availability of laser-based experimental data, these models have largely been assessed in flows in two-dimensional axisymmetric model engines. The main such data was obtained from transparent model engines, operating at low speeds and low compression ratios (1), (2). Perhaps the major work undertaken in this area was that of Ahmadi-Befrui and Gosman (3). They applied a number of such models to non-swirling flows in two model engines. Both engines had a flat cylinder head and piston face, thus ensuring no squish effects during compression. The engine speed was 200 rpm, and compression ratios were 3.5 and 6.7. They found that all three models studied gave very similar predictions of the mean flow and turbulence intensities throughout the intake and compression strokes. Only when they examined the turbulence length scale behaviour did they find significant differences between the performances of the models.

A quite different approach to modifying the k-ϵ model was taken by Gul (4). He solved an additional transport equation for τ, the turbulent time scale, and extended the model to account for thin shear flows. Gul computed the same test cases as Ahmadi-Befrui and Gosman (3). Like them, he employed a first order upwind scheme for convection. Overall the k-ϵ-τ model gave a small improvement in predictive ability, particularly for the turbulence intensity.

In many engineering applications the differential stress model (DSM) (5) has become the favoured approach to modelling the turbulence of the flow because of its ability to predict the anisotropy of the flow, and its superior response to such features as streamwise curvature and adverse pressure gradients. There appears to be every reason to suppose that such an approach would bring benefits for the calculation of in-cylinder flows too, for here all the above mentioned flow phenomena appear. The work of El-Tahry (6) is the only one to date which uses a DSM for in-cylinder flows. He concentrated on model engine flows, but without compression. The inlet valve was stationary and non-protruding. His calculations employed the skew upwind differencing scheme for convection, with both the DSM and k-ϵ turbulence model. Overall, he concluded that the DSM and k-ϵ gave similar results for the computed mean flows, with fair agreement with the experimental data. The DSM gave significantly better agreement with data on turbulence intensities.

El-Tahry's work contains several limitations. The bulk-compression effects are very important. It is of considerable interest to ascertain how a second-moment closure model such as the DSM can handle this phenomenon. Clearly El-Tahry was not able to throw any light on the DSM's ability to predict the turbulence of the flow near TDC of compression, which is a crucial test of the engine model.

In the remainder of this paper a differential stress model for the turbulence is applied to three different reciprocating model engine test cases. These are the two flat-faced piston cases investigated by Ahmadi-Befrui and Gosman (3), and a swirling flow case (7). Parallel calculations have also been made employing the k-ϵ turbulence model, in its standard form, and in a form modified for swirling flows. Thus direct comparisons can be made between the performances of the two models for compressing flows. In this analysis numerical errors are reduced by employing the second-order accurate QUICK scheme in the momentum equations. This is the first instance of this scheme being used for reciprocating engine flows.

2 MATHEMATICAL MODELS

The flow in the axisymmetric cylinder is expressed by the unsteady compressible conservation equations for mass(continuity), axial, radial and circumferential momentum, energy, and a turbulence model to account for the Reynolds stresses. These equations are too lengthy to be detailed here. For non-swirling flows they are given in (8), (9), for both the k-ϵ turbulence model and the DSM. Both models include the velocity divergence term in the ϵ equation. The particular DSM employed for the non-swirling cases is that due to Gibson and Launder (10). This was employed without the wall reflection terms.

A much wider exploration of variants of the DSM has been made for swirling flows. These include the new wall reflection terms due to Craft and Launder (11) and the cubic IPCM for the pressure-strain interactions (5). In addition the τ equation has been extracted from Gul's (4) k-ϵ-τ model to form a DSM-τ model.

In the DSM the normal stresses are stored at the same nodal points as other scalars whereas the shear stresses are stored at the corners of scalar control volumes. The velocity components are stored halfway between scalar nodes.

Finite volume equations are formed by integrating the relevant form of the transport equation over a control volume surrounding each of the velocity and scalar locations. Cell face values are expressed in terms of nodal values through the QUICK scheme, in the momentum equations. The QUICK scheme was not used for the turbulence model equations, as this introduces instability and convergence difficulties. Instead the HYBRID scheme was generally used. However, for some of the swirling flow calculations the TVD-based MUSCL scheme was also employed. The non-swirling cases employ the Euler implicit method for temporal advancement, however, it was necessary to switch to the second-order accurate Crank-Nicolson scheme for the swirling flow case.

The general equations need to be amended near boundaries to account for boundary conditions. For both turbulence models two-layer wall functions are used to bridge the viscous sub-layer (5). At inlet an assumed profile is given for each of the dependent variables, details of which are given with the description of each of the test cases in the next section. At the symmetry axis, the normal velocity and the gradients of all the other properties are set to zero.

The published valve lift diagram was discretised to obtain the valve motion. The number of grid lines covering the valve curtain area was fixed for a given grid. These grid lines then expanded and contracted with the valve motion. The grid lines between the valve face and the piston face also expanded and contracted in the axial direction with the piston motion.

The same number of lines was used in this volume throughout the intake and compression strokes. The transport equations were modified to account for the translation of the grid lines by a coordinate transformation in the axial direction. Further details can be found in Lea (8).

3 TEST CASES

Two experimental cases, without swirl, have been simulated of a flat-piston geometry with 3.5 and 6.7 compression ratios, (1) (2). A further case (7) in which the inlèt flow is swirled by inlet vanes, giving a swirl number of 1.2 and having a compression ratio of 3.5, is also simulated. All engines ran at 200 rpm. These geometries, although idealised, still exhibit many of the flow phenomena existing in real engines, and thus allow turbulence models to be assessed. The lower compression ratio cases provide experimental data throughout the intake and compression strokes. The majority of the assessments are therefore made using these test cases. The other test case provides data only towards the end of the compression stroke. The computations presented here have been guided by the results of steady inflow calculations (8). The effects of grid density, the differencing scheme for convection, and time-step increment were investigated for the low-compression non-swirling case and for both turbulence models. The best of these numerical practises were then employed for the other cases, again using both the turbulence models.

A 45 x 45 line grid was used for the majority of the calculations. A further grid of 60 x 60 lines was also employed for some calculations. This showed that, in conjunction with the QUICK scheme, the 45 x 45 grid results give a sufficiently accurate representation of the turbulence model's performance, to allow conclusions to be drawn.

The computations were initially marched through one full 720° engine cycle, to provide more realistic initial conditions for the start of the intake stroke. The second cycle simulations showed that the results became independent of the cycle during the intake stroke of the second cycle. The time step used was equivalent to ¼° crank angle.

The axial and radial components of the mean velocity at inlet were set by assuming that the flow enters at the valve seat angle, 60° in these cases, and with a uniform profile. This latter assumption is based on steady-flow calculations (8). There it was shown that experimentally-determined inlet flow profiles give a much better description of the subsequent in-cylinder flow. It was also shown however, that one such profile measured on one flow rig could not be transferred to another flow case. Indeed, the second case was calculated more accurately if a uniform profile was assumed at inlet. Because of this result, and because there are no measured inlet profiles for these reciprocating cases, it was decided to compromise and accept the uniform profile as best.

In the absence of data, the levels of k at inlet were set to 2% of the incoming local mean kinetic energy, and the Reynolds stress distributions were set isotropic. Inlet ϵ was prescribed from an assumed turbulence mixing length equal to one third of the valve lift.

For the swirling case the inlet swirl velocity W = 1.2VL/R, where V is the radial velocity at inlet, L is the valve lift and R is the valve radius.

4 RESULTS

4.1 Non-swirling, compression ratio = 3.5

A comparison of predictions given by the k-ϵ model and the DSM for this case is shown in Fig. 1, for intake, and Fig. 2 for compression. The mean flow predictions during intake are quite similar, with a slightly superior prediction of the peak jet velocity by the DSM, and a better reproduction of the flow behind the valve by the k-ϵ model. For the turbulence intensities the k-ϵ model gives larger values in the jet region than the DSM and as a consequence better matches the experimental measurements. This is in contrast to steady flow calculations reported in (8) which indicate that in this region the DSM gives larger values of turbulence intensities. For the k-ϵ model, all the turbulence intensity values quoted here are obtained from the Boussinesq relationship expressing the normal stresses in terms of the appropriate mean flow gradients, and not from k. The turbulence intensities behind the valve are also better captured by the k-ϵ model. This is again in contrast to the steady flow calculations. There also the k-ϵ model gives larger values than does the DSM, but the measured turbulence intensity levels are so low that the DSM results are in better agreement. This raises the question as to why the reciprocating engine turbulence intensity values are measured so much higher than for the steady flow case, in which the valve lift was 6 mm and hence the inflow rate was as high as for much of the intake stroke here. One possibility is that the measured turbulence intensities for the reciprocating case include the effects of cycle-to-cycle variation.

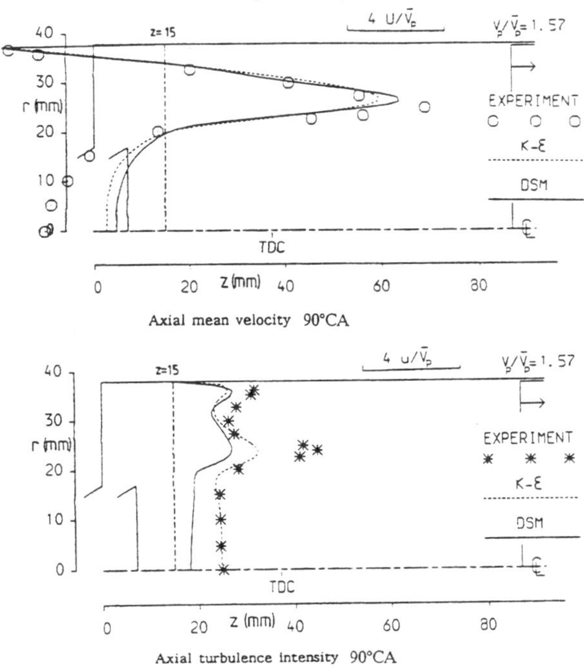

Fig. 1 Predictions during intake, non-swirling flow, compression ratio = 3.5

The results shown in Fig. 2 clearly demonstrate the superiority of the DSM to capture the dissipative effects of compression on the flow. The mean flow vortex which is created during the intake stroke is quickly suppressed, due to higher magnitudes of the internal stresses at the end of the intake stroke/beginning of compression, and excellent agreement is found with the experimental data, apart from near the cylinder head/wall corner where the experiment indicates a small vortex exists. As for the turbulence intensities the agreement given by the DSM is everywhere superior to that given by the k-ϵ model. This is particularly marked near the piston face. The uni-directional nature of the flow during compression will result in cycle-to-cycle effects having less importance and effecting the measured turbulence intensities to a much smaller degree than during intake. The features shown here persist up until TDC of compression, (8), (9).

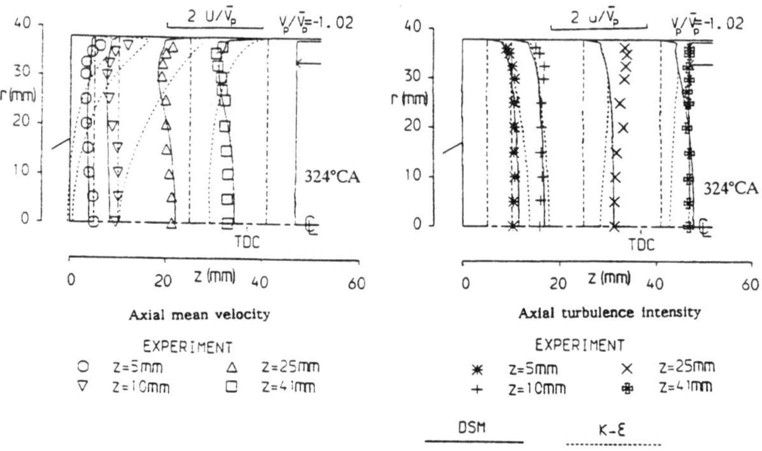

Fig. 2 Predictions during compression, non-swirling flow, compression ratio = 3.5

4.2 Non-swirling, compression ratio = 6.7

The primary effect of the reduced volume available to the inflow during the intake stroke is that the main vortex generated behind the valve is increased in strength. This vigorous vortex persists throughout the intake stroke, and into compression. Fig. 3 shows the results

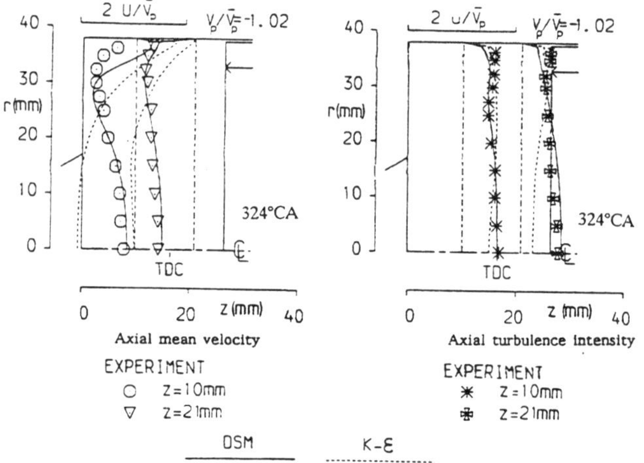

Fig. 3 Predictions during compression, non-swirling flow, compression ratio = 6.7

during compression, for both the DSM and k-ϵ models. The DSM results show the persistence of the intake generated vortex which had been suppressed in the low compression case. Unlike the k-ϵ results, however, the vortex is restricted to the cylinder head/wall corner. Apart from the corner vortex, the DSM gives excellent predictions of the axial mean flow. The k-ϵ model gives very poor results, in a similar manner as for the low compression case. The DSM also returns superior predictions of the turbulence intensities.

Again these features persist up to TDC of compression, (8), (9).

4.3 Swirling, compression ratio = 3.5

To avoid excessive CPU requirements, the Euler implicit method had to be replaced by the second-order accurate Crank-Nicolson scheme. Time-step independent results were then obtained using ¼° crank angle steps.

The results obtained using the variants of the DSM, outlined above, will be published elsewhere. Here just one set are shown. These make use of the τ model equation, which increases the predictions of turbulence intensity during intake (4), the wall reflection model of Craft and Launder (11), which is designed to damp the normal components of the Reynolds stress tensor for impinging flows, as found during the intake stroke on the cylinder wall, and the IPCM for the pressure-strain correlation (5).

A parallel set of calculations were also made using two versions of the k-ϵ model. The first was a standard model, augmented by the velocity divergence term for compression and expansion. In the second the Richardson number modification to the ϵ equation for swirling flows (12) was implemented.

Figures 4 and 5 show the predictions, for all the models, during intake. There are no substantial differences between the two k-ϵ model results. The flow in the jet is best captured by the DSM, due to its less diffusive nature. This also results in the backflow in the corner vortex being better predicted by the DSM. The suppression of turbulent diffusion by streamline curvature in the DSM results in the wall jet being narrower and stronger, than for the k-ϵ model. At z=40mm the DSM therefore predicts better agreement with the data. However, by z=65mm the data suggests that another mechanism causes the wall jet to spread. This is not captured by either model, but the increased turbulent diffusion given by the k-ϵ model results in better agreement here. The backflow of the main vortex is better predicted by the DSM, over the middle of the flow region. However, the suppression of the axial flow behind the valve is overpredicted by both models. The DSM performs worse here, despite use of the new wall reflection model which suppresses all the Reynolds stresses as the wall is approached, leading to reduced turbulent diffusion.

In the annular jet region the k-ϵ model predicts higher and more accurate values of the axial turbulence intensities. However, as the wall jet develops, the higher levels of turbulent diffusion produced by the k-ϵ model result in the turbulence intensities diminishing in value. The augmentation of turbulence intensities by use of the τ equation, results in the DSM predicting larger values of u as the wall jet develops. Thus near the piston face the experimental levels have almost been reached. Despite the DSM predicting the mean flow better in the backflow region of the main vortex, the turbulence intensity levels are suppressed. This is in agreement with the non-swirling flow case. The high levels of measured turbulence intensity near the centreline may partly be due to cycle-to-cycle variations in the mean flow possibly caused by swirl-centre precession.

Figure 6 presents the predictions for all the models at 324° crank angle. The experimental data indicates that the vortical structure set up during intake has been suppressed by this stage of compression. Uni-axial flow predominates with larger velocities found near the cylinder wall. All the turbulence models essentially give this result also. There is some backflow predicted in the middle region of the cylinder by the DSM, and near the cylinder centreline by the modified k-ϵ model, but all the variations are very small.

The axial turbulence intensities are accurately predicted by the DSM, except near the centreline. The standard k-ϵ model, however, overpredicts by about 100% throughout most of the chamber. The standard k-ϵ model is well-known to overpredict turbulence

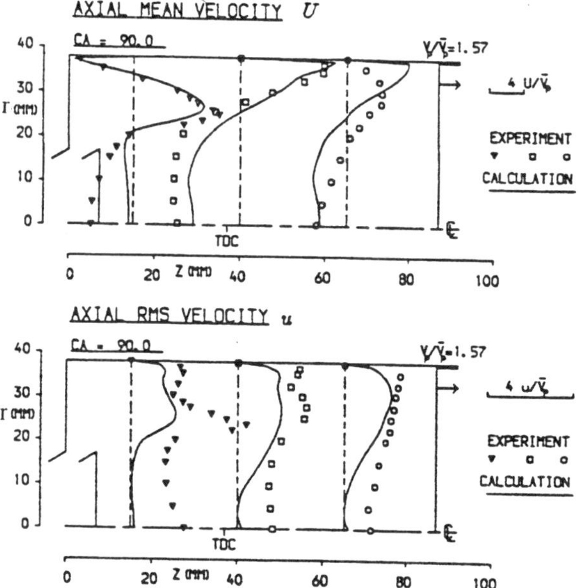

Fig 4 DSM predictions during intake, swirling flow

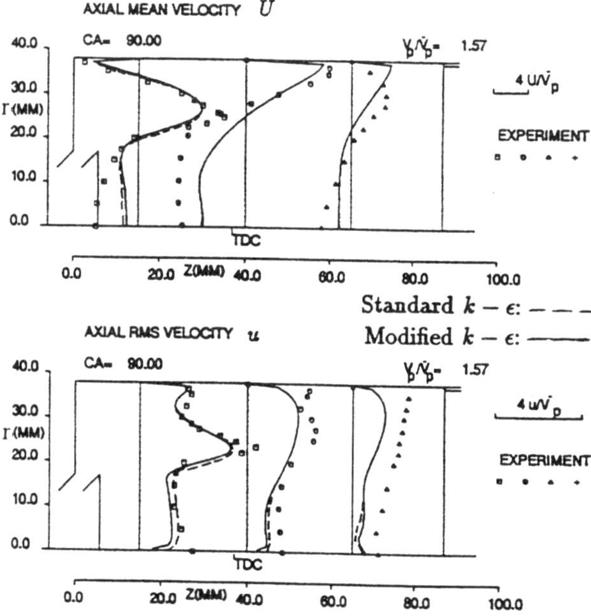

Fig 5 k-ϵ predictions during intake, swirling flow

intensities in swirling flows, and it can be clearly seen that the Richardson number modification to the model has been successful in suppressing the turbulence intensities to values close to the experimental data. Even so the agreement is not as good as for the DSM.

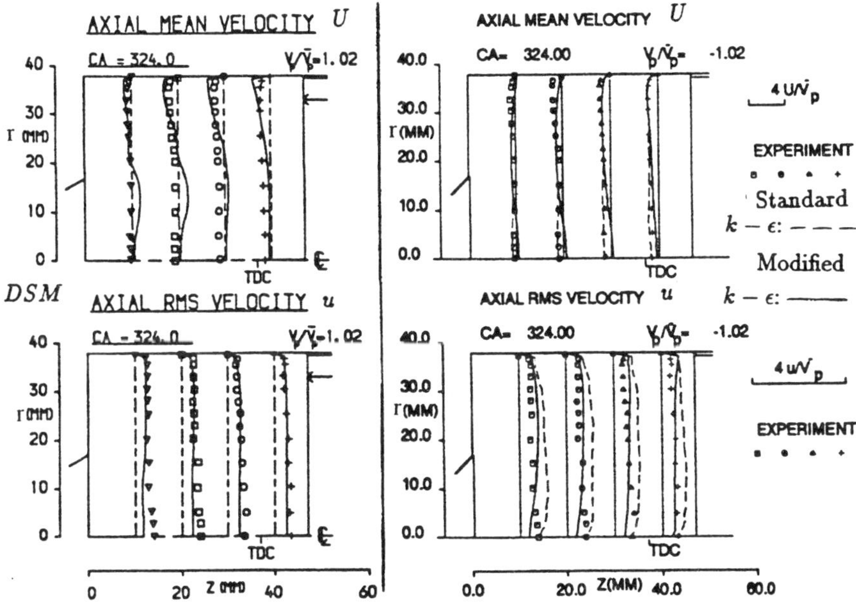

Fig 6 Predictions during compression, swirling flow

5 CONCLUSIONS

During the intake stroke, all the cases examined showed that the DSM predicts a more vigorous vortical structure than does the k-ϵ model. This behaviour is driven by two main effects. Firstly, the intake jet is less diffused by the DSM. Secondly, the DSM responds to stabilising streamline curvature and therefore suppresses turbulent diffusion of momentum. As the clearance height was decreased, the turbulence suppression with the DSM persisted until later in the stroke, because the increased geometric constraint ensures that stabilising secondary strain remains significant until later. However, comparison of mean flow simulations with experiment shows that the two models return very similar predictions for the intake stroke. This was partly because most of the data relates to near valve positions, where the flow is dominated by pressure field and geometry effects, and partly because experimental data are only available during the first half of the stroke. For the swirling flow case where more data exists, differences appear close to the piston face, due to the higher turbulent diffusion given by the k-ϵ model.

The corresponding turbulence intensities for the non-swirling cases were poorly predicted, but some of the discrepancy is thought to be because of cycle-to-cycle variations in the mean flow contributing to the intensity levels. For the swirling flow case the DSM gave poor results in the near jet region but improving results as the flow proceeds downstream along the wall. The k-ϵ results gave the inverse performance. The k-ϵ results were superior in the main backflow region. Again, it is possible that the experimental intensity data are being augmented by cycle-to-cycle variations, in this case caused by swirl centre precession.

The DSM mean motion for the non-swirling cases tended to decay towards a state of uni-axial compression at TDC. This was in good agreement with measurements. The k-ϵ model predicted the continued presence of a mean vortical structure occupying the whole of the in-cylinder space. For the swirling case all the models predicted the tendency to uni-axial flow during compression, with the standard k-ϵ model giving marginally the best agreement with data.

The measured decay of turbulence during the compression stroke was well matched by all turbulence models, particularly so for the DSM. At TDC of compression for the non-swirling cases predicted turbulence intensities were generally in agreement with not only the present data but also more widely reported correlations of open-chamber data. For the swirling flow case, the standard k-ϵ model greatly overpredicted the turbulence levels. These were suppressed by the modified model, but the DSM still gave the best agreement with data.

Acknowledgement
The research has been supported by the SERC under grants GR/D/78237 and GR/H/48606.

REFERENCES

(1) BICEN, A.F. Air flow characteristics of model internal-combustion engines, Ph.D. Thesis, University of London, 1983.

(2) AHMADI-BEFRUI, B., ARCOUMANIS, C., BICEN, A.F., GOSMAN, A.D., JAHANBAKHSH, A. and WHITELAW, J.H. Calculations and measurements of the flow in a motored model engine and implications for open-chamber direct-injection engines, in Three dimensional turbulent shear flows, Eds. S. Carmi et al, 1982, ASME, 1-9.

(3) AHMADI-BEFRUI, B. and GOSMAN, A.D. Assessment of variants of the k-ϵ turbulence model for engine flow applications, Int. J. for Numerical Meth. in Fluids, 1989, 9, 1073-1086.

(4) GUL, M.Z. Prediction of in-cylinder flow by a multiple time-scale turbulence model, Ph.D. Thesis, Faculty of Technology, University of Manchester, 1994.

(5) LAUNDER, B.E. Second-moment closure: present ... and future?, Int. J. of Heat and Fluid Flow, 1989, 10, 282-300.

(6) EL-TAHRY, S.F. A comparison of three turbulence models in engine-like geometries, in Proc. of Int. Symp. on Diagnostics and modelling of combustion in reciprocating engines: COMODIA 85, Tokyo, 1985, 203-213.

(7) ARCOUMANIS, C., BICEN, A.F. and WHITELAW, J.H. Squish and swirl-squish interaction in motored model engines, J. of Fluids Engng., Trans. of the ASME, 1983, 105, 105-111.

(8) LEA, C.J. Second-moment closure computations of in-cylinder flows in idealised reciprocating engines, Ph.D. Thesis, Faculty of Technology, University of Manchester, 1994.

(9) LEA, C.J. and WATKINS, A.P. Differential stress modelling of turbulent flows in model reciprocating engines, Submitted to Proc. I. Mech. E., 1995.

(10) GIBSON, M.M. and LAUNDER, B.E. Ground effects on pressure fluctuations in the atmospheric boundary layer, J. Fluid Mech., 1978, 86, 491-511.

(11) CRAFT, T. and LAUNDER, B.E. New wall-reflection model applied to the turbulent impinging jet, AIAA J., 1992, 30, 12, 2970-2972.

(12) RODI, W. Influence of buoyancy and rotation on equations for the turbulent length scale, Proc. 2nd Symp. on Turbulent Shear Flows, 1979, 10.37-10.42.

C499/027/96

Zonal modelling of diesel engine smoke emission

S WELCH BSc, MSc and **J B MOSS** BSc, MSc, PhD, CEng, MRAeS, FIMechE
School of Mechanical Engineering, Cranfield University, Bedford, UK

SYNOPSIS

A zonal phenomenological model has been developed and validated as a basis for soot emissions modelling. The behaviour and sensitivities of various soot formation and oxidation expressions were examined using the simulation. Results were compared with the extensive parametric data-set due to Kamimoto (1). It was found that further calibration of soot models is generally necessary and that there is no marked difference in performance despite their very different origins. Sensitivity to soot model constants is generally low, but predictions may be very sensitive to certain model dependencies which affect the variation in the rates during the combustion process.

NOTATION

c_{min} - smaller of the fuel concentration and the air concentration divided by the stoichiometric air-fuel ratio [kg m^{-3}]
E - activation energy [kcal / mol]
k - turbulent kinetic energy [m^2 s^{-2}]
m - mass [kg]
$\dot{m}$ - mass transfer rate [kg s^{-1}]
N - engine running speed [r / min]
p_{noz} - injection pressure [Pa]
V_f - volume of fuel injected per stroke [mm^3]
γ - mass fraction in fine structure (Magnussen EDC model) [-]
ε - turbulent energy dissipation rate [m^2 s^{-3}]
θ_{ip} - injection period [degrees crank angle]
υ - kinematic viscosity [m^2 s^{-1}]
ϕ - equivalence ratio [-]
χ - mass fraction burnt [-]

Subscripts

for - formation
o - zero convection conditions
ox - oxidation

1 INTRODUCTION

Whilst the diesel engine has become increasingly popular as an automotive powerplant in recent years, serious concern is now being expressed over the health effects of exhaust particulates. In particular, mortality and morbidity have been shown to correlate well with atmospheric fine-particle concentrations (PM10) and this has led to a tightening of legislated limits. Computer models can give an insight into sooting processes and cut engine development work. There is therefore great interest in the development of reliable models for the prediction of smoke emission. Despite much previous work, no consensus has been reached and the most popular models in current use lack a fundamental basis.

The main focus of this work is the prediction of exhaust smoke levels, rather than detailed in-cylinder distributions. This problem is best addressed using combustion models of the zonal phenomenological type, and one such model has been developed as a basis for the soot modelling study. However, the findings are not limited to combustion models of this type, and an attempt has been made to assess the general applicability of the soot models examined.

Expressions used in previous work to describe soot formation and burn-out derive from a wide variety of sources, and usually relate to operating conditions far removed from those pertaining to diesel combustion. Moreover, development of these soot models has often been undertaken in conjunction with a particular combustion model. Thus, their applicability in different combustion models remains uncertain. The analysis reported in this paper includes a comparative study of a number of soot models. The critical controlling parameters are identified and model sensitivities are examined.

2 COMBUSTION MODELLING METHODOLOGY

The approach taken to combustion modelling is of great importance in the prediction of emissions. Any useful emissions model must be underpinned by an accurate description of the combustion process. The combustion model must supply the detailed information relating to mixture composition and thermodynamic state which are essential inputs to the soot model mechanisms.

The processes leading to the emission of smoke from a diesel engine are extremely complex. The soot formation and oxidation processes are closely coupled and the former, in particular, is intimately related to the chemistry of the combustion process. The combustion regime is turbulent and reactants are often highly localised in turbulent eddies. Moreover, the sooting processes themselves are highly non-linear, so they are very sensitive to variation in conditions, on micro and macroscopic scales.

Thus, it is necessary that the combustion model can be integrated consistently with the emissions model. Also, any useful model must contain some representation of the different states existing in the mixture. Finally, if the model is to be used for development work involving parametric studies, it is essential that the strategy adopted is relatively undemanding in computational terms.

2.1 Modelling techniques

Many computational models have been developed for diesel-engine simulation, though not all are well-suited to emissions work. The two main classes are phenomenological models and multi-dimensional or computational fluid dynamic (CFD) models. The former consist of a set of independent sub-models describing the important combustion phenomena, structured around a thermodynamic analysis. They can be further classified into zero-dimensional models, which compute only chamber averages, and quasi-dimensional and stochastic models, which attempt to characterise the different regions of the mixture using zones and a statistically-based mixing model respectively. CFD models solve the governing conservation equations within a computational grid, such that the ensemble-averaged state of the mixture is fully described at every point in space.

The zero-dimensional models are the least useful, since they do not have the required spatial resolution and heat-release calculations usually lack a fundamental basis. The strength of the CFD model is that detailed spatially-resolved predictions are given. Thus, this type of model has a role in the design of the combustion chamber to give advantageous gas flows. However, such models are poorly suited to the prediction of exhaust smoke, and little success has been reported in this area. Additionally, computational limitations do not allow the full potential of the soot models to be exploited.

The above requirements of a realistic emissions model are best met by phenomenological models of the quasi-dimensional (multi-zone) or stochastic variety. There are several advantages in using this type of model. Because the component sub-models (spray, evaporation, heat transfer etc.) are fairly independent, any level of detail can be built in, within the bounds of experimental knowledge. Together with the fact that some model parameters may be tuned to give a good match to experiment, this ensures that the combustion process may be represented with sufficient accuracy for the purposes of emissions modelling. This contrasts with the CFD approach, where there is limited scope for tuning model constants.

In phenomenological models, there is no restriction on the complexity of the mixture description which can be fed into the emissions model. Thus, the emissions model can take full advantage of the composition and temperature information obtained from the combustion model, for example, in using reactants and combustion-products temperatures and a representative range of mixture compositions. Though a CFD model supplies similar information in greater detail, it cannot be used in practice. If a presumed pdf approach is adopted, mixture fraction and soot parameters must be assumed to be uncorrelated, since these relationships are inaccessible. This is a major simplification, since it is known that high soot concentration will in fact be strongly correlated with high temperature and high fuel concentration. Alternatively, by using Monte-Carlo methods these problems may be overcome, but such models are very demanding computationally. Thus, though a CFD model provides much greater spatial detail, in practice soot model inputs are confined to local averages.

In the class of interest, multi-zone or stochastic models, much work has already been done. From independent origins, at least eighteen such models have been developed for the simulation of diesel engine combustion. The majority are of the multi-zone type and have been applied to direct-injection diesel engines; only three are stochastically-based, including one of only two models which have been specifically applied to an indirect-injection diesel. A number of these have been applied to the prediction of smoke emission (see section 4).

3 COMBUSTION MODEL DESCRIPTION

This section describes some of the key components of the combustion model developed in the current work (a fuller description is given in ref (2)). Most sub-models adopt the approaches used in previous work, but special attention has been given to certain topics. The structure of the model is illustrated in Fig 1.

The model follows the combustion process by calculating the composition and temperature of a number of zones and sub-zones. Zones are injected in successive sets and each represents a region of the spray, introducing an element of dimensionality. Magnussen's eddy dissipation concept (EDC) is used to provide a consistent framework for the computation of fast combustion processes and the relatively slower sooting processes (3). This requires that the mixture is partitioned into two regions, the surrounding fluids and the heated fine structure. Mixture is allowed to burn only when intimately mixed in the latter region and the combustion rate may be set by the rate of transfer of mixture through this region. Building upon Magnussen's model, a further level of partitioning is introduced, in which separate sub-zones are used for mixture which has passed through the heated fine structure and the remaining pure reactants. This differs from the approach of many existing two-zone models, in that unburnt fuel or air can be found in the burnt-mixture region whenever combustion is rich or lean respectively. This modification is clearly of great significance for soot modelling, since unburnt fuel in high temperature regions gives high soot formation rates.

3.1 Model details

The spray model is based upon the penetration equations given by Hiroyasu (4). Air entrainment is calculated by applying conservation of momentum in the radial direction. Progressive breakup of the spray is accommodated and velocity in the break-up region is assumed to vary linearly between the break-up velocity and that of the fully-developed jet. The effects of swirl and impingement are calculated by parallel computations using Hiroyasu's scaling factors.

Evaporation is described using representative sets of droplets. The droplet size is described using Hiroyasu's SMD correlation (5) and size-distribution expression (6). Various convection correlations were examined and the most satisfactory performance was obtained using that for evaporative mass transfer rate due to Ranz (7):

$$\frac{\dot{m}}{\dot{m}_o} = 1 + 0.3\,Re^{\frac{1}{2}}\,Pr^{\frac{1}{3}} \qquad (1)$$

In this expression, the Reynolds number is obtained simply from the size and velocity of each representative droplet.

The mass transfer rate between the sub-zones is given directly by Magnussen's eddy dissipation concept expression for the rate of transfer through the heated fine structure region:

$$\dot{m} = 23.6\left(\frac{\nu\varepsilon}{k^2}\right)^{\frac{1}{4}}\frac{\varepsilon}{k} \qquad (2)$$

The turbulence parameters are calculated using a k-ε turbulence model. Various correlations for k and ε were investigated. The most satisfactory performance was achieved using Dent's expression for the turbulent energy dissipation rate (assumed constant) (8):

$$\varepsilon = \frac{3 p_{noz} N V_f}{\theta_{ip}} \tag{3}$$

In describing the combustion rate, a distinction is made between the initial premixed combustion of mixture prepared during the ignition-delay period and the subsequent diffusion burning. The fraction of combustible mixture allowed to burn during the premixed phase was set using an empirical constant selected to match experimental data over a range of operating conditions. During diffusion burning, the combustion rate is expressed in terms of the mass transfer rate given in equ. (2):

$$\frac{dm_{fuel}}{dt} = \dot{m} \frac{\chi}{1 - \gamma\chi} c_{\min} \tag{4}$$

Convective heat-transfer calculations use the Woschni correlation. The overall heat transfer is distributed between the model zones according to the temperature difference with the wall and the zonal surface area (9). Radiative heat transfer is highly dependent on the soot concentration in the mixture. In this case, heat transfer is calculated assuming the spray soot cloud acts as a grey body. The emissivity is expressed in terms of the soot density and the cloud size (10). In reality, the distribution of the radiative heat transfer is complex and for simplicity, transfers between zones and sub-zones are neglected. Partitioning between zones is taken to be proportional to the emissivity and the difference between zonal and wall temperatures each raised to the fourth power.

Internal energies and specific heat capacities are obtained using the polynomial expressions of Krieger and Borman (11). Temperatures are followed in each sub-zone allowing for the effect of chemical heat-release (due to combustion, soot formation and soot oxidation), compressive work, heat transfer and mixing.

3.2 Validation

The combustion model was validated by comparison with existing experimental measurements from the literature. In particular, the extensive dataset published by Kamimoto et al. (1) for a 0.7 litre DI diesel engine has been used to establish satisfactory performance of all model predictions over a range of operating conditions. The final values of certain model parameters were chosen to give best agreement with this data.

A key performance parameter is the heat-release rate. Very good agreement was obtained between predicted and measured values as illustrated in Fig 2 for the standard operating conditions of $\phi = 0.5$, $N = 1250\,\mathrm{r/min}$. Since combustion is usually rich during the main diffusion-burn phase, heat release is approximately proportional to the air entrainment rate. In the case of light-duty engine (0.52 litre), Kuo recommended a reduction of the scaling factor for air entrainment into burning mixture from the original value of 0.7 given by Hiroyasu to a value of 0.4 (12). This choice was adopted as standard.

Only in the case of swirl ratio variation did the simulation show clearly different trends to the experimental data. Using Hiroyasu's correlation for the reduction of penetration due to swirl, air entrainment rates rose rapidly at high penetrations. This is contrary to the empirical finding, where air entrainment falls off at higher penetrations because deflection of the jet reduces the velocity difference relative to the swirling air. Kamimoto's data show that the effect of increased swirl levels on entrainment is minor, and the best match with the experimental data was obtained by ignoring the swirl effect.

Heat release will also be affected by the combustion rate calculated according to equ. (4). Using this expression in its basic form led to a serious underestimation of the heat-release rate during the early part of the diffusion burning. At this point, the heat release rate scales directly with the burn rate of equ. (4) and it was found necessary to include an additional multiplicative factor of 50. This approach is also used in CFD applications of the Magnussen combustion-rate expression, and a wide range of values have been used for the constant factor. Finally, a value of 0.27 was selected for the premixed burn factor, this giving a satisfactory agreement with the empirical data over the full range of the dataset.

4 EXHAUST SMOKE MODELLING

4.1 Soot formation and oxidation

Information relating to soot formation and oxidation is available from a variety of combustion systems and conditions, often in the form of correlations. These are not necessarily appropriate for use in the unique combustion environment of the diesel engine, and the source of the expressions must be carefully considered.

The key to accurate prediction of smoke levels is identification of the important controlling processes and faithful representation of these in models. Diesel-engine studies have reported that exhaust soot may be well-correlated by both temperature (Ahmad (13), Iida (14)) and turbulent mixing timescale (Dent (8)). Thus, both pure chemical and turbulent processes are of interest. A number of the expressions previously applied to the diesel derive from laminar-flow conditions in simple combustion systems, e.g. those due to Farmer (15), Tesner (16), Lee (17), Nagle (18) and Magnussen (19). Magnussen also demonstrated that such expressions will over-estimate rates under turbulent combustion conditions (19) and this must be taken into account in the case of the diesel.

Other expressions have been derived in conjunction with combustion models, e.g. those due to Khan (20) and Hiroyasu (4). In this case, though the correlations used have the typical form of a chemical-kinetic expression, the effects of turbulence have been introduced in calibration. Moreover, the rate is expressed in terms of accessible model parameters, rather than exact local values. In these cases, averaged species concentration and temperatures are used, and the error introduced in using these representative parameters is also accounted for in calibration. However, use of such calibration procedures is highly undesirable since the resulting expressions are not fundamentally-based and their general applicability is questionable. In fact, use of the pair of expressions due to Hiroyasu has given mixed results in prediction of exhaust smoke trends (4,12). The generality of the Khan expression is even less well established, since both constants and the equivalence-ratio exponent were fitted using only exhaust data, and oxidation was neglected.

Finally, the experimental conditions relating to the derivation of a particular correlation must be taken into account in assessing its applicability in the diesel engine. Virtually all correlations available were determined using simple fuels at atmospheric conditions. Exceptions include the Khan and Hiroyasu correlations mentioned above. With oxidation, the partial pressure of oxygen is more significant than the total pressure. This value may exceed one atmosphere in the diesel engine, and only the study of Nagle (18), as extended by Park (21), covers these conditions.

Some of these issues can be addressed by examining model performance in the context of diesel combustion simulations. Much previous work has been done and reasonable results have been reported in specific simulations in the literature. However, little attempt has been made to examine sensitivity to combustion-model performance and soot-model parameters. Also, there is very limited information regarding applicability in different types of combustion model and in different combustion regimes and engines. These issues provided motivation for the current study.

4.2 Implementation and validation

In the current work, the soot-formation expressions due to Hiroyasu (4), Harmadi (6), Khan (20), Farmer (15) and Tesner (16) were implemented, together with the oxidation expressions of Hiroyasu (4), Nagle (18), Lee (17), Magnussen (laminar) (19), Feugier (22) and Magnussen (EDC) (3). Yields are predicted according to any pair of expressions, though the Tesner and Magnussen EDC expressions must be used as a pair. As an example of the soot correlations, the Hiroyasu expressions are given below:

$$\left.\frac{dm_{soot}}{dt}\right|_{for} = A_{for}\, m_{fuel}\, p^{0.5}\, e^{-\frac{E_{for}}{RT}} \tag{5}$$

$$\left.\frac{dm_{soot}}{dt}\right|_{ox} = -A\, m_{soot}\, p_{O_2}\, p^{0.8}\, e^{-\frac{E_{ox}}{RT}} \tag{6}$$

In the interests of simplicity, it was chosen to use zonal compositions and temperatures for evaluation of these rates, rather than the values for each thermodynamic subzone. However, it was assumed that soot oxidation occurred preferentially from the combustion products region, and soot and species concentrations were followed in each subzone (see Fig 1). All other expressions are computed in the same way, though the Tesner-Magnussen model receives special treatment with the rate also being calculated separately for each subzone. Also, in this model, both soot and radical nuclei concentrations are followed in each subzone.

In order to provide a standardised basis for the comparative study, the Hiroyasu expressions were fitted to in-cylinder and exhaust data from the experimental study of Kamimoto (1) and related works. The values of the pre-exponential constants in equs. (5) and (6) were chosen to give the best match considering all cases of the parametric study. All soot oxidation rates were calculated using the resulting soot concentration as an input, thus allowing direct comparison.

4.3 Results and discussion

Illustrative results are shown in Figs 2 - 5. Fig 2 includes the total soot yield obtained using the Hiroyasu expressions, which is in good agreement with the experimental result for this case. The pre-exponential constants in the soot formation and oxidation correlations were set as follows: $A_{for} = 2$, $A_{ox} = 6 \times 10^{-6}$. Figs 3 and 4 shows the calculated overall rates obtained using each of the formation and oxidation expressions. The maximum possible formation rate is also included on Fig 3 to provide an upper bound. On each figure, the rates from the Hiroyasu expressions represent the basis for comparison since they have been fitted to match experimental yield data.

The figures show that the calculated rates are generally quantitatively poor compared with the Hiroyasu standards. Thus, if the expressions are to be used for emissions prediction, calibration will usually be necessary, and this will tend to obscure the expected reduction of rates due to turbulent effects. In light of this, it is not important to account for turbulence effects explicitly.

For formation, the expressions due to Harmadi and Tesner give rates which are comparable with the Hiroyasu standard during early diffusion burning. However, in the latter case, it was found that the number of radical nuclei drops off rapidly since the Magnussen EDC oxidation rate does not fall with temperature. Thus, the Tesner formation rate also drops and the exhaust level using this pair of expressions is negligible. The Khan correlation gives very high levels during the main phase of diffusion burning, but also falls more rapidly than the Hiroyasu rate. This behaviour results from an explicit dependence on equivalence ratio. Combustion is initially rich such that unburnt air concentrations fall to very low levels, giving high values of equivalence ratio; however, once more air has been entrained, the equivalence ratio falls to very low values. Thus, in practice the expression is switched on only during rich combustion. Finally, the Farmer expression is quantitatively poor.

The oxidation-rate correlations give a similar spread of magnitudes. The Magnussen EDC expression stands out in giving a different variation with crank angle to the quasi-chemical rates. This is because the EDC correlation is controlled by turbulence parameters which remain relatively constant and involves no explicit dependence on temperature. The Lee and Feugier rates are very high. Though the Magnussen laminar expression was fitted to the same experimental data as the Lee expression, the calculated rate is much lower because of the different dependency on oxygen partial pressure. In the simulation, the calculated oxygen partial pressures are much higher than the value of 0.05 - 0.1 atm relating to the original experimental work. Thus, the danger of applying correlations beyond the range of experimental conditions for which they were derived is clearly demonstrated. In fact, amongst the expressions examined, only that due to Nagle attempts to account for the variation of oxidation rate over a wide range of oxygen partial pressures.

It is instructive to plot the oxidation rate against the reciprocal temperature, as in Fig 5. The form of this plot follows the widely cited comparative plot of Park (21); the data points represent calculated rates in randomly-selected zones in the period between ignition and exhaust valve opening. The plot reveals interesting differences in behaviour. The Magnussen laminar expression is shown to be very insensitive to variation in temperature and fuel concentration. The rate drops only very late in the cycle when the temperature is relatively low. The Hiroyasu and Lee data points are each grouped in two arms, which both show an increase in the rate with temperature. The upper arms correspond to early combustion, when the oxygen partial pressure is relatively high, whilst the lower arms relate to the period when the temperature is falling and oxygen concentration and pressures are low, later in the cycle. In each case, the slope of the Hiroyasu-expression curve is lower, due to the use of a much lower activation energy. With the Nagle expression, all data points are grouped on the same curve, and the rate remains relatively high later in the cycle. This is because the expression gives an independence from oxygen partial pressure at temperatures below about 2000 K.

The sensitivity of rate predictions to the values of certain model parameters was examined. It was found that virtually identical results could be obtained using a value of around 40 kcal / mol for the Hiroyasu expression activation energies (such a value is much more typical of experimental data than the original values of 12.5 and 14 kcal / mol). Using the Khan correlation, it was also found that very similar trends could be obtained using values of unity or zero for the equivalence ratio exponent.

Overall, it was found that there are no distinct differences in behaviour between expressions derived from different sources. The most important factor affecting emissions is rather the overall sensitivity of an expression to composition and temperature, and in this respect, the behaviour of the oxidation expressions is more significant. The different sensitivities to oxygen partial pressure, as shown in Fig 5, give different trends when the rates are plotted against crank angle. This has a serious impact on emissions prediction because the calculated yields are very sensitive to the matching of the formation and oxidation expressions throughout the combustion period. To illustrate this problem, calculated smoke concentrations obtained with two different oxidation expressions are given in Table 1. The Hiroyasu and Nagle expressions were chosen in view of their different sensitivities to oxygen partial pressure (see Fig 5). In both cases, Hiroyasu's expression has been used for formation (equ. 5). Three different factors have been used to scale the oxidation rate to investigate the sensitivity of the yield predictions. The results for the combination of both Hiroyasu expressions with unscaled rate were fitted to the experimental value for the standard case of the experimental data-set ($0.362\ g\ m^{-3}$).

The results show that it is difficult to achieve a good balance between peak and exhaust value using the Nagle oxidation rate. Also, there is a very high sensitivity to the value of the scaling factor in contrast to the Hiroyasu oxidation expression results. Thus, the importance of using well-matched soot formation and oxidation expressions is demonstrated.

5 CONCLUSIONS

Zonal models present many advantages for the prediction of exhaust smoke from diesel engines. A comparative study of soot models suggests that in general, similar predictions may be obtained using local chemical rates and quasi-chemical expressions which have been calibrated in models. The Nagle and Magnussen EDC expressions are exceptions which show rather different dependencies. In other cases, the mismatch between the experimental conditions of the expression derivation and the simulation values leads to some quantitative inaccuracies. Thus, calibration will generally be necessary, and this tends to obscure the expected reduction in chemical rates due to turbulence. Sensitivity to model constants is generally low and similar results can be obtained by calibration. Improved predictions of exhaust emission may be achieved by upgrading certain features of the combustion model representation and paying particular attention to the matching of the soot formation and oxidation expressions.

REFERENCES

1. KAMIMOTO,T., AOYAGI,Y., MATSUI,Y., MATSUOKA,S.: "Some effects of engine variables on measured rate of air entrainment and heat release in a DI diesel engine", SAE 800253, 1980
2. WELCH,S.: "Computational modelling of diesel engine smoke emission", PhD thesis, Cranfield University, 1995
3. MAGNUSSEN,B.F.: "Heat transfer in gas turbine combustors", Heat transfer and cooling in gas turbines, CP-390, paper no. 23, Bergen, Norway, 1985, pp. 1-17, AGARD
4. HIROYASU,H., KADOTA,T.: "Development and use of a spray combustion modeling to predict diesel engine efficiency and pollutant emissions (Part 1: Combustion Modeling)", Bull. JSME, Apr., 1983, vol. 26, no. 214, paper no. 214-12, pp. 569-575
5. HIROYASU,H., ARAI,M., TABATA,M.: "Empirical equations for the Sauter Mean Diameter of a diesel spray", SAE 890464, 1989
6. HIROYASU,H., YOSHIMATSU,A., ARAI,M.: "Mathematical model for predicting the rate of heat release and exhaust emissions in IDI diesel engines", IMechE, C102/82, 1982
7. RANZ,W.E., MARSHALL,W.R.: "Evaporation from drops", Chemical. Eng. Prog., 1952, vol. 48, no. 3, pp. 141-146, no. 4, pp. 173-180
8. DENT,J.C.: "Turbulent mixing rate - its effect on smoke and hydrocarbon emissions from diesel engines", SAE 800092, 1980
9. XIAO,Y., BORGNAKKE,C.: "A stochastic combustion model of direct injection diesel engines", SAE 912354, 1991
10. BAZARI,Z.: "A DI diesel combustion and emission predictive capability for use in cycle simulation", SAE 920462, 1992
11. KRIEGER,R.B., BORMAN,G.L.: "The computation of apparent heat release for internal combustion engines", ASME paper no. 66-WA/ DGP-4, 1966
12. KUO,T-W.: "Evaluation of a phenomenological spray-combustion model for two open-chamber diesel engines", SAE 872057, 1987
13. AHMAD,T., PLEE,S.L., MYERS,J.P.: "Diffusion flame temperature - its influence on diesel particulate and hydrocarbon emissions", IMechE, C101/82, 1982, pp. 267-275
14. IIDA,N. "Surrounding gas effects on soot formation and extinction - observation of diesel spray combustion using a rapid compression machine", SAE 930603, 1993
15. FARMER,R., EDELMAN,R., WONG,E.: "Modeling soot emissions in combustion systems", in Particulate carbon formation during combustion, eds. SIEGLA,D.C., SMITH,G.W., 1981, Plenum Press, New York, pp. 299-320
16. TESNER,P., SNEGIRIOVA,T.D., KNORRE,V.G.: "Kinetics of dispersed carbon formation", Comb. & Flame, 1971, vol. 17, pp. 253-260
17. LEE,K.B., THRING,M.W., BEÉR,J.M.: "On the rate of combustion of soot in a laminar soot flame", Comb. & Flame, 1962, vol. 6, pp. 137-145
18. NAGLE,J., STRICKLAND-CONSTABLE,R.: "Oxidation of Carbon between 1000-2000°C", Proc. Fifth Carbon Conf., 1962, vol. 1, pp. 154-164
19. MAGNUSSEN,B.F.: "The rate of combustion of soot in turbulent flames", Thirteenth Symposium (Int.) on Combustion, 1971, pp. 869-877, The Combustion Institute
20. KHAN,I.M., GREEVES,G.: "A method for calculating the formation and combustion of soot in diesel engines", in Heat transfer in flames, eds. AFGAN,N.H., BEER,J.M., 1974, Scripta Book Co., Washington, pp. 391-404
21. PARK,C., APPLETON,J.P.: "Shock-tube measurements of soot oxidation rates", Comb. & Flame, 1973, vol. 20, pp. 369-379
22. FEUGIER,A.: "Soot oxidation in laminar hydrocarbon flames", Comb. & Flame, 1972, vol. 19, pp. 249-256

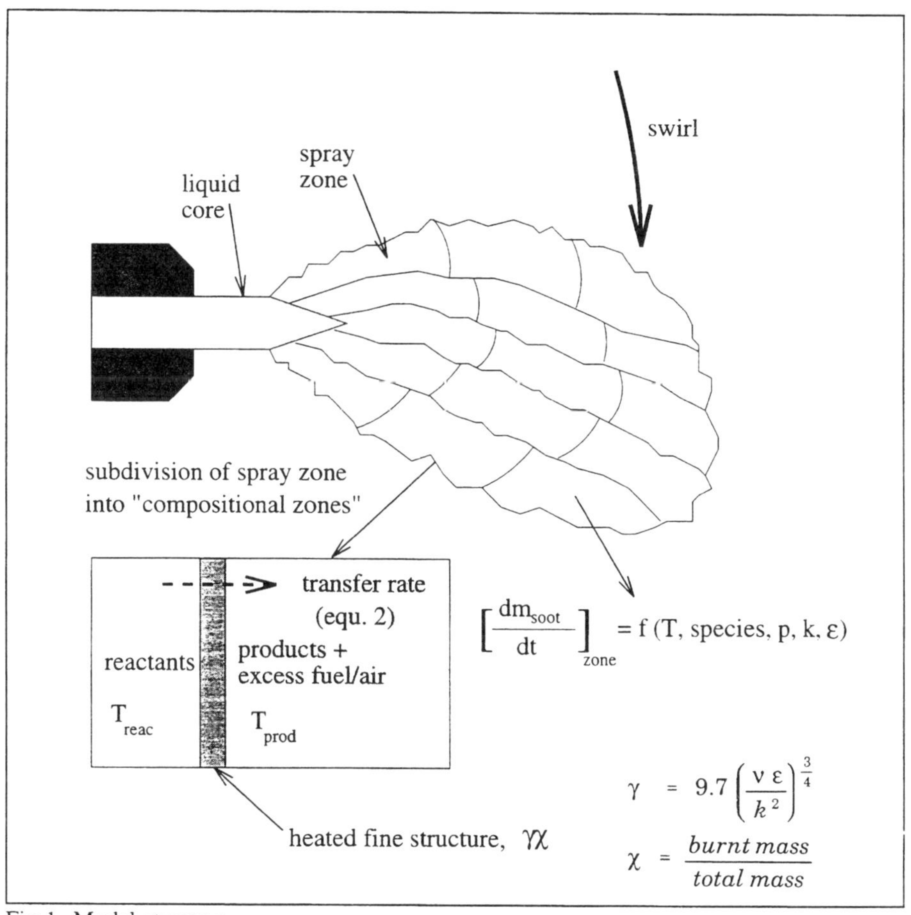

Fig 1 Model structure

Predicted smoke concentrations with Hiroyasu formation expression [g m^{-3}]	Oxidation expression			
	Hiroyasu		Nagle	
	Peak	Exhaust	Peak	Exhaust
oxidation rate x 100%	1.579	0.362	2.669	0.189
oxidation rate x 95%	1.575	0.391	2.920	0.317
oxidation rate x 90%	1.608	0.424	3.178	0.476

Table 1 - Predicted exhaust smoke concentration with different oxidation rate combinations

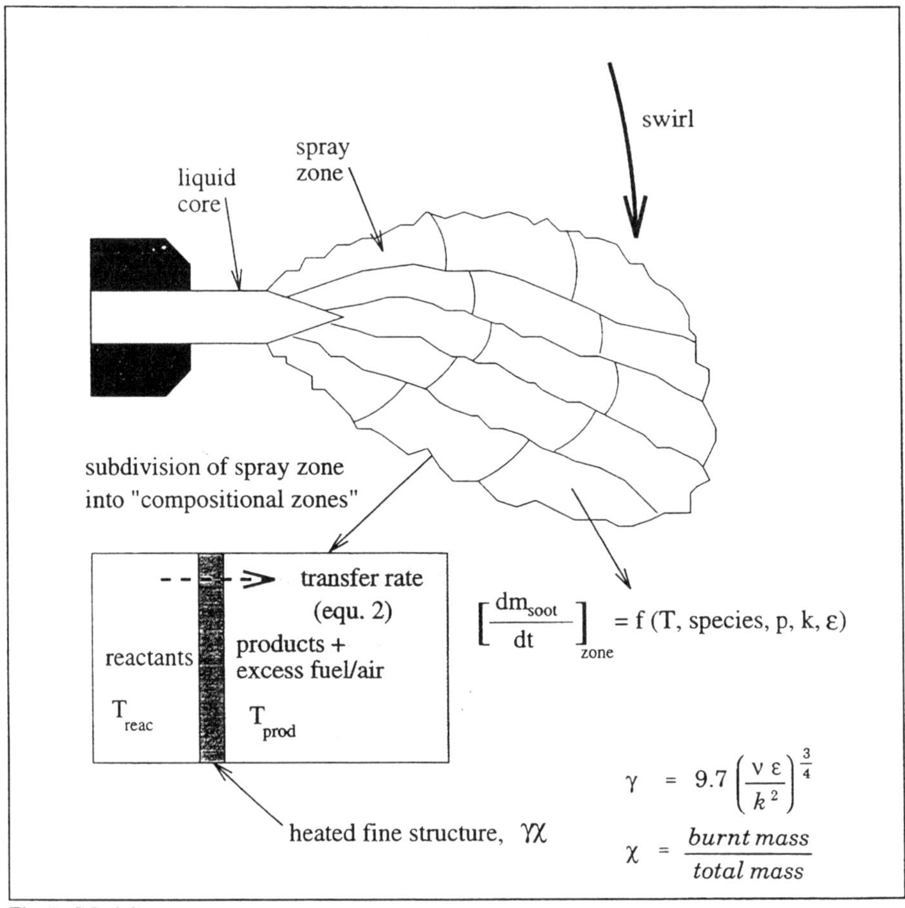

Fig 1 Model structure

Predicted smoke concentrations with Hiroyasu formation expression [g m^{-3}]	Oxidation expression			
	Hiroyasu		Nagle	
	Peak	Exhaust	Peak	Exhaust
oxidation rate x 100%	1.579	0.362	2.669	0.189
oxidation rate x 95%	1.575	0.391	2.920	0.317
oxidation rate x 90%	1.608	0.424	3.178	0.476

Table 1 - Predicted exhaust smoke concentration with different oxidation rate combinations

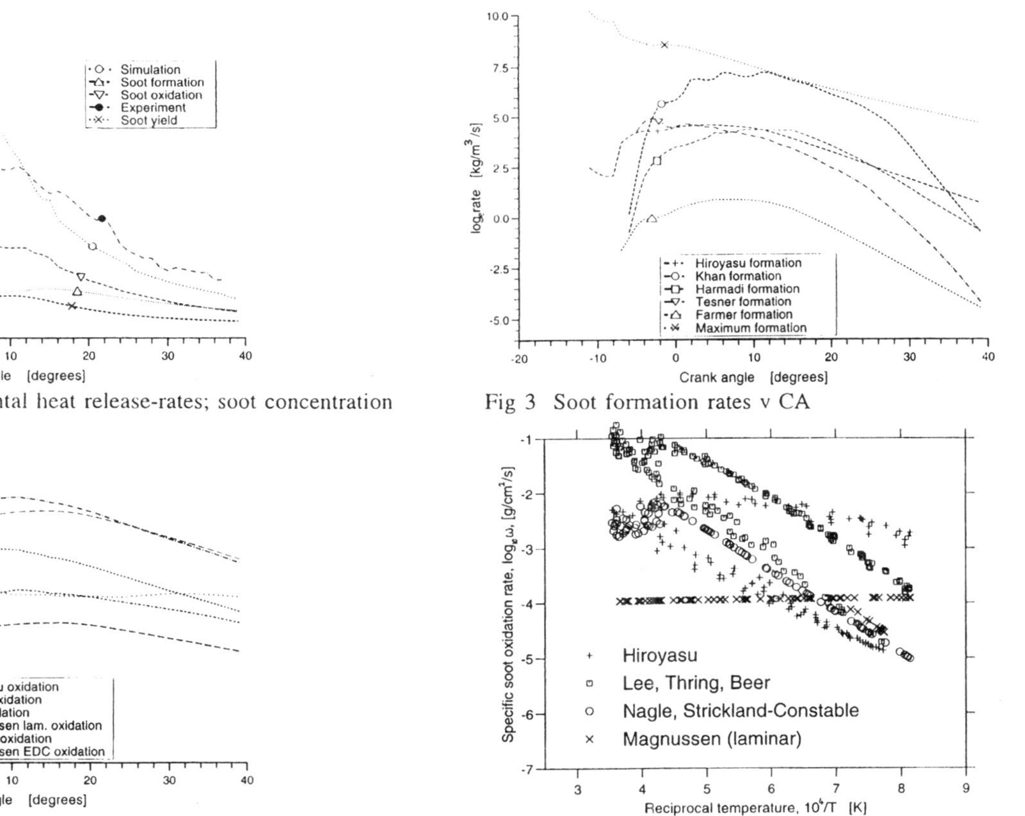

Fig 2 Predicted and experimental heat release-rates; soot concentration

Fig 3 Soot formation rates v CA

Fig 4 Soot oxidation rates v CA

Fig 5 Soot oxidation rate v reciprocal temperature

C499/047/96

A simple auto-ignition model for three dimensional diesel combustion computations

P BELARDINI, C BERTOLI MSAE, and **N DEL GIACOMO**
Istituto Motori - CNR, Napoli, Italy
M C CAMERETTI
DIME Università Federico II, Napoli, Italy

SYNOPSIS

Multidimensional Diesel combustion calculations require the simulation of the low temperature chemistry during the ignition delay period.
The correct prediction of the ignition delay time is very important to properly reproduce the relative weight of the premixed and diffusive phases of the diesel combustion process.On the other hand this is very significant to evaluate the NOx and HC emissions.
In the present paper different formulations were used to simulate the low temperature chemistry, using multi-step schemes, as well as a simplified phenomenological approach, based on an experimental correlation. The different methods investigated were implemented in the KIVA II code and used to perform 3D combustion calculations in a number of test cases. The tests are referred to a direct injection diesel engine running in different operating conditions and burning fuels with different cetane number. The numerical results are analysed in comparison with the corresponding experimental measurements.

INTRODUCTION

In diesel engine's combustion the injected fuel ignites spontaneously. The ignition delay period is commonly defined as the period between the start of injection and the start of combustion.
Usually the start of injection, from an experimental point of view, is clearly defined using the needle lift law measurement ($5 \div 10\%$ of the maximum lift). Vice versa different definitions of the start of combustion can be assumed [1]: the pressure rise, or the first positive value of the heat release rate, or the first luminosity of the flame is assumed by different authors as start of combustion.
During the ignition delay time the fuel vapour mixes with oxygen and kinetic chemical reactions take place producing free radicals deriving from the partial fuel oxidation.
Because the rate of these chain-type reactions increases exponentially, also the heat release rises, leading to the auto ignition in different places of the combustion chamber.
The ignition delay duration in diesel engines controls a number of relevant phenomena. In particular, once settled the injection law, the quantity of fuel injected during the ignition delay period controls the NOx emissions [2]. Also the HC emissions are well linked to the cetane number of the fuel, and therefore to the ignition delay time, as assessed for example by Heaton et al. [3].The ignition delay influences also the main fractions (premixed or diffusive) of the

diesel combustion affecting also the soot formation process, as outlined by Barbella et al. [4], who measured the main species concentrations in the different phases of diesel combustion by the fast sampling valve technique.
Finally the ignition delay has a hard impact on the temporal rise of the cylinder pressure, that is the main source of the diesel engine noise.
From the previous considerations emerges how important is the correct estimation of the ignition delay time in diesel combustion calculations. In any case, because of the computational cost of 3D CFD calculations, a strong simplified approach is highly desirable, on condition that the accuracy is acceptable. Thus, in the present paper a simplified mechanism of ignition delay time computing, implemented in the KIVA II code, is presented and discussed.
Anyway, to evaluate if more advantageous results may be achieved, also more sophisticated ignition delay models are considered and tested.For each model tested comparisons between computed and measured ignition delay times are also presented for a D.I. diesel engine combustion system.

REVUE OF THE PAST WORK

Starting from pioneering work of Wolfer [5] in 1938 a number of formulas were proposed by the different authors to compute the ignition delay time in Diesel Engines. A comprehensive review of these formulas was done by Hiroyasu [6].
However, Belardini et al. [1] showed that there is a large scatter of the computed ignition delay values when using the different formulations. More recently in 3D Diesel combustion computations some of these formulas were employed and updated. Clearly also different approaches can be adopted.
Hiroyasu and his colleagues, in a number of works as well as in 3D computations [7], adopted the following approach. In each computational cell the ignition happens when:

$$\int_0^{\tau_{ig}} \frac{1}{\tau_d} d\tau_d = 1$$

Where τ_{ig} is the ignition time and 0 is the start of injection. τ_d is computed as:

$$\tau_d = 4*10^{-3} p^{-2.5} \varphi^{-1.04} exp\frac{6000}{T_g}$$

In this expression p is the cylinder pressure [pascal]; T_g is the gas temperature of the computational cell; ϕ is the equivalence ratio of the computational cell.
Adopting the previously outlined method to compute the ignition delay and describing combustion with a Magnussen type combustion model, Hiroyasu obtained a good correspondence between computed and measured heat release rates.
More recently Yoshiraki et al. [8] used the previous approach but slightly modified: they chose a different value for the activation temperature (5000 K) and introduced for the n-dodecane mass fraction lean and rich limits as cut off values for the combustion calculations.
Pinchon [9] performed three dimensional computations of combustion in a swirl chamber diesel engine. To represent the low temperature chemistry he adopted an extension of the well known Shell method, proposed by Teobald et al. [10] to predict the ignition of n-dodecane.
The model takes into account four reactions, in which the generalised radical R represents the pool of radicals involved in the auto ignition process, M is any molecule and ΔQ is the energy released by the reaction:

Initiation stage:	fuel	$\Rightarrow$	2R
Branching stage:	fuel + R	$\Rightarrow$	3R
Termination stage:	2R + M	$\Rightarrow$	products +M
Propagation stage:	fuel + O2 + R	$\Rightarrow$	comb. products + R + ΔQ

For each equation the corresponding reaction rate is:

$$\omega_i = A_i T^{n_i} e^{\frac{-T_i}{T}} \prod_j \left(\frac{\rho_j}{w_j}\right)^{a'm_{i,j}}$$

Where T_i is the activation temperature, ρ_j and w_j are respectively the density and the molecular weight of the j specie. A_i and $a'm_{i,j}$ are constants, for which Pinchon didn't give any values, as well as no comparisons with the experimental data were performed.
Using the same auto ignition and combustion models as Pinchon, Zellat et al. [11] did some comparisons between computed and measured combustion pressure cycles, in the pre chamber as well as in the main chamber, in different engine operating conditions. The agreement was quite acceptable, but no information about the values used for the model constants is given. In both works the switch criterion between the low and the high temperature combustion models is roughly based on a predetermined temperature threshold in the computing domain.
Also Gorokhovski and Borghi [12] adopted the Shell model [13] in a coupled ignition - soot formation global mechanism. They modified the original mechanism replacing three of the eight reactions (the propagation reactions) with a modified mass balanced reaction, so obtaining for the auto ignition a six step's mechanism. In [12] no comparison between computed and measured data is reported.
Abraham and Bracco [14] modelled in a simplified way the low temperature chemistry. They considered a new specie R, representative of the evolution of the radicals, that lead to auto ignition.
For all the species involved in the process, the specie conservation equations are solved. For the specie R the diffusion velocity in the j direction V_{Rj} and the corresponding formation rate $\dot{\omega}_R$ are expressed as:

$$V_{Rj} = -\frac{1}{Y_R}\left(D_e \frac{\partial Y_R}{\partial x_j}\right), \qquad \dot{\omega}_R = \frac{1}{\tau_R} \qquad \text{with} \qquad \tau_R = A\left(\frac{p}{p_0}\right)^B e^{(D(1+C|1-\phi|))/T}$$

In these expressions Y_R is the mass fraction of the specie R: the ignition occurs when Y_R equals 1; D_e is the effective diffusivity, P the cell pressure [atm], T the cell temperature [K] and Φ the equivalence ratio. A, B, C, D are the model constants, to be tuned with the experiments: they were settled respectively to 0.0845, -1.5, 2.5 and 4350 [K].
Abraham and Bracco, simulating bomb experiments, compared the computed ignition delay times with those of literature, obtaining a globally acceptable fit. They performed also 3D computations of D.I. diesel combustion, obtaining reasonable trends in the numerical results.
In a recent work Patterson et al. [15] have been performed 3D computations for a quiescent chamber D.I. diesel engine using an improved version of the KIVA-II code. To describe the low temperature kinetic the Shell model was adopted. No details were furnished about the model constants values. Vice versa many comparisons were given between computed and measured heat release patterns in a number of test cases. The numerical results in normal engine operation (without split injection) are very close to the experimental ones, also in the case of high injection pressure. Moreover the NOx predictions match with the measured values

at the exhausts. Therefore it is possible to deduce that the ignition delay periods are properly computed. However it must be noted that Patterson et al.[15] use tetradecane to simulate the properties of the diesel fuel: probably the tuning of the ignition delay model was performed taking into account the differences in cetane number for the two fuels, but in the paper this is not noticed.
Anyway whatever model used, even if simplified, needs the tuning of the model constants and this may not be the same if using different fuels, characterised by different volatility and cetane number.
In the present paper a simplified approach is discussed, similar to those used by Nishida and Hiroyasu [7], but adopting as ignition delay correlation the Hardenberg and Hase formulation [16], as reported by Belardini et al. in [17], [18]. With respect to the previous papers, the experimental data base, used for the model testing, has been enlarged including secondary reference fuels with quite different cetane number.
In alternative to this simple model, just for the n-heptane ignition, a more fundamental approach was investigated: it is the model elaborated by Muller et al. in [19] , also used in the version developed by Griffiths in [22].

EXPERIMENTAL TESTS

In the present work a single cylinder D.I diesel engine was used to perform measurements of the ignition delay time: its characteristics are reported in table 1.
The engine was equipped with combustion and line pressure transducers as well as with a Hall current needle lift transducer. To evaluate the ignition delay angles, the crank angle resolution was of about 0.2 C.A. degrees: the heat release curves were computed with an integration pass of 0.5 C.A. degrees to smooth the oscillations.

Compression ratio	18.0
Bore	10.0 [cm]
Stroke	9.5 [cm]
Connecting rod length	17.8 [cm]
Combustion chamber	Toroidal
Equivalent Aspect Ratio	4.6
Speed	1250 [rpm]
Injector	4 holes ϕ=0.28
Spray cone angle	160
Injection velocity	190 [m/s]

Table 1: The engine characteristics

To have more reliable data, the measurements were performed over a mean of 128 engine cycles: in order to reduce the injection law cyclic variation, the injection pump (a Bosch P type) was used without the mechanical speed governor (one of tha main source of injection law variations) and with a proper choice of the plunger diameter and of the inner bore to lenght ratio of the injection line. So doing the variation coefficient of the injection start was maintained below the 1%.
To evaluate the combustion start both equivalent criteria reported in [1] were used: the pressure rise and the first positive going of rate of heat release (ROHR) curve. For these

parameters the variation coefficient falls in the range 3-6 % depending on the injection advance (at retarded timings the cyclic variation rises).
Finally the engine head was also equipped with a small quartz window to measure the first appearance of the luminous flame.
As concerns the experimental test cases, each used fuel was tested at two engine speed (1250 and 2000 RPM), at two different load values (0.02 and 0.035 grams/cycle) and at three different injection timings.
Because the aim of the present work is an accurate evaluation of ignition delay data, blends of secondary reference fuels were used: actually the T18, the U11 and the check fuel for the ASTM D613 procedure. As resulting value of cetane number CN=73, CN=55, CN=49.9 and CN=35 were obtained. In addition, just at lower engine speed to avoid wear problems to the injection apparatus, the pure tetradecane (CN=93) and the n-heptane fuel (CN=56) were tested.
Therefore the experimental matrix comprises a total number of 32+12 test points.

THE SINGLE STEP IGNITION DELAY MODEL

As concern's auto ignition in the approach we have used the auto ignition starts when it is reached the condition:

$$\int_0^{t_{ig}} \frac{1}{t_d} dt_d = 1$$

In the present work, for τ_d [sec] the Hardenberg and Hase [17] formula was used:

$$t_d = \frac{1}{6*n} * \left(0.36+0.22S_p\right) * exp\left(E_A * \left(\frac{1}{RT} - \frac{1}{17190}\right) + \left(\frac{21.2}{p-12.4}\right)^{0.63} \right)$$

where 0 is the time at which injection starts, n is the engine speed [rpm], T is the cell temperature [K], p is the cell pressure [bar], S_p is the mean piston speed [m/sec]. This formula is sensitive to the fuel cetane number via the term:

$$E_A = \frac{618840}{CN+8} [J/mol]$$

The energy activation E_A was originally expressed by Hardenberg and Hase as:

$$E_A = \frac{618840}{CN+25} [J/mol]$$

But, to obtain a better overall fitting of the experimental data, it was modified as above reported: in this way a higher sensitivity to the fuel cetane number was achieved.
This model was used with the whole set of test cases: in the numerical runs the blended fuels were simulated adopting the n-dodecane fuel as from the standard fuel library of the KIVA-II code, but using the true value of cetane number. The original code was modified to incorporate an improved version of the beak-up model for fuel jets, the interaction between fuel jets and walls and an eddy break-up theory based high temperature combustion model . The details of the different sub- models can be found in the ref. [24]. Anyway all the numerical test cases were performed with the same tuning of the constants of these sub- models
To properly compare numerical and measured values of start of combustion, the numerical combustion pressure curve was processed as in the experiments.

In fig.1 the measured values of start of combustion are plotted versus the computed ones for all test conditions, showing a good agreement between numerical and experimental data. The proposed model, without any retuning of the model constants, is able to reproduce the significant variations in the combustion evolution, confirming that a proper sensitivity to cetane number was achieved.

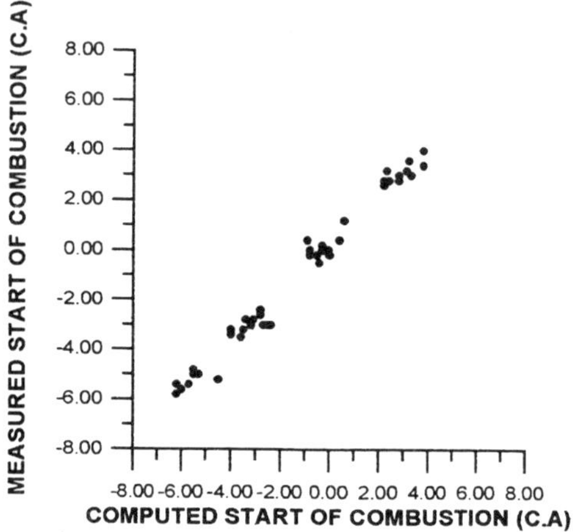

Fig.1: Single step ignition delay model: measured values of start of combustion versus the computed ones for all test cases.

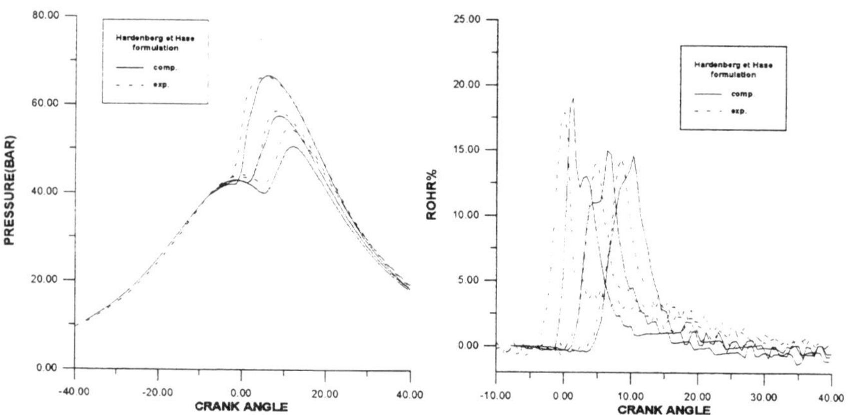

The Hardenberg and Hase correlation: computed and measured heat release and combustion patterns for the n-heptane at three different injection timings (-7.4CA, -4.4CA, -1.4CA).
Fig.2

Also at a more detailed analysis the model seems to give numerical results well related with the experimental ones. The fig.2 shows the combustion pressure cycles and the heat release

patterns obtained at the same engine speed (1250 RPM), at air/fuel ratio equal to 35 (injected mass = 0.02 g/cycle), but at three different injection timings (-7.4CA, -4.4 CA, -1.4 CA) burning n-heptane fuel: the same diagram shows predicted as well measured data, sufficiently close one each other.
This ignition delay model, without any change in the model constants, gives good predictions of the combustion and heat release patterns also with the other fuels of the previously described set of test cases.

THE MULTISTEPS IGNITION DELAY MODELS

The Muller et al. model for the n-heptane combustion.

Muller et al. in [19] proposed a reduced kinetic model to represent the oxidation of n-heptane over the temperature range of 700-1500 K: the experimental validation was performed comparing the simulation of the adiabatic ignition delay with experimental results by shock tube experiments. The model is very attractive because of the low number of reactions involved.
In particular four steps are proposed, here reported together with the expressions used for the corresponding reaction rates:

$$F \stackrel{1}{\Rightarrow} X \qquad k_1 = A_1 e^{-\frac{E_1}{RT}}$$

$$X + 11O_2 \stackrel{2}{\Rightarrow} P \qquad k_2 = \frac{p}{RT} A_2 e^{-\frac{E_2}{RT}}$$

$$F + 2O_2 \underset{3b}{\overset{3f}{\Leftrightarrow}} I \qquad k_{3f} = \frac{p}{RT} A_{3f} e^{-\frac{E_{3f}}{RT}} \qquad k_{3b} = A_{3b} e^{-\frac{E_{3b}}{RT}}$$

$$I + 9O_2 \stackrel{4}{\Rightarrow} P \qquad k_4 = \frac{p}{RT} A_4 e^{-\frac{E_4}{RT}}$$

Here F is the fuel, X and I represent the combined intermediates and P a combination of products:

$$F = C_7H_{16}$$
$$X = 3C_2H_4 + CH_3 + H$$
$$I = OC_7H_{13}OOH + H_2O$$
$$P = 7CO_2 + 8H_2O$$

The preexponential factors and the activation temperatures are reported in the tab.2.

$A_1 = 2*10^{10}$ (s^{-1})	$E_1/R = 21650$ K
$A_2 = 2*10^{12}$ ($cm^3\ mol^{-1} s^{-1}$)	$E_2/R = E_1/R$
$A_{3b} = 4*10^{22}$ (s^{-1})	$E_{3f}/R = E_1/R$
$A_{3f} = 3*10^{18}$ ($cm^3\ mol^{-1} s^{-1}$)	$E_{3b}/R = 37285$ K
$A_4 = 5*10^{13}$ ($cm^3\ mol^{-1} s^{-1}$)	$E_4/R = 13230$ K

Table 2: Muller model constants

The model was implemented in the standard kinetic routine of KIVA II and used together with the high temperature combustion model.

The switch criterion between the low temperature and the high temperature single step mechanism was based on a temperature threshold in the cell (1200 K).
This ignition delay model was applied to the same test cases used for fig.2: with the chosen setting of the model constants the results obtained in terms of predicted ignition delay time are well related with the measured values.

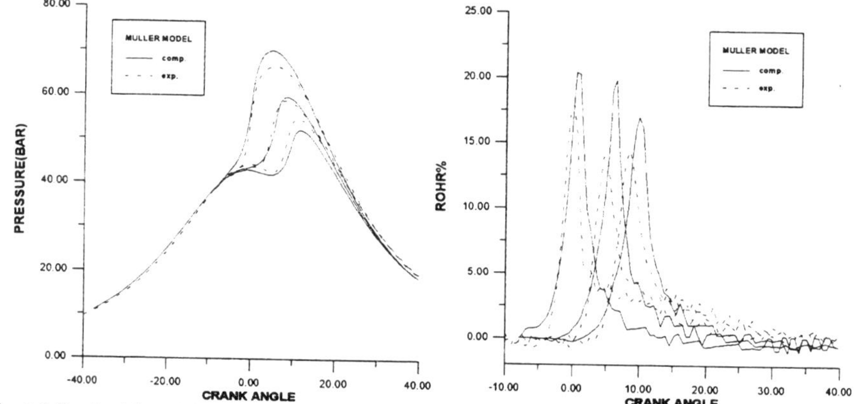

The Muller ignition delay model: computed and measured heat release and combustion patterns for the n-heptane at three different injection timings (-7.4CA, -4.4CA, -1.4CA).
Fig.3

Also the combustion and the heat release patterns seem to be acceptable: they reported in fig.3 in comparison with the experimental cycles obtained for the same tests used for fig.2.
But for this ignition delay model, when used with fuel different from the n-heptane, it is necessary to retune the model constants in order to obtain combustion and heat release patterns well related with the experimental ones.
Obviously this is a strong limitation and makes the single step model previously described more convenient when the modelling aims to describe a wide range of test case, more than to follow in details a single test case.

The Griffiths model for the n-heptane combustion

In [23] Griffiths proposed a revised form of the Muller model to take into account the negative temperature dependence of the ignition delay time not described by the original formulation of the model . We applied the model in a form slightly modified, that is here reported:

1- $F \Rightarrow X$
2- $X + 11O_2 \Rightarrow P$
3- $F + 2O_2 \Rightarrow RO_2$
4- $RO_2 \Rightarrow QO_2H$
5+6- $QO_2H + O_2 \Leftrightarrow O_2QO_2H$
7- $O_2QO_2H + 12O_2 \Rightarrow 7CO_2 + 16OH$
8- $F + 8OH \Rightarrow RO_2 + 4H_2O$
9- $QO_2H \Rightarrow I$
10- $I + 9O_2 \Rightarrow P$

The steps from 3 to 9 replace the equation 3 in the Muller mechanism: the equation 7 introduces the auto catalysis, that occurs in competition with the reverse step 5; reaction 4 signifies the isomerisation of the alkylperoxy radical. To match with the experimental results, the model was used with the setting of the reaction rate constants that is reported in table 3.
Also in this case it was used the same test condition used for fig.2, using a temperature threshold in the cell of 1200 K as switch value between the low and the high temperature mechanism.
Also in this case the results obtained in terms of predicted ignition delay time, and combustion and heat release patterns, that are reported in fig.4, are well related with the measured values. But it was necessary to change the original setting of the model constants, that also in this case must be retuned whenever changing fuel.

$A_1 = 2*10^{10}$ (s^{-1})	$E_1/R = 21650$ K
$A_2 = 2*10^6$ ($cm^3\ mol^{-1}s^{-1}$)	$E_2/R = 7220$ K
$A_3 = 5*10^7$ (s^{-1})	$E_3/R - 20000$ K
$A_4 = 1*10^{11}$ ($cm^3\ mol^{-1}s^{-1}$)	$E_4/R = 10500$ K
$A_5 = 5*10^6$ ($cm^3\ mol^{-1}s^{-1}$)	$E_5/R = 4000$ K
$A_6 = 5*10^{11}$ ($cm^3\ mol^{-1}s^{-1}$)	$E_6/R = 6500$ K
$A_7 = 5*10^{10}$ ($cm^3\ mol^{-1}s^{-1}$)	$E_7/R = 12000$ K
$A_8 = 1*10^6$ ($cm^3\ mol^{-1}s^{-1}$)	$E_8/R = 1500$ K
$A_9 = 1*10^{11}$ ($cm^3\ mol^{-1}s^{-1}$)	$E_9/R = 8400$ K
$A_{10} = 5*10^7$ ($cm^3\ mol^{-1}s^{-1}$)	$E_{10}/R = 13230$ K

tab 3. Griffiths model constants

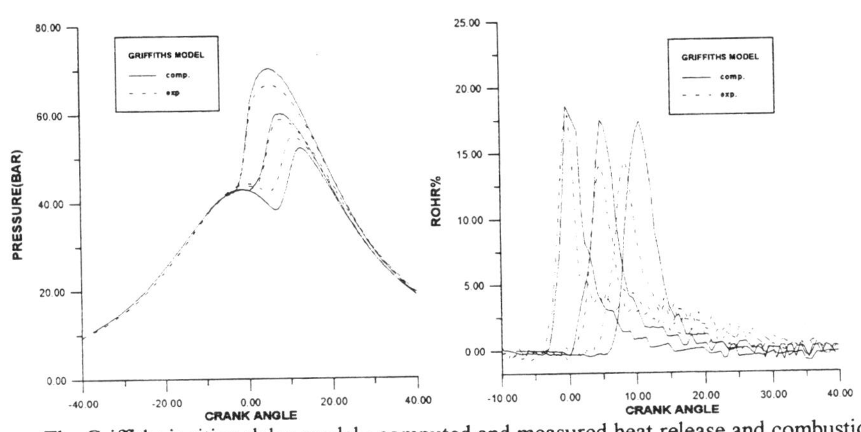

The Griffiths ignition delay model.: computed and measured heat release and combustion patterns for the n-heptane at three different injection timings (-7.4CA, -4.4CA, -1.4CA).
Fig.4

CONCLUSIONS

The right use of a 3D code to optimise the engine combustion system design requires that the computational cost remains acceptable. Thus simplified formulations for some physical problems are strongly wanted if the overall precision of computations is acceptable.

To perform computations of the auto ignition phenomenon in Diesel engines a simplified formulation, based on an experimental correlation, was introduced and tested in the KIVA II code. Experiments performed in a single cylinder diesel engine in a number of test cases confirm the good accuracy of the computations also changing the cetane number of the fuel. Finally a 4 steps model as well as a 9 steps model for low temperature combustion of n-heptane were implemented in the code. Computations showed that in any case the retuning of the model constants is necessary to obtain more realistic predictions.

REFERENCES

[1] P. Belardini, C. Bertoli, F.E. Corcione, G. Police: "Ignition delay measurement in a D.I. Diesel Engine", IMEchE c/86/83, vol.2, pp.1-7, Oxford 1983;

[2] P. Zalenka, V. Kriegler, P.L. Herzog, V.P. Cartellieri: "Ways toward to the clean heavy duty diesel", SAEpaper 900602;

[3] D.M. Heaton, B. Martin, C. Bertoli, F. Giavazzi: "Fuel Composition Effects on Diesel Engine Emissions. A Joint European Study", IMEchE Seminar on fuels for automotive and industrial Diesel Engines, 1993 MEP, London 1993;

[4] R. Barbella, C. Bertoli, A. Ciajolo, A. D'Anna: "In Cylinder Sampling of High Molecular Weight Hydrocarbons from a D.I. Light Duty Diesel Engine", SAEpaper 890437;

[5] H.H. Wolfer: "Der Zunderzug im Diesel Motor", CDE - Forrshungsheft, vol 392, pp.15-24, 1938;

[6]. H. Hiroyasu: "Diesel Engine Combustion and its Modeling", I COMODIA, Tokyo 1985;

[7]. K.Nishida, H. Hiroyasu: "Simplified three Dimensional Modeling of Mixture Formation and Combustion in a D.I. Diesel Engine", SAEpaper 890269;

[8] T. Yoshiraki, K. Nishida, H. Hiroyasu: "Approach to low NOx and Smoke Emission Engines by using Phenomenological Simulation", SAEpaper 930612;

[9] P. Pinchon: "Three-dimensional Modelling of Combustion in a prechamber Diesel Engine", SAEpaper 890660;

[10] M.A. Theobald, W.K. Cheng: "A Numerical Study of Diesel Ignition", ASME, 87 - FE2, 1987;

[11] M. Zellath, T.H. Rolland and F. Poplow: "Three- Dimensional Modelling of Combustion and Soot Formation in an Indirect Injection Diesel Engine", SAEpaper 900254;

[12] M. Gorokhovski, R. Borghi: "Numerical Simulation of Soot Formation and Oxidation in Diesel Engines", SAEpaper 930075;

[13] M.P. Helsted, L.J. Kirsh, C.P. Quinn: "The Autoignition and Hydrocarbons Fuels at High Temperatures and Pressure-Fitting of Mathematical Model", Combustion and Flame, Vol.30, pp.45-66, 1977;

[14] J.C. Abraham and F.V. Bracco: "Simple Modeling of Autoignition in Diesel Engines for 3D Computations", SAEpaper 932656;

[15] M.A. Patterson, S.C. Kong, G.J. Hampson, R.D. Reitz: "Modeling the Effect of Fuel Injection Characteristics on Diesel Engine Soot and NOx Emissions", SAEpaper 940523;

[16] H.O. Hardenberg, F.W. Hase: " An Empirical formula for Computing Pressure rise Delay of a Fuel from its Cetane number and from Relevant Parameters of D.I. Diesel Engines", SAEpaper 790493;

[17] P. Belardini, C. Bertoli, A. Ciajolo, A. d'Anna, N. Del Giacomo: "Three Dimensional Calculations of D.I. Diesel Engine Combustion and Comparison with in Cylinder Sampling Valve Data", SAEpaper 922225;

[18] P. Belardini, C. Bertoli, N. Del Giacomo, B. Iorio: "Soot Formation and Oxidation in a D.I. Diesel Engine: A Comparison between Measurements and three Dimensional Computations", SAEpaper 932650;

[19] U.C. Muller, N. Peters, A. Linon: "Global Kinetics for n-heptane Ignition at High Pressures", Twenty-Fourth Symposium (International) on Combustion, The Combustion Institute Pittsburgh 1992;
[20] P. Belardini, C. Bertoli, F. Corcione, G. Valentino: "In Cylinder Flow Measurements by LDA and Numerical Simulation by KIVA II Code", SAEpaper 920155;
[21] P. Belardini, C. Bertoli, M.C. Cameretti, N. Del Giacomo: "A Coupled Diesel Combustion and Soot Formation Model for KIVA II Code: Characteristics and Experimental Validation", submitted to COMODIA '94 Symposium;
[22] A.A. Amsden, P.J. O'Rourke, T.D. Butler: "KIVA-II: A Computer Program for Chemically Reactive Flows with Sprays", Los Alamos National Laboratory Report No. LA-11560-MS, 1989.
[23] J.F. Griffiths: "Kinetic Fundamentals of Alkane Autoignition at Low Temperatures", COMBUSTION AND FLAME 93:202-206,1993.
[24] C. Beatrice, P. Belardini, C. Bertoli, M.C. Cameretti, and N.C. Cirillo: "Fuel Jet Models for Multidimensional Diesel Combustion Calculation: An Update", SAEpaper 950086.